Applied and Numerical Methods

Applied and Numerical Methods

Dr Maria Tattaris

ARCLER
PRESS

www.arclerpress.com

Applied and Numerical Methods

Dr Maria Tattaris

Arcler Press

2010 Winston Park Drive,

2nd Floor

Oakville, ON L6H 5R7

Canada

www.arclerpress.com

Tel: 001-289-291-7705
001-905-616-2116
Fax: 001-289-291-7601
Email: orders@arclereducation.com

ISBN: 978-1-77361-217-1 (Hardcover)

ABOUT THE AUTHOR

Dr Maria Tattaris

Maria Tattaris has a background in Mathematics and a PhD in remote sensing within the area of atmospheric sciences, awarded in 2012. Her interests include field spectroscopy, image processing, spectral analysis, field data collection, and investigating new data analysis methods for remote sensing techniques, all within the areas of environmental, plant, atmospheric and related sciences.

TABLE OF CONTENTS

PREFACE

In many respects, this book is an ideal text book for beginners. It combines a clear approach and well-rounded explanations with plenty of practice. It makes an ideal companion for those students beginning a course in applied and numerical methods, differential equations and mechanics at school, or for undergraduate courses in mathematics, engineering or physics.

The study of differential equations is an extensive subject in pure and applied mathematics, physics, and engineering. All of these disciplines deal with the properties of differential equations of different types. Pure mathematics has an emphasis on the life and incomparability of solutions, while applied mathematics emphasizes the rigorous methods of approximating solutions. Differential equations are an important tool in modeling virtually every physical, technical, or biological process, from celestial motion, to bridge design, to interactions between neurons. Differential equations, such as those used to solve real-life problems, may not necessarily be directly solvable, i.e. do not have closed example solutions. Instead, solutions can be approximated using numerical methods. Many fundamental laws of physics can be formulated as differential equations. In biology and economics, differential equations are used to model complex systems. Likewise, mechanics and numerical methods are of much importance. They play an important role in mathematical problems.

Author

CHAPTER 1

DIFFERENTIAL EQUATIONS

INTRODUCTION

Mathematical models for real world phenomena often take the form of equations relating various quantities and their rates of change (derivatives), For example, the motion of a particle involves the distances covered in time t and velocity v and/or acceleration a. The rate of change $\frac{ds}{dt}$ of s with respect to t is the velocity v, and the rate of change $\frac{dv}{dt}$ of the velocity with respect to t is the acceleration a. A particle moving in a straight line has an equation of motion $s = f(t)$, where t is given in seconds and s in meters. Its velocity satisfies the equation:

$$v = \frac{ds}{dt} = 4t^2 + 5t - 3.$$

This leads us to the definition of a differential equation (D.E).

A differential equation is a mathematical equation that relates some function with its derivatives. Functions typically represent physical quantities, the by - product represent their rates of change, and the equation defines an association between the two. As such relations are exceptionally common, differential equations play a prominent role in many disciplines, including engineering, physics, economics, and biology.

In pure mathematics, differential equations are studied from several different perspectives, mostly concerned with their solutions-the set of functions that satisfy the equation. Only the simplest differential equations are solvable by explicit formulas. However, an estimation of the solutions of a given differential equation may be determined without determining their exact form.

Differential Equations and their Classification

An equation involving one dependent variable and its derivatives with respect to one or more independent variables is called a differential equation. For example:

- $\frac{dy}{dx} + y\cos x = \sin x,$
- $\frac{d^2y}{dx^2} + xy\left(\frac{dy}{dx}\right)^2 = 0$,
- $\left[1 + \left(\frac{dy}{dx}\right)^2\right]^{\frac{3}{2}} = \frac{d^2y}{dx^2}.$

Ordinary Differential Equation (O. D. E)

A differential equation that features ordinary derivatives of the dependent variable with respect to a single independent variable is called an ordinary differential equation (O. D. E). The above equations (i), (ii) and (iii) are examples of ordinary differential equations.

Partial Differential Equation

A differential equation involving partial derivatives of the dependent variable with respect to more than one independent variable is called a partial differential equation.

$$x\frac{\partial z}{\partial x} + y\frac{\partial z}{\partial y} = nx,$$

$$\frac{\partial^2 u}{\partial x^2} + \frac{\partial^2 u}{\partial y^2} + \frac{\partial^2 u}{\partial z^2} = 0.$$

Order of a Differential Equation

The order of a differential equation is the order of the highest derivative that occurs in the equation.

Degree of a Differential Equation

The degree of a differential equation is the greatest exponent of the highest order derivative that appears in the equation. (The dependent variable and its derivatives should be expressed in a form free from radicals and fractions).

- $\frac{dy}{dx} + y\cos x = \sin x,$ (order 1, degree 1)
- $\frac{d^2y}{dx^2} + xy\left(\frac{dy}{dx}\right)^2 = 0,$ (order 2, degree 1)
- $\left[1+\left(\frac{dy}{dx}\right)^2\right]^{\frac{3}{2}} = \frac{d^2y}{dx^2}.$ (order 2, degree 2)

The exponent of $\frac{d^2y}{dx^2}$ is 2 after removing the radical by squaring, and subsequently take the cubic root of both sides of the equation.

- $x\frac{\partial z}{\partial x} + y\frac{\partial z}{\partial y} = nx,$ (order 1, degree 1)
- $\frac{\partial^2 u}{\partial x^2} + \frac{\partial^2 u}{\partial y^2} + \frac{\partial^2 u}{\partial z^2} = 0.$ (order 2, degree 1)

In this chapter, we shall study ordinary differential equation only.

Recall that a function $T: U \to V$, where U, V, are vector spaces over the same field F, is called linear if, for $\alpha, \beta \in F, x, y \in U$, $T(\alpha x + \beta y) = \alpha T(x) + \beta T(y)$.

Linear Differential Equations

An ordinary differential equation $F\left(x, y, \frac{dy}{dx}, \frac{d^2y}{dx^2}, \ldots, \frac{d^ny}{dx^n}\right)=0$ is said to be linear if F is a linear function with respect to the variables $x, \frac{dy}{dx}, \frac{d^2y}{dx^2}, \ldots, \frac{d^ny}{dx^n}$.

A similar definition applies to partial differential equations.

Thus, the general form of a linear ordinary differential equation of order n is given as:

$$a_0(x)\frac{d^ny}{dx^n}+a_1(x)\frac{d^{n-1}y}{dx^{n-1}}+\ldots+a_{n-1}(x)\frac{dy}{dx}+a_n(x)y=F(x),$$

where $a_0(x)$ is not identically zero.

The equation $\frac{d^3y}{dx^3}+2e^x\frac{d^2y}{dx^2}+y\frac{dy}{dx}=x^3$ is not linear because of the term $y\frac{dy}{dx}$.

It should be carefully noted that in a linear ordinary differential equation:

- The dependent variable y and its derivative are all of degree one.
- No product of y or any of its derivative appear.
- No transcendental function of y and / or its derivative appear.

A differential equation that is not linear is called a nonlinear differential equation. Differential equations occur in the mathematical formulation of many problems in science and engineering. Some such problems are:

- Determining the motion of projectiles, rockets, satellites or planets.
- Finding the charge or current in an electric circuit.
- Study of chemical reactions.
- Derivation of curves with given geometrical properties.

Solution of a Differential Equation

A solution (or integral) of a differential equation is a relationship between variables not containing derivatives, such that the relationship and the derivatives obtained satisfy the given differential equation identically. For example, the equation

$\frac{dy}{dx} = -\mu y$ has a solution $y = c.e^{-\mu x}$, where c is an arbitrary constant.

The differential equation $\frac{d^2y}{dx^2} + y = 0$ has solutions $y = A\cos x$, $y = B\sin x$ and $y = A\cos x + B\sin x$, where A and B are arbitrary constants.

A solution of a differential equation that contains as many arbitrary constants as the order of the equation is called a general solution (or integral) of the differential equation. A solution obtained from the general solution by giving particular values to the constants is called a particular solution. The graph of a particular integral is called an integral curve of the differential equation.

Formation of a Differential Equation

Given an equation $f(x, y, c_1, c_2, \ldots, c_n) = 0$ (1)

between variables x, y and containing n constants $c_1, c_2, \ldots, c_n$, it is always possible to form a differential equation of order n such that the given equation (1) is the general solution of the differential equation. This is done by differentiating (1) n times, thereby obtaining equations and then eliminating the n constants from the original equation and n derived equations. The method can be illustrated by means of examples.

Example: The equation $y = x + a$ (1)

Represents a family of parallel straight lines for different values of a. Elimination of one constant 'a' requires two equations. The second equation is obtained by differentiating (1).

Thus $\frac{dy}{dx}=1$ is the differential equation of (1), with a eliminated.

Example: Form a differential equation by eliminating the two constants A and B from the equation $y = A\sin x + B\cos x.$
(1)

Solution: It is clear that three equations are required to eliminate the two unknowns A and $B.$ We obtain the two other needed equations by successive differentiation of (1). Thus from (1), we have

$$\frac{dy}{dx} = A\cos x - B\sin x, \quad (2)$$

$$\frac{d^2y}{dx^2} = -A\sin x - B\cos x = -y, \text{ using (1).} \quad (3)$$

So $$\frac{d^2y}{dx^2} + y = 0 \quad (4)$$

is the required differential equation and (1) is its general solution.

Example: Find the differential equation of all parabolas whose axes are parallel to the $y-axis.$

Solution: The general equation of a parabola whose axis is parallel to the $y-axis$ is

$$y = ax^2 + bx + c. \quad (1)$$

In order to obtain its differential equation, we have to eliminate a,b,c from (1). For that, we need three more equations. Differentiating (1) successively, we have

$$\frac{dy}{dx} = 2ax + b,$$

$$\frac{d^2y}{dx^2} = 2a,$$

$$\frac{d^3y}{dx^3} = 0.$$

The last equation does not contain any of the constants a,b and c. Thus

$$\frac{d^3y}{dx^3}=0$$

is the differential equation of all parabolas whose axes are parallel to the $y-axis$.

Initial and Boundary Conditions

We have observed that the general solution of a differential equation contains the same number of arbitrary constants as the order of the equation. Sometimes we need to find the solutions of differential equations subject to supplementary conditions. Two types of conditions will be often encountered.

Initial Condition

It is often required to find the solution of a differential equation subject to certain conditions. Conditions related to a value of the independent variable, such as $y=y_0$ at $x=x_0$ (written as $y(x_0)=y_0$) and $\frac{dy}{dx}=y'(x_0)$ at $x=x_0$, where x_0 belongs to some interval $[\alpha,\beta]$, are known as initial conditions (or one- point boundary conditions), with x_0 as the initial point. An initial value problem consists of a differential equation (of any order) together with a collection of initial conditions that must be satisfied by the solution of the differential equation and derivatives at the initial point.

Boundary Conditions

The problem of finding the solution of a differential equation such that all the associated constraints relate to two different values of the independent variable is called a two-point boundary value problem (or simply a boundary value problem). The associated supplementary boundary conditions are called two-point boundary conditions.

Example: Solve $\frac{dy}{dx}=2x,$ (1)

such that $y(1)=4.$

Solution: This is an initial value problem. We note that $y=x^2+c,$ c being an arbitrary constant, is the general solution of (1). Since $y(1)=4,$ we have

$y(1)=1^2+c.$

Therefore, $4=1+c,$ or $c=3.$

Thus $y=x^2+3$ is the solution of the initial value problem (1).

Note that the general solution represents a family of parabolas for different values of $c.$ The solution $y=x^2+3$ is a particular member of the family that passes through $(1,4).$

Example: Solve $\dfrac{d^2y}{dx^2}+y=0,$ (1)

subject to the conditions $y\left(\dfrac{\pi}{4}\right)=\sqrt{2},\ y'\left(\dfrac{\pi}{4}\right)=\dfrac{1}{\sqrt{2}}.$

Solution: Since both the conditions relate to one value of $x,$ namely $x=\dfrac{\pi}{4},$ this is an initial value problem. We have already noted in a previous example that:

$$y=A\sin x+B\cos x \tag{2}$$

is the general solution of (1). Differentiating (2) w. r. t $x,$ we get

$$y'=A\cos x-B\sin x. \tag{3}$$

By the given conditions, we have, from (2) and (3) respectively,

$$y\left(\frac{\pi}{4}\right)=\sqrt{2}=A\sin\frac{\pi}{4}+B\cos\frac{\pi}{4}=\frac{A}{\sqrt{2}}+\frac{B}{\sqrt{2}},$$

$$y'\left(\frac{\pi}{4}\right)=\frac{1}{\sqrt{2}}=A\cos\frac{\pi}{4}-B\sin\frac{\pi}{4}=\frac{A}{\sqrt{2}}-\frac{B}{\sqrt{2}}.$$

Hence $A+B=2$ and $A-B=1$.

Solving for A and B we find that

$$A=\frac{3}{2}, B=\frac{1}{2}.$$

With these values of A *and* B, the particular solution of (1) is given as

$$y=\frac{3}{2}\sin x+\frac{1}{2}\cos x.$$

Example: Verify that $y=c_1\cos x$ and $y=c_2\sin x$ are solutions of $\frac{d^2y}{dx^2}+y=0.$ Find a particular solution of the equation satisfying the boundary conditions

$$y(0)=1, y\left(\frac{\pi}{2}\right)=2.$$

Solution: We have $y=c_1\cos x.$

Differentiating twice, we get

$$\frac{dy}{dx}=-c_1\sin x,$$

$$\frac{d^2y}{dx^2}=c_1\cos x=-y.$$

Or $\frac{d^2y}{dx^2}+y=0.$

Thus $y=c_1\cos x$ is a solution of $\frac{d^2y}{dx^2}+y=0.$ (1)

Similarly, we can check that $y=c_2\sin x$ is also a solution of (1)

The general solution of (1) is

$$y=c_1\cos x+c_2\sin x. \quad (2)$$

Applying the boundary conditions, we obtain from (2),

$$y(0)=1=c_1+0$$

and $$y\left(\frac{\pi}{2}\right)=2=0+c_2.$$

Thus $c_1=1$ and $c_2=2.$ Hence the particular solution of (1) satisfying the given conditions is

$$y=\cos x+2\sin x.$$

Example: Solve the boundary value problem $\frac{d^2y}{dx^2}+y=0,\ y(0)=1, y(\pi)=5.$

Solution: Applying the boundary conditions to the general solution

$$y=A\sin x+B\cos x$$

of the given equation, we have

$$y(0)=1=B,$$

$$y(\pi)=5=-B.$$

Thus, we obtain two values of B and we are unable to determine any definite value of $A.$ Hence the boundary value problem has no solution.

It follows that a boundary value problem need not always have a solution.

In the above section, we classified differential equations and saw how a differential equation can be formed by eliminating constants from a given functional equation. The problem of finding a general solution (integral) of a given differential may or may not exist. Even if the integral of a given equation exists, it may not be easy to find. We shall now discuss methods of solutions of special types of differential equations.

Separable Equations

Definition. A differential equation of the type:

$$F(x)G(x)dx+f(x)g(y)dy=0 \tag{1}$$

is called an equation with separable variables, or simply a separable equation. Equation (1) may be written as:

$$\frac{F(x)}{f(x)}dx+\frac{g(y)}{G(y)}dy=0,$$

which can be easily integrated.

Example: Solve $\frac{dy}{dx}=\frac{x^2}{y}$. (1)

Solution: Equation (1) can be written as

$$y\,dy=x^2dx.$$

Integrating both the sides, we get

$$\frac{y^2}{2}=\frac{x^3}{3}+c_1,$$

or $3y^2=2x^2+c$

which is the required solution of (1).

Example: Solve $\frac{dy}{dx}=\frac{1}{x\tan y}$.

Solution: The given equation is separable and can be written as

$$\frac{dx}{x}=\tan y\,dy.$$

Integrating, we have

$$\ln|x|=-\ln|\cos y|+c$$

or $\ln|x|+\ln|\cos y|=c,$

i.e., $\ln|x\cos y|=c,$

or $|x\cos y|=e^c=a(say)$

$x\cos y=a$ because $a>0.$

Example: Solve $x\sin y\,dx+\left(x^2+1\right)\cos y\,dy=0.$ (1)

Solution: Dividing (1) by $(x^2+1)\sin y$, we have:

$$\frac{x}{x^2+1}dx + \cot y\, dy = 0,$$

which is a separable equation. Therefore,

$$\int\frac{x\,dx}{x^2+1} + \int\cot y\, dy = c_1 = \ln|c_2|,$$

i.e., $$\frac{1}{2}\ln\left(x^2+1\right) + \ln|\sin y| = \ln|c_2|$$

or $$\ln\left(x^2+1\right) + 2\ln|\sin y| = 2\ln|c_2|.$$

Now, we can write $2\ln|\sin y| = \ln|\sin y|^2 = \ln(\sin y)^2 = \ln\sin^2 y,$

and $2\ln|c_2| = \ln|c_2|^2 = \ln c_2^2 = \ln c,$ where $c_2^2 = c > 0.$

Hence $\ln(x^2+1)\sin^2 y = c$ is the required solution.

Example: Solve the initial value problem

$$\frac{dy}{dx} = \frac{2x}{y+x^2y},\ y(0) = -2.$$

Solution: We have $$\frac{dy}{dx} = \frac{2x}{y\left(1+x^2\right)}$$

or $$y\,dy = \frac{2x}{1+x^2}dx.$$

Integrating, we obtain

$$\frac{y^2}{2} = \ln(1+x^2) + \ln c,$$ c being a constant.

$$= \ln\left(c\left(1+x^2\right)\right)$$

$$y^2 = \ln(c^2\left(1+x^2\right)^2) \qquad \text{(i)}$$

Now setting $y(0)=-2$ into (1), we have:

$4=\ln c^2\ or\ c^2=e^4.$

So (i) becomes

$$y^2=\ln\left(e^4\left(1+x^2\right)^2\right).$$

Which is the required solution.

Homogenous Equations

Definition. A function $f(x,y)$ is homogeneous of degree n if:

$f(tx,ty)=t^n f(x,y),$

where t is a non-zero real number. Thus $\sqrt{xy},\frac{x^{10}+y^{10}}{x^2+y^2}\ and\ \sin\left(\frac{x}{y}\right)$ are homogeneous functions of degree 1,8 and 0 respectively.

A first order differential equation

$$\frac{dy}{dx}=f(x,y) \qquad (1)$$

is said to be a homogeneous function of any degree. If (1) is written in the form

$M(x,y)dx+N(x,y)dy=0,$

it is a homogeneous function of the same degree.

The equation $\frac{dy}{dx}=\ln x-\ln y+\frac{x+y}{x-y}=\ln\frac{1}{\frac{y}{x}}+\frac{1+\frac{y}{x}}{1-\frac{y}{x}}$ is homogenous, but

$\frac{dy}{dx}=\frac{y^3+2xy}{x^2}=y\left(\frac{y}{x}\right)^2+2\left(\frac{y}{x}\right)$ is not homogeneous.

Theorem. A homogeneous equation $\frac{dy}{dx}=g\left(\frac{y}{x}\right)$ (1)

can be transformed into a separable equation (in terms of variables v and x) by the substitution $y=vx$.

Proof: Substitute $y=vx$ into (1). Then $\frac{dy}{dx}=v+x\frac{dv}{dx}$ and (1) becomes

$$v+\frac{xdv}{dx}=g(v),$$

or $$v-g(v)+x\frac{dv}{dx}=0,$$

or $$\left[v-g(v)\right]dx+x\,dv=0.$$

This equation is separable and can be solved as in the previous section.

Example: Solve $\frac{dy}{dx}=\frac{x^3+y^3}{x^2y+xy^2}$.

Solution: We have $$\frac{dy}{dx}=\frac{1+\left(\frac{y}{x}\right)^3}{\left(\frac{y}{x}\right)+\left(\frac{y}{x}\right)^2}.$$

So that the equation is homogeneous. Setting $y=vx$ into (1), we obtain

$$v+x\frac{dv}{dx}=\frac{1+v^3}{v+v^2}=\frac{1+v^3}{v(1+v)}=\frac{v^2-v+1}{v},$$

$$x\frac{dv}{dx}=\frac{v^2-v+1}{v}-v=\frac{1-v}{v}.$$

Or $$\frac{v}{1-v}dv=\frac{dx}{x}.$$

Or $$\left(-1+\frac{1}{1-v}\right)dv=\frac{dx}{x},$$

integrating, we get:

$$-v-\ln|1-v|=\ln|x|+\ln|c|,$$

or $$ln|cx(1-v)|+v=0.$$

Replacing $v by \frac{y}{x}$ in the above equation, we have:

$$\ln|c(x-y)|+\frac{y}{x}=0,$$

which is the required solution.

Example: Solve the initial value problem $(x^2+3y^2)dx-2xy\,dy=0$, $y(2)=6$.

Solution: Here $$\frac{dy}{dx}=\frac{x^3+3y^2}{2xy}=\frac{3\left(\frac{y}{x}\right)^2+1}{2\left(\frac{y}{x}\right)},\qquad (1)$$

which shows that the equation is homogeneous. Putting $y=vx$ into (1), we have

$$v+x\frac{dv}{dx}=\frac{3v^2+1}{2v},$$

$$x\,.\frac{dv}{dx}=\frac{3v^2+1}{2v}-v=\frac{v^2+1}{2v}.$$

Or $$\frac{2v}{1+v^2}dv=\frac{dx}{x},$$

on integrating, we get

$$\ln(1+v^2)=\ln|x|+\ln|c|.$$

Or $$1+v^2=|cx|,$$

replacing we obtain:

$$1+\frac{y^2}{x^2}=|cx|,$$

$$or\, x^2+y^2=|cx|x^2.$$

Now if $x\geq 0,$ we can write this as

$$x^2+y^2=cx^3.$$

Or $\quad y^2=cx^3-x^2.$

Or $\quad y=\pm\sqrt{cx^3-x^2}.$

Applying the initial condition, we get:

$$y(2)=6=\pm\sqrt{8c-4},$$

or $\quad 8c-4=36\, i.e., c=5.$

Hence $y=\sqrt{5x^3-x^2}$ is the required solution. We take the plus sign in the radical, since $y(2)$ is positive.

Differential Equations Reducible to Homogeneous Form

Some differential equations are not homogeneous, however, they can be reduced to homogeneous form.

The differential equation $(a_1x+b_1y+c_1)dx+(a_2x+b_2y+c_2)dy=0$ is not homogeneous. However, it can be reduced to a homogeneous form as illustrated below.

Case 1. If $\frac{a_1}{a_2}\neq\frac{b_1}{b_2}$, then apply the transformations

$$x=X+h, y=Y+k.$$

The given equation then becomes:

$$(a_1X+b_1Y+a_1h+b_1k+c_1)dX+(a_2X+b_2Y+a_2h+b_2k+c_2)dY=0 \quad (1)$$

Let $h\,and\,k$. be the solution of the system of equations:

$$\left.\begin{array}{l} a_1h+b_1k+c_1=0 \\ a_2h+b_2k+c_2=0 \end{array}\right\}.$$

Then for these values of $h\,and\,k,$ (1) reduces to the homogenous equation

$(a_1X + b_1Y)\,dX+(a_2X + b_2Y)dY=0,$

for the variables $X\,and\,Y.$

Case II. If $\frac{a_1}{a_2}=\frac{b_1}{b_2},$ then take $z=a_1x+b_1y$ and the given equation will reduce to a separable equation in the variables x and $z.$

Example: Solve $\frac{dy}{dx}=\frac{2y-x+5}{2x-y-4}.$ (1)

Solution: Use $x=X+h, y=Y+k.$

Hence, the given equation becomes:

$$\frac{dY}{dX}=\frac{2Y-X+2K-h+5}{2X-Y+2h-k-4}. \tag{2}$$

The solution of the system of equations

$$\left.\begin{array}{l} -h+2k+5=0 \\ 2h-\ k-4=0 \end{array}\right\}$$

is therefore $h=1,k=-2.$

For these values of h and $k,$ (2) reduce to:

$$\frac{dY}{dX}=\frac{2Y-X}{2X-Y}, \tag{3}$$

which is homogeneous.

Substituting $Y=vX$ into (3), we have

$v+X\frac{dv}{dX}=\frac{2v-1}{2-v},$ which is separable. Therefore,

$$\left(\frac{2-v}{v^2-1}\right)dv=\frac{dX}{X}.$$

Integrating, we obtain

$$\ln\left|\frac{v-1}{v+1}\right|-\frac{1}{2}\ln\left|v^2-1\right|=\ln|X|+\ln|c_0|,$$

$$\ln\left|\frac{v-1}{v+1}\right|^2=\ln|X|^2+\ln|c_0|^2+\ln\left|v^2-1\right|,$$

$$\ln\left|\frac{v-1}{v+1}\right|^2=\ln\left\{|c_0X|^2.\left|v^2-1\right|\right\},$$

$$\left|\frac{v-1}{v+1}\right|^2=c|X|^2.|v-1|.|v+1|,\ c=c_0^2.$$

Or $|v-1|=c|X|^2.|v+1|^3$.

Replacing $v\,by\,\frac{Y}{X}$, this becomes:

$$\left|\frac{Y-X}{X}\right|=c|X|^2.\left|\frac{Y+X}{X}\right|^3.$$

Or $\quad |Y-X|=c|Y+X|^3 \quad (4)$

But $X=x-1,Y=y+2$. Hence (4) takes the form

$$|y-x+3|=c\,|y+x+1|^3.$$

Which is the required solution.

Example: Solve $(2x+y+1)dx+(4x+2y-1)dy=0$. (1)

Solution: Since $\frac{a_1}{a_2}=\frac{2}{4}=\frac{b_1}{b_2}=\frac{1}{2}$ we apply $2x+y=z$, and (1) becomes

$$(z+1)dx+(2z-1)(dz-2dx)=0,$$

or $(z+1-4z+2)dx+(2z-1)dz=0,$

i.e., $3(1-z)dx+(2z-1)dz=0$ which is separable.

Dividing by $1-z,$ we have

$$3\,dx+\frac{2z-1}{1-z}dz=0.$$

Or
$$3\,dx+\left(-2+\frac{1}{1-z}\right)dz=0,$$

integrating, we obtain:

$$3x-2z-\ln|1-z|=c_0$$

Replacing z by $2x+y,$ this reduces to:

$$3x-2(2x+y)-\ln|1-2x-y|=c_0.$$

Or
$$-x-2y-\ln|2x+y-1|=c_0,$$

or $\ln|2x+y-1|+x+2y=c,\ \ c=-c_0$,

which is the required solution.

Exact Equation

Definition. The expression $M(x,y)dx+N(x,y)dy$ (1)

is called an exact differential if there exists a continuously differentiable function $f(x,y)$ of two real variables x and $y,$ such that the expression equals the total differential $df.$ We know from calculus that:

$$df=\frac{\partial f}{\partial x}dx+\frac{\partial f}{\partial y}dy.$$

Thus, if (1) is exact then

$$M(x,y)=\frac{\partial f}{\partial x}=f_x \text{ and } N(x,y)=\frac{\partial f}{\partial y}=f_y.$$

If (1) is an exact differential then the differential equation

$$M(x, y)dx + N(x, y)dy = 0$$

is called an exact equation.

Theorem. The differential equation $M(x, y)dx + N(x, y)dy = 0$ (1)

is an exact equation if and only if $\frac{\partial M}{\partial y} = \frac{\partial N}{\partial x}$,

where the functions $M(x, y)$ and $N(x, y)$ have continuous first order partial derivatives.

Proof. Suppose that equation (1) is exact, so that $M\,dx + Mdy$ is an exact differential. By definition, there exists a function $f(x, y)$ such that:

$$M(x, y) = \frac{\partial f}{\partial x} = f_x \text{ and } N(x, y) = \frac{\partial f}{\partial y} = f_y.$$

Then $M_y = \frac{\partial M}{\partial y} = f_{xy} = \frac{\partial^2 f}{\partial y\,\partial x}$ and $N_x = \frac{\partial N}{\partial x} = f_{yx} = \frac{\partial^2 f}{\partial x\,\partial y}$.

Since M and N contain continuous first order partial derivatives, we have $f_{xy} = f_{yx}$ and, therefore

$\frac{\partial M}{\partial y} = \frac{\partial N}{\partial x}$, as desired.

Solution of an Exact Equation.

If $M(x, y)dx + N(x, y)dy = 0$ (1)

is an exact equation, then there exists a function $f(x, y)$ such that:

$$df = \frac{\partial f}{\partial x}dx + \frac{\partial f}{\partial y}dy$$

$$= M(x, y)dx + N(x, y)dy$$

Therefore $\frac{\partial f}{\partial x}=M$ and $\frac{\partial f}{\partial y}=N$.

Integrating $\frac{\partial f}{\partial x}=M$ with respect to x, we have

$f(x,y)=\int M\,dx+hs(y).$ (2)

The constant of integration $h(y)$ is an arbitrary function of y since it must vanish under differentiation w.r.t x.

Differentiating (2) $w.r.t.\,y$, we get

$$\frac{\partial f}{\partial y}=\frac{\partial}{\partial y}\left(\int Mdx\right)+h'(y),$$

i.e., $\frac{\partial f}{\partial y}=N=\frac{\partial}{\partial y}\left(\int Mdx\right)+h'(y),$

so $h'(y)=N-\frac{\partial}{\partial y}\left(\int Mdx\right).$

Integrating the above equation $w.r.t\,y,$ we obtain h and hence $f(x,y)=c$ is the required solution of (1).

Example: Solve $\left(3x^2y+2\right)dx+\left(x^3+y\right)dy=0$. (1)

Solution: Here $M=3x^2y+2$ and $N=x^3+y$

- $\frac{\partial M}{\partial y}=3x^2,\ \frac{\partial N}{\partial x}=3x^2.$

Thus $\frac{\partial M}{\partial y}=\frac{\partial N}{\partial x}$ and so the equation is exact.

To find the solution of (1), we note that the left-hand side of the equation is an exact differential. Therefore, there exists a function $f(x,y)$ such that:

$\frac{\partial f}{\partial x}=3x^2y+2$ (2)

and $\frac{\partial f}{\partial y} = x^3 + y$ (3)

Integrating (2) $w.r.t\, x$, we have

$$f(x,y) = x^3 y + 2x + h(y),$$

where $h(y)$ is the constant of integration. Differentiating the above equation $w.r.t\, y$ and using (3), we obtain:

$$\frac{\partial f}{\partial y} = x^3 + h'(y) = x^3 + y,$$

or $h'(y) = y.$

Integrating, we have:

$$h(y) = \frac{y^2}{2}.$$

Thus $f(x,y) = x^3 y + 2x + \frac{y^2}{2}:$

Hence the general solution of (1) is:

$$x^3 y + 2x + \frac{y^2}{2} = c.$$

Alternative Method:

Integrating (2) and (3) $w.r.t\, x$ and y respectively, we have:

$$f(x,y) = x^3 y + 2y + h(y)$$

and $f(x,y) = x^3 y + \frac{y^2}{2} + g(x).$

Thus $h(y) = \frac{y^2}{2}$ and $g(x) = 2x$.

Hence, the general solution is:

$$x^3 y+2x+\frac{y^2}{2}=c.$$

Example: Solve the initial value problem

$$\left(2y\sin x\cos x+y^2\sin x\right)dx+\left(\sin^2 x-2y\cos x\right)dy=0, \quad y(0)=3.$$

Solution: Here $M=2y\sin x\cos x+y^2\sin x$

and $N=\sin^2 x-2y\cos x$.

$$\frac{\partial M}{\partial y}=2\sin x\cos x+2y\sin x$$

$$\frac{\partial N}{\partial x}=2\sin x\cos x+2y\sin x.$$

Thus $\frac{\partial M}{\partial y}=\frac{\partial N}{\partial x}$ showing that the given equation is exact.

Hence there exists a function $f(x,y)$ such that:

$$\frac{\partial f}{\partial x}=2y\sin x\cos x+y^2\sin x \qquad (1)$$

$$\frac{\partial f}{\partial y}=\sin^2 x-2y\cos x. \qquad (2)$$

Integrating (1) $w.r.t\ x$, we have:

$$f(x,y)=y\sin^2 x-y^2\cos x+h(y).$$

Differentiating this equation $w.r.t\ y$ and using (2), we get:

$$\sin^2 x-2y\cos x+h'(y)=\sin^2 x-2y\cos x,$$

i.e., $h'(y)=0$ and so $h(y)=c_1$.

Hence the general solution of the given equation is:

$$f(x,y)=c_2,$$

i.e., $y\sin^2 x - y^2 \cos x + c_1 = c_2$

$y.\sin^2 x - y^2.\cos x = c_2 - c_1 = c.$

Applying the initial condition of $x = 0, y = 3,$ we have,

$-9 = c.$

Hence $y^2 \cos x - y\sin^2 x = 9$ is the required solution.

Integrating Factors

Definition. Consider the non-exact differential equation

$$M(x, y)dx + N(x, y)dy = 0 \qquad (1)$$

If it is multiplied by a function $\mu(x, y)$ and the resulting equation $\mu(x, y)M(x, y)dx + \mu(x, y)N(x, y)dy = 0$ is exact, then $\mu(x, y)$ is called an integrating factor (I.F) of the differential equation (1). The number of integrating factors of an equation may be infinite.

We list below (without proofs) several rules that can be used to find the integrating factors of special types of equations.

Rule 1. If $M(x, y)dx + N(x, y)dy = 0$ (1)

is not exact and $\frac{M_{y - N_x}}{N} = P,$ where P is a function of x only, then (1) has an integrating factor $\mu(x)$ which also depends on x. $\mu(x)$ is solution of the differential equation

$$\frac{d\mu}{dx} = P\mu$$

i.e., $\mu(x) = \exp \int P\, dx.$

Note that $M_y = \frac{\partial M}{\partial y}, N_x = \frac{\partial N}{\partial x}.$

Rule II. If $\dfrac{N_x - M_y}{M} = Q,$ where Q is a function of y only, then the differential equation

$Mdx + Ndy = 0$ has an integrating factor

$\mu(y) = \exp \int Q\,dy.$

Rule III. If $M\,dx + Ndy = 0$ (1)

is homogeneous and $xM + yN \neq 0,$ then $\dfrac{1}{xM + yN}$ is an I.F of (1).

Rule IV. If $M\,dx + Ndy = 0$ is of the form $y\,f(x, y)\,dx + x\,g(x, y)\,dy = 0$

and $xM - yN \neq 0,$ then $\dfrac{1}{xM - yN}$ is an I.F of (1).

The following differential formulas are useful in the calculation of certain exact equations:

$$d\left(\frac{y}{x}\right) = \frac{x\,dy - y\,dx}{x^2}$$

$$d\left(\frac{x}{y}\right) = \frac{y\,dx - x\,dy}{y^2}$$

$$d(xy) = x\,dy + y\,dx$$

$$d\left(x^2 + y^2\right) = 2\left(x\,dx + y\,dy\right)$$

$$d\left(\ln\frac{x}{y}\right) = \frac{y\,dx - x\,dy}{xy}$$

$$d\left(arc\tan\frac{x}{y}\right) = \frac{y\,dx - x\,dy}{x^2 + y^2}$$

Example: Solve $y\,dx + \left(x^2 y - x\right)dy = 0$.

Solution. The equation is not exact. Rearranging the equation, we have:

$y\,dx - x\,dy + x^2 y\,dy = 0.$

Or $$\frac{y\,dx - x\,dy}{x^2} + y dy = 0.$$

Now it is an exact equation and may be written as $-d\left(\frac{y}{x}\right) + y\,dy = 0.$

Integrating, we have:

$$-\frac{y}{x} + \frac{y^2}{2} = c,$$

or $xy^2 - 2y = cx$ is the general solution.

Example: Solve $\left(x^2 - 2x + 2y^2\right)dx + 2xy\,dy = 0$. (1)

Solution. Here $$\frac{M_y - N_x}{N} = \frac{4y - 2y}{2xy} = \frac{1}{x}.$$

Therefore, the I.F $\mu(x, y)$ is the solution of $\frac{d\mu}{dx} = \frac{\mu}{x}$,

or $\mu = x$ is an I.F.

Multiplying the equation by x, we have:

$\left(x^3 - 2x^2 + 2xy^2\right)dx + 2x^2 y = 0.$

This equation is exact. Hence, we can easily determine its solution as

$$\frac{x^4}{4} - \frac{2x^3}{3} + x^2 y^2 = c_0,$$

or $3x^4 - 8x^3 + 12x^2 y^2 = c$ as the solution of (1).

Example: Solve $dx + \left(\frac{x}{y} - \sin y\right)dy = 0$.

Solution. Here, by rule II, we have:

$$\frac{Nx - M_y}{M} = \frac{\frac{1}{y} - 0}{1} = \frac{1}{y},$$

which is a function of y only. Therefore, $\mu(y) = \exp \int \frac{dy}{y} = e^{\ln y} = y$

is an I.F. Multiplying the original equation by y, we have:

$y\,dx + (x - y\sin y)\,dy = 0,$

or $y\,dx + x\,dy - y\sin y\,dy = 0,$

or $d(xy) - y\sin y\,dy = 0.$

Integrating, we get $xy + y\cos y - \sin y = c,$

which is the required solution.

Example: Solve $(x^2y - 2xy^2)\,dx - (x^3 - 3x^2y)\,dy = 0.$ (1)

Solution: The equation is homogeneous but not exact. We have:

$xM + yN = x^3y - 2x^2y^2 - x^3y + 3x^2y^2 = x^2y^2 \neq 0.$

So, using Rule III, we have

$\frac{1}{xM + yN} = \frac{1}{x^2y^2}$ is an I.F. Therefore, multiplying (1) by $\frac{1}{x^2y^2}$, we obtain

$$\left(\frac{1}{y} - \frac{2}{x}\right)dx - \left(\frac{x}{y^2} - \frac{3}{y}\right)dy = 0.$$

This equation is exact. Integrating, we get:

$$\frac{x}{y} - 2\ln|x| + 3\ln|y| = c,$$

is the required solution.

Note that, here the term $\frac{x}{y}$, involving both x and y has been taken only once.

Example: Solve $y(xy + 2x^2y^2)\,dx + x(xy - x^2y^2)\,dy = 0$ (1)

Solution: The equation is of the form $yf(xy)dx + x\,g(xy)dy = 0.$

Now, $xM - yN = x^2y^2 + 2x^3y^3 - x^2y^2 + x^3y^3 = 3x^3y^3 \neq 0.$

Therefore, $\frac{1}{x^3y^3}$ is an I.F.

Multiplying (1) by $\frac{1}{x^3y^3},$ we have

$$\left(\frac{1}{x^2y} + \frac{2}{x}\right)dx + \left(\frac{1}{xy^2} - \frac{1}{y}\right)dy = 0.$$

This is an exact equation. Integrating, we get:

$$-\frac{2}{xy} + 2\ln|x| - \ln|y| = c$$

as the required solution.

Linear Equations

A first order ordinary differential equation (ODE) is linear in both the dependent variable y and the independent variable x if it can be written in the form:

$$\frac{dy}{dx} + P(x)y = Q(x), \tag{1}$$

where $P\,and\,Q$ are functions of x.

Solution of a Linear Equation.

The linear equation $\frac{dy}{dx} + P(x) = Q(x)$ may be written as

$$\left[P(x)y - Q(x)\right]dx + dy = 0. \tag{2}$$

This is of the form $M\,dx + N\,dy = 0,$ where

$M = P(x)y - Q(x)\,and\,N = 1.$

Now $\frac{\partial M}{\partial y} = P(x)$ and $\frac{\partial N}{\partial x} = 0$.

Thus (2) is not exact unless $P(x) = 0,$ in which case (1) is separable. However, an integrating factor (depending only on x) of (2) may be easily found. Let $\mu(x)$ be an I.F. of (2). Multiplying (2) by $\mu(x),$ we get:

$$\left[\mu(x)P(x)y - \mu(x)Q(x)\right]dx + \mu(x)dy = 0. \tag{3}$$

Now (3) is an exact equation if and only if

$$\frac{\partial}{\partial y}\left[\mu(x)P(x)y - \mu(x)Q(x)\right] = \frac{\partial}{\partial x}\left[\mu(x)\right].$$

This condition reduces to $\mu(x)P(x) = \frac{d}{dx}\left[\mu(x)\right],$

i.e., $\mu P(x) = \frac{d\mu}{dx},$

or $\frac{d\mu}{\mu} = P(x)dx.$

Integrating, we obtain

$\ln|\mu| = \int P(x)dx$

or $\mu = \exp[\int P(x)dx\ ] > 0.$

Thus $e^{\int P\,dx}$ *or* $[\exp \int P(x)dx\ ]$ is an I.F. of the linear equation (1).

Multiplying (1) by $e^{\int P\,dx}$, we have $e^{\int P\,dx}\frac{dy}{dx} + e^{\int P\,dx}P(x)y = Q(x)e^{\int P\,dx},$

or $\frac{d}{dx}\left[ye^{\int P\,dx}dx\right] = Q(x)e^{\int P\,dx}.$

Integrating this we obtain the solution of (1) in the form:

$ye^{\int P\,dx} = \int[Q(x)e^{\int P\,dx}]dx + c.$

Example: Solve $(x-1)^3 \frac{dy}{dx} + 4(x-1)^2 y = x+1.$ (1)

Solution: We write the equation in the standard form:

$$\frac{dy}{dx} + \frac{4}{x-1} y = \frac{x+1}{(x-1)^3}.$$

Here $P(x) = \frac{4}{x-1}.$ Therefore, an I.F. of (1) is

$$\exp[\int \frac{4\,dx}{x-1}] = \exp\left[\ln(x-1)^4\right] = (x-1)^4.$$

Multiplying (1) by this I.F., we get

$$(x-1)^4 \frac{dy}{dx} + 4(x-1)^3 y = x^2 - 1,$$

or $\frac{d}{dx}\left[y(x-1)^4\right] = x^2 - 1.$

Integrating, we obtain

$$y(x-1)^4 = \frac{x^3}{3} - x + c,$$

which is the required solution.

Example: Solve $\left(x + 2y^3\right)\frac{dy}{dx} = y.$

Solution: We have $\frac{dy}{dx} = \frac{y}{x + 2y^3}.$

This equation is clearly not linear in y. However, in the first order differential equation, the roles of x and y are interchangeable in the sense that either variable may be regarded as the dependent variable. Let us regard x as dependent variable and y as independent variable. The equation may be written as

$$\frac{dx}{dy} - \frac{1}{y} x = 2y^2, \qquad (1)$$

which is linear in x with

$$I.F. = \exp\left[\int\left(-\frac{1}{y}\right)dy\right] = \exp\left[\ln\frac{1}{y}\right] = \frac{1}{y}.$$

Multiplying (1) by $\frac{1}{y}$, we get

$$\frac{1}{y}\frac{dx}{dy} - \frac{1}{y^2}x = 2y,$$

or $\frac{d}{dx}\left(\frac{x}{y}\right) = 2y.$

Integrating, we have $\frac{x}{y} = y^2 + c$ or $x = y\left(y^2 + c\right),$ the required solution.

Example: Solve the initial value problem

$$\frac{dr}{d\theta} + r\tan\theta = \cos^2\theta, \quad r\left(\frac{\pi}{4}\right) = 1.$$

Solution. The equation is linear in r with

$$I.F. = \exp\left[\int\tan\theta\, d\theta\right] = \exp\left[\ln\sec\theta\right] = \sec\theta.$$

Multiplying the given equation by $\sec\theta$, we have

$$\sec\theta\frac{dr}{d\theta} + r\sec\theta\tan\theta = \cos\theta$$

$$\frac{d}{d\theta}\left[r\sec\theta\right] = \cos\theta.$$

Integrating, we obtain

$$r\sec\theta = \sin\theta + c.$$

Applying the initial conditions, we have:

$$1.\sec\left(\frac{\pi}{4}\right) = \sin\left(\frac{\pi}{4}\right) + c,$$

or $c=\sqrt{2}-\dfrac{1}{\sqrt{2}}=\dfrac{1}{\sqrt{2}}$.

Therefore, $r\sec\theta=\sin\theta+\dfrac{1}{\sqrt{2}}$,

$$r=\sin\theta\cos\theta+\frac{\cos\theta}{\sqrt{2}},$$

or $2r=\sin 2\theta+\sqrt{2}\cos\theta$, the required solution.

The Bernoulli Equation

Definition: An equation of the form $\dfrac{dy}{dx}+P(x)y=Q(x)y^n$ (1)

is defined as a Bernoulli differential equation. This equation is linear if $n=0$ or 1. If n is not zero or 1, then (1) is reducible to a linear equation. Dividing by y^n, (1) becomes

$$y^{-n}\frac{dy}{dx}+P(x)y^{1-n}=Q(x). \quad (2)$$

Substituting $v=y^{1-n}$ into (2) reduces (1) to

$$\frac{dv}{dx}+(1-n)P(x)v=(1-n)Q(x),$$

which is linear in v.

Note. Consider the equation

$$f'(y)\frac{dy}{dx}+P(x)f(y)=Q(x).$$

Letting $v=f(y)$, this equation becomes $\dfrac{dv}{dx}+P(x)v=Q(x)$, which is linear in v.

Example: Solve $\dfrac{dy}{dx}+\dfrac{xy}{1-y^2}=xy^{\frac{1}{2}}$. (1)

Solution: Dividing by $y^{\frac{1}{2}}$, (1) becomes

$$y^{-\frac{1}{2}}\frac{dy}{dx}+\frac{x}{1-x^2}y^{\frac{1}{2}}=x. \qquad (2)$$

Putting $y^{\frac{1}{2}}=v,$

or $\frac{1}{2}y^{-\frac{1}{2}}\frac{dy}{dx}=\frac{dv}{dx}$,

then (2) reduces to

$$\frac{dv}{dx}+\frac{x}{2\left(1-x^2\right)}v=\frac{x}{2}. \qquad (3)$$

This is linear in v.

$$I.F.=\exp\left[\int\frac{x}{2\left(1-x^2\right)}dx\right]=\exp\left[-\frac{1}{4}\ln\left(1-x^2\right)\right]=\left(1-x^2\right)^{-\frac{1}{4}}.$$

Multiplying (3) by $\left(1-x^2\right)^{-\frac{1}{4}}$, we get

$$\left(1-x^2\right)^{-\frac{1}{4}}\frac{dv}{dx}+\frac{x}{2\left(1-x^2\right)^{\frac{5}{4}}}v=\frac{x}{2\left(1-x^2\right)^{\frac{1}{4}}}$$

or $\frac{d}{dx}\left[\left(1-x^2\right)^{-\frac{1}{4}}v\right]=-\frac{1}{4}\left[-2x\left(1-x^2\right)^{-\frac{1}{4}}\right]$.

Integrating, we have

$$v\left(1-x^2\right)^{-\frac{1}{4}}=-\frac{1}{4}\frac{\left(1-x^2\right)^{\frac{3}{4}}}{\frac{3}{4}}+c,$$

or $v=c\left(1-x^2\right)^{\frac{1}{4}}-\frac{1-x^2}{3}$

or $y^{\frac{1}{2}}=c\left(1-x^2\right)^{-\frac{1}{4}}-\frac{1-x^2}{3},$

the required solution of (1).

Equation Solvable for *p*

Example: Solve $x^2p^2+xp-y^2-y=0.$ (1)

Solution: We factorize the left-hand side of (1) to obtain:

$$\left(x^2p^2-y^2\right)+\left(xp-y\right)=\left(xp-y\right)\left(xp+y+1\right)=0.$$

Therefore, either

$xp-y=0$ (2)

or $xp+y+1=0.$ (3)

(2) gives $x\frac{dy}{dx}=y$ or $\frac{dy}{dx}=\frac{dx}{x}$

or $y=cx.$ (4)

From (3), we have $xp=-\left(y+1\right)$

or $x\frac{dy}{dx}=-\left(y+1\right)$ or $\frac{dy}{y+1}=-\frac{dx}{x}$

which yields $x\left(y+1\right)=c.$ (5)

Combining (4) and (5), the required solution of (1) is

$$\left(y-cx\right)\left(xy+x-c\right)=0.$$

Example: Solve $xp^3-\left(x^2+x+y\right)p^2+\left(x^2+xy+y\right)p-xy=0.$ (1)

Solution: By inspection, we find that $p-1$ is a factor of left hand side of (1). Thus, the given equation is

$$\left(p-1\right)\left[xp^2-\left(x^2+y\right)p+xy\right]=0,$$

or $\left(p-1\right)\left(xp-y\right)\left(p-x\right)=0.$

Therefore, either

$p-1=0$ (2)

or $xp-y=0$ (3)

or $p-x=0$ (4)

From (2), we have $\frac{dy}{dx}=1 or \ y=x+c.$ (5)

From (3), we get $x\frac{dy}{dx}=y \ or \ y=cx.$ (6)

0From (4), we obtain $\frac{dy}{dx}=x$.

or $y=\frac{x^2}{2}+c \quad or \ x^2+2(y-c)=0.$ (7)

The general solution of (1) is obtained by combining (5), (6) and (7). Thus

$$(y-x-c)(y-cx)\left(x^2-2y+2c\right)=0.$$

is the general solution of (1).

Equations solvable for *y*

Example: Solve $y=p^2x+p.$ (1)

Solution. Differentiating (1) $w.r.t.x,$ we have:

$$\frac{dy}{dx}=p=p^2+2xp\frac{dp}{dx}+\frac{dp}{dx},$$

or $\frac{dp}{dx}(2px+1)+p^2-p=0,$

$$\frac{dp}{dx}=\frac{-p(p-1)}{2px+1}$$

$$\frac{dx}{dp}=\frac{-2px-1}{p(p-1)},$$

i.e., $$\frac{dx}{dp}+\frac{2x}{p-1}=-\frac{1}{p(p-1)},\tag{2}$$

which is linear in x.

$$I.F.=\exp\left[\int\frac{2}{p-1}dp\right]=\exp\left[\ln(p-1)^2\right]=(p-1)^2.$$

Multiplying (2) by $(p-1)^2$ and integrating, we get

$$x(p-1)^2=c-p+\ln p,\tag{3}$$

$$x=\frac{c-p+\ln p}{(p-1)^2}.$$

Substituting this value of x into (1), we have

$$y=p^2\frac{c-p+\ln p}{(p-1)^2}+p$$

$$y(p-1)^2=p^2(c-p+\ln p)+p(p-1)^2.\tag{4}$$

Thus, (3) and (4) constitute the required solution of (1) with p as a parameter.

Example: Solve $y+px=p^2x^4$. (1)

Solution. From (1), we get

$$y=p^2x^4-px.$$

Differentiating the above equation $w.r.t.x,$ we have

$$\frac{dy}{dx}=p=4x^3p^2+2px^4\frac{dp}{dx}-p-x\frac{dp}{dx},$$

$$2p-4x^3p^2-2x^4p\frac{dp}{dx}+x\frac{dp}{dx}=0,$$

$$2p\left(1-2px^3\right)+x\left(1-2px^3\right)\frac{dp}{dx}=0,$$

$$\left(1-2px^3\right)\left(2p+x\frac{dp}{dx}\right)=0.$$

Hence, either $1-2px^3=0,$

or $2p+x\dfrac{dp}{dx}=0.$ (2)

Equation (2) gives

$$\frac{dp}{p}+2\frac{dx}{x}=0.$$

Or $\ln p+2\ln x=\ln c.$

Or $px^2=c$ or $p=\dfrac{c}{x^2}.$

Substituting this value of p into (1), we obtain

$$y=c^2-\frac{c}{x}.$$

Or $xy-c^2x+c=0,$ the required solution.

We have yet to deal with relation (2). If we eliminate p from (1) and (2), we get

$y=-\dfrac{1}{4x^2}$ (3)

It is easy to check that (3) is also a solution of (1) and this solution does not involve any constant.

Equations Solvable for x

Example: Solve $xp=1+p^3.$ (1)

Solution: We have $x=\dfrac{1}{p}+p$ (2)

Differentiating (1) $w.r.t.y,$ we have:

$$\frac{dx}{dy} = \frac{1}{p} = -\frac{1}{p^2}\frac{dp}{dy} + \frac{dp}{dx}.$$

Or $\frac{1}{p} = \left(1 - \frac{1}{p^2}\right)\frac{dp}{dy}.$

Or $1 = \left(p - \frac{1}{p}\right)\frac{dp}{dy}.$

Which is a separable equation. Therefore,

$$dy = \left(p - \frac{1}{p}\right)dp.$$

Integrating, we get $c + y = \frac{p^2}{2} - \ln p$,

i.e., $2y = p^2 - 2\ln p - 2c$. (3)

Thus, (2) and (3) constitute the solution of (1).

Clairaut's Equation

Definition. The equation $y = xp + f(p)$ (1)
is known as Clairaut's equation.

Differentiating both sides of (1), $w.r.t.x,$ we get

$$p = p + xp' + f'(p).p'$$

Cancelling like terms, we have:

$$xp' + p'f'(p) = 0.$$

Or $$p'\left[x + f'(p)\right] = 0.$$

Since one of the factors must be zero, two different solution arise.

- If $p' = 0,$ then $p = c$ and substitution of this value into (1) yields the general solution: $y = cx + f(c).$

- If $x+f'(p)=0$, then $x=-f'(p)$ and (1) can be rewritten as $y=-pf'(p)+f(p)$.

Thus x and y are both expressed as functions of p, and we obtain the parametric equations

$$\left.\begin{aligned} x &= -f'(p) \\ y &= f(p)-pf'(p) \end{aligned}\right\} \tag{2}$$

of a curve, representing a solution of (1). The solution (2) is known as the singular solution. This solution is not deducible from the general solution, p may be eliminated between the two equations in (2) to obtain an equation with x and y involving no constant.

Example: Find the general solution and singular solution of $y=xp+\frac{1}{4}p^4$. (1)

Solution. The general solution of the equation is $y=cx+\frac{1}{4}c^4$.

Differentiating (1) $w.r.t.x$, we have

$p=p+xp'+p^3p'$,

Or $$x=-p^3. \tag{2}$$

We eliminate p from (1) and (2) and get

$$y=x(-x)^{\frac{1}{3}}+\frac{1}{4}(-x)^{\frac{4}{3}}=-\frac{3}{4}x(-x)^{\frac{1}{3}}.$$

Or $64y^3+27x^4=0$ is the singular solution.

Example: Solve $x^2(y-px)=yp^2$. (1)

Solution: This does not have the form of Clairaut's equation. Let us write

$x^2=u$ and $y^2=v$.

Then, $2x\,dx = du$ and $2y\,dy = dv$.

Hence $\frac{y}{x}\frac{dy}{dx} = \frac{dv}{du}$,

$$p = \frac{dy}{dx} = \frac{x}{y}\frac{dv}{du}.$$

Substituting this value of p into (1), we get

$$x^2\left(y - \frac{x^2}{y}\frac{dv}{du}\right) = \frac{x^2}{y}\left(\frac{dv}{du}\right)^2.$$

Or $y^2 - x^2\frac{dv}{du} = \left(\frac{dv}{du}\right)^2.$

Or $v - u\frac{dv}{du} = \left(\frac{dv}{du}\right)^2.$

Or $v = u\frac{dv}{du} + \left(\frac{dv}{du}\right)^2.$ (2)

This is Clairaut's equation. Hence its general solution is:

$$v = cu + c^2.$$

Or $y^2 = cx^2 + c$ is the general solution of (1).

Differentiating (2) $w.r.t\, u$, we get:

$$\frac{dv}{du} = u\frac{d^2v}{du^2} + \frac{dv}{du} + 2\frac{dv}{du}\frac{d^2v}{du^2}.$$

Or $\frac{d^2v}{du^2}\left(u + 2\frac{dv}{du}\right) = 0$, or $v = cu + c^2$, which is the general solution.

If $u + 2\frac{dv}{du} = 0,$ then

$$u = -2\frac{dv}{du} \qquad (3)$$

and $$v = u\frac{dv}{du} + \left(\frac{dv}{du}\right)^2 \qquad (4)$$

give the singular solution.

Eliminating $\frac{dv}{du}$ from (3) and (4), we obtain

$$v = -\frac{u^2}{4}.$$

Or $u^2 + 4v = 0$ is the singular solution.

Replacing u, v by x^2 and y^2 respectively, we have:

- $y^2 = cx^2 + c^2$ as the general solution.
- $x^4 + 4y^2 = 0$ as the general solution.

Envelope

Let $f(x, y, c) = 0$ be a one-parameter family of curves. Suppose all members of the family of curves are drawn for various values of the parameter *c,* arranged in order of magnitude. The two curves that correspond to two consecutive values of *c* will be denoted as neighboring curves. The locus of the ultimate points of intersection of neighboring curves is known as the envelope of the family $f(x, y, c) = 0.$

Let A, B, C represents three neighboring intersecting members of the family. Let *P* be the point of intersection of A and B, and Q be the point of intersection of B and C. By definition $P and Q$ are points on the envelope. Thus, the curve B and the envelope have two contiguous points in common, and therefore, a common tangent. Hence B and the envelope touch each other. In the same way, we can show that the envelope touches any other member of the family. Thus, a curve E, which, at each of its points, touches some of the curves of the family, is the envelope of the family.

Consider the family of circles with its center on the $x-axis$ and a radius of 1, that is, the family of circles with the equation $(x-1)^2+y^2=1$. The pair of lines $y=\pm 1$ is the envelope. It is easy to see that the lines $y=\pm 1$ touch each member of the family of the circles.

Equation of the Envelope

To find an equation of the envelope of the family $f(x,y,c)=0,$ consider two neighboring curves

$$\left.\begin{aligned} f(x,y,c)=0 \\ \text{and } f(x,y,c+h)=0 \end{aligned}\right\}. \qquad (1)$$

Find the intersection of these two curves and let $h \to 0$. The point of intersection must approach the point of contact of the curve $f(x,y,c)=0$ with the envelope. At the point of intersection, we have the equation

$$\frac{f(x,y,c+h)-f(x,y,c)}{h}=0, \qquad (2)$$

as well as (1). Letting $h \to 0,$ we have from (2)

$$f_c(x,y,c)=0 \qquad (3)$$

and from (1) $f(x,y,c)=0$. (4)

If we eliminate c from (3) and (4), we obtain an equation of the envelope. The eliminant is called the c - discriminant of the family $f(x,y,c)=0$.

The c-discriminant may contain loci other than the envelope.

Example: The family of parabolas

$(x-c)^2-2y=0$ has the x-axis as an envelope.

Singular Solution

Let $f(x,y,p)=0$ (1)

be a nonlinear first order differential equation in which the left-hand side is a polynomial in p. The general solution of this differential equation will be a one parameter family

$$f(x, y, c) = 0. \tag{2}$$

The envelope E of (2) is a curve which, at each of its points, touches some member of the family (2). At a point of contact P of the envelope and a member of (2), the values of x, y, p are the same. The values of x, y, p for the curve at P satisfy (1). Thus, the envelope of family (2) is a solution of the differential equation (1).

Example: Consider $p^2 - xp + y = 0.$

This is Clairaut's equation and its general solution (replacing p with c) is:

$$y = cx + c^2. \tag{1}$$

Now we find the envelope of the family (1)

$$f(x, y, c) \equiv y - cx + c^2 = 0 \tag{2}$$

$$f_c = \frac{\partial f}{\partial c} - x + 2c - 0. \tag{3}$$

Substituting the value of c from (3) into (2), we have:

$$y - x\left(\frac{x}{2}\right) + \left(\frac{x}{2}\right)^2 = 0$$

i.e., $y - \frac{x^2}{2} + \frac{x^2}{4} = 0$

or $4y = x^2,$

and (4) is an equation of the envelope of the family (1).

We check whether (4) is a solution of the given differential equation. Differentiating (4) $w.r.t.x,$ we get:

$$4\frac{dy}{dx} = 2x,$$

or

$2p = x$ (5)

Substituting (4) and (5) into $p^2 - px + y$, we have:

$$\frac{x^2}{4} - \frac{x^2}{2} + \frac{x^2}{4} = 0.$$

Thus (4) is a solution of the given differential equation. This solution does not involve any constant and it cannot be derived from the general solution by giving particular values to c. Such a solution is known as a singular solution (S.S).

Definition A solution of a differential equation $f(x, y, p) = 0$ is called a singular solution (S.S) if:

- It is not derived from the general solution by giving any particular value to the arbitrary constants.
- At each of its points, it is tangent to some member of the one parameter family of curves represented by the general solution.

We have seen that the envelope, of the one parameter family of curves represented by the general solution of a differential equation is a singular solution. Thus, the c- discriminant equation may contain a singular solution.

The singular solution may also be obtained from the differential equation directly without finding the general solution. At the ultimate point of intersection of the neighboring curves, the $p's$ for the intersecting curves become equal, and thus the locus of the points where the $p's$ have equal roots will include the envelope. If we eliminate p from: $f(x, y, p) = 0,\ \dfrac{\partial f}{\partial p} = 0.$

The resulting equation is called the p-discriminant. The p-discriminant represents the locus for each point of which $f(x, y, p) = 0$ has equal roots.

If the envelope of the general solution of $f(x, y, p) = 0$ exists, it will be contained in the p-discriminant. Thus, the p-discriminant equation may contain (i) a singular solution or (ii) solutions that are not singular and (iii) loci that are not solutions at all.

Example: Solve $xp^2 - 2yp + 4x = 0.$ (1)

Solution: $2yp = xp^2 + 4x$,

$$2y = xp + \frac{4x}{p}.$$

Differentiating $w.r.t.x,$ we get:

$$2p = x\frac{dp}{dx} + p + \frac{4}{p} - \frac{4x}{p^2}\frac{dp}{dx}.$$

Or $p - \frac{4}{p} = \left(x - \frac{4x}{p^2}\right)\frac{dp}{dx}$,

$$\frac{p^2-4}{p} = \frac{x\left(p^2-4\right)}{p^2}\frac{dp.}{dx}$$

Or $\frac{dp}{p} = \frac{dx}{x}.$

Or $p = cx$. (2)

Eliminating p from (1) and (2), we obtain:

$$x\left(c^2x^2\right) - 2ycx + 4x = 0.$$

Or $c^2x^2 - 2yc + 4 = 0.$ (3)

From (3), we obtain the c - discriminant as:

$4y^2 - 16x^2 = 0$ or $y^2 = 4x^2.$

The p-discriminant is from (1):

$4y^2 = 16x^2$ or $y^2 = 4x^2.$

Since the c-discriminant and p-discriminant are the same viz., $y^2 = 4x^2$, which satisfies the given differential equation, it is also the singular solution.

Example: Solve $x^2p^2 + yp\left(2x + y\right) + y^2 = 0,$ (1)

by making substitution $y=u, xy=v$ and find the singular solutions.

Solution: Let $y=u, x=\frac{v}{y}=\frac{v}{u}$.

Thus $\frac{dy}{du}=1, \frac{dx}{du}=\frac{u\,.\frac{dv}{du}-v}{u^2},$

$$p=\frac{dy}{dx}=\frac{dy}{du}.\frac{du}{dx}=\frac{u^2}{u.\frac{dv}{du}-v}.$$

Substituting into (1), we have

$$\frac{v^2}{u^2}\left[\frac{u^2}{u.\frac{dv}{du}-v}\right]^2+\left(2v+u^2\right).\frac{u^2}{u.\frac{dv}{du}-v}$$

Or $v^2+\left(2v+u^2\right)\left(u\frac{dv}{du}-v\right)+\left(u.\frac{dv}{du}-v\right)^2=0$

$$v^2+2uv\frac{dv}{du}-2v^2+u^3.\frac{dv}{du}-u^2v+u^2\left(\frac{dv}{du}\right)^2-2uv\frac{dv}{du}+v^2=0$$

$$v=u\frac{dv}{du}+\left(\frac{dv}{du}\right)^2.$$

Which is a Clairaut's equation. Its solution is:

$v=cu+c^2,$

i.e., $xy=cy+c^2,$ (2)

is the solution of (1).

- From (2), the c-discriminant is:

$y^2+4xy=0$ i.e., $y(y+4x)=0$.

Therefore, $y=0, y+4x=0$. (3)

Differentiating (3), we have $p=-4$. Substituting $p=-4$ into the left-hand side of (1) and using (3), we obtain:

$$16x^2-4x(-4)(-2x)+(-4x)^2=32x^2-32x^2=0.$$

Thus $y+4x=0$ is also a singular solution.

Example: Solve $\left(x^2-1\right)p^2-2xyp-x^2=0$ (1)

Moreover, find the singular solution, if any.

Solution: Solving the equation for y, we have

$$2y=\frac{\left(x^2-1\right)p^2-x^2}{xp}=xp-\frac{p}{x}-\frac{x}{p}.$$

Differentiating $w.r.t.x$, we get:

$$2p=xp'+p-\frac{xp'-p}{x^2}-\frac{p-xp'}{p^2},\quad \text{where} \quad p'=\frac{dp}{dx}.$$

Or $$p=p'\left(x-\frac{1}{x}+\frac{x}{p^2}\right)+\frac{p}{x^2}-\frac{1}{p}$$

Or $$p\left(1-\frac{1}{x^2}+\frac{1}{p^2}\right)=xp'\left(1-\frac{1}{x^2}+\frac{1}{p^2}\right).$$

Therefore, $$\frac{dp}{p}=\frac{dx}{x} or\ p=cx.$$

Substituting this value of p into (1), we get:

$$\left(x^2-1\right)c^2x^2-2x^2yc-x=0.$$

Or $\left(x^2-1\right)c^2-2yc-1=0.$ (2)

This is the general solution of (1).

From (2), the c-discriminant is

$$4y^2+4\left(x^2-1\right)=0,$$

i.e., $x^2+y^2=1,$

which is also the envelope of the family (2). We also note that the p-discriminant is:

$$4x^2y^2 + 4x^2\left(x^2 - 1\right) = 0,$$

i.e., $x^2 + y^2 = 1$.

Since the envelope and the p-discriminant are same, the singular solution is:

$$x^2 + y^2 = 1.$$

Example: Find the singular solution of

$$x^3p^2 + x^2yp + a^3 = 0. \qquad (1)$$

Solution: The p-discriminant of (1) is:

$$x^4y^2 - 4x^3a^3 = 0 \, or \, x^3\left(y^2x - 4a^3\right) = 0 \qquad (2)$$

The parts of (2) that satisfy (1) are singular solutions of the given equation. From (2), we have

$x = 0$, or

$$y^2x - 4a^3 = 0 \qquad (3)$$

We rewrite (1) as:

$$x^3 + x^2y\frac{dx}{dy} + a^3\left(\frac{dx}{dy}\right)^2 = 0. \qquad (4)$$

Clearly, $x = 0$ satisfies (4).

Thus $x = 0$ is a singular solution of (1).

Differentiating $y^2x - 4a^3 = 0 \ \ w.r.t\, x$, we have:

$$2xyp + y^2 = 0 \, or \, y\left(2xp + y\right) = 0.$$

i.e., $p = -\dfrac{y}{2x}$.

Putting this value of p into the left-hand side of (1) and using (3), we obtain:

$$x\left(-\frac{y}{2x}\right)^2+x^2y\left(-\frac{y}{2x}\right)+a^3=-\frac{xy^2}{4}+a^3=-\left(xy^2-4a^3\right)=0.$$

Thus $xy^2-4a^3=0$ and its derivative satisfy (1) and so it is a solution of (1).

Hence $x=0$ and $xy^2-4a^3=0$ are singular solutions of (1).

The Ricatti Equation

Definition. We have already studied first order linear differential equations of the form:

$$y'+P(x)y=R(x) \tag{1}$$

If we add the term $q(x)y^2$ to the left-hand side of (1), we obtain the nonlinear differential equation $y'+P(x)y+Q(x)y^2=R(x)$. (2)

Equation (2) is denoted the Ricatti equation.

Note: In many cases, the solution of (2) cannot be expressed in terms of elementary functions. However, the Ricatti equation

$$y'+Py+Qy^2=R \tag{1}$$

can be reduced to a linear equation by the substitution $y=y_1+\frac{1}{u}$, where y_1 is a particular solution of (1) and u is an unknown nonzero function of x.

Proof: Let $y=y_1+\frac{1}{u}$ be as above. Differentiating $w.r.t.x$, we have:

$$y'=\frac{dy}{dx}=\frac{dy_1}{dx}-\frac{1}{u^2}\frac{du}{dx}.$$

Substituting for $y\,and\,y'$ into (1), we get:

$$\frac{dy_1}{dx}-\frac{1}{u^2}\frac{du}{dx}+P\left(y_1+\frac{1}{u}\right)+Q\left(y_1^2+\frac{2y_1}{u}+\frac{1}{u^2}\right)=R.$$

Or $\frac{dy_1}{dx}+Py_1+Qy_1^2-R-\frac{1}{u^2}\left(\frac{du}{dx}-Pu-2Qy_1u-Q\right)=0$. (2)

Since y_1 is a solution of (1), we have

0 $\frac{dy_1}{dx}+Py_1+Qy_1^2-R=0,$ andhence,(2)reducesto $\frac{du}{dx}-(P+2Qy_1)u=Q,$

w0hich is a linear equation. Thus, if a particular solution of (1) is known then its general solution can be found.

Example: Solve $\frac{dy}{dx}-y^2=-1, y(0)=3,$ given that $y_1=1$ is a particular solution of the given equation.

Solution: Here $P=0, Q=-1, R=-1.$

Writing $y=1+\frac{1}{u},$ the given equation reduces to:

$$\frac{du}{dx}-(0+2(-1)(1))=-1,$$

i.e., $\frac{du}{dx}+2u=-1,$ (1)

which is a linear equation.

The I.F. of (1) is $e^{\int 2\,dx}=e^{2x}.$

Multiplying (1) by the I.F., we get

$$\frac{du}{dx}e^{2x}+2u\,e^{2x}=-e^{2x}.$$

Or $\frac{d}{dx}\left(ue^{2x}\right)=-e^{2x}.$

Integrating, we obtain

$$u\,e^{2x}=-\int e^{2x}\,dx+c=-\frac{e^{2x}}{2}+c.$$

Or $u=-\frac{1}{2}+\frac{c}{e^{2x}}.$ (2)

From $y = 1 + \frac{1}{u}$, we get, by the initial condition,

$$y(0) = 3 = 1 + \frac{1}{u(0)} \; or \; u(0) = \frac{1}{2}.$$

Hence from (2), we have:

$$u(0) = \frac{1}{2} = -\frac{1}{2} + c \;, c = 1.$$

Thus $u = -\frac{1}{2} + \frac{1}{e^{2x}} = \frac{2 - e^{2x}}{2e^{2x}}$,

and the required solution is $y = 1 + \frac{1}{u} = 1 + \frac{2e^{2x}}{2 - e^{2x}} = \frac{2 + e^{2x}}{2 - e^{2x}}$.

Example: Solve $\frac{dy}{dx} - \frac{y}{x} - x^3 y^2 = -x^5$, (1)

by finding a particular solution.

Solution: It is a Ricatti equation with $P = -\frac{1}{x}, Q = -x^3 \; and \; R = -x^5$.

An obvious solution of (1) is $y_1 = x.$ Substituting $y = x + \frac{1}{u}$ into (1), we have

$$\frac{du}{dx} - \left[-\frac{1}{x} + 2\left(-x^3\right)x \right] u = -x^3.$$

Or $\frac{du}{dx} + \left(\frac{1}{x} + 2x^4 \right) u = -x^3$, (2)

which is a linear equation with an I.F. $e^{\int \left(\frac{1}{x} + 2x^4 \right) dx}$

$$= e^{\ln x} . e^{\frac{2}{5}x^5} = xe^{\frac{2}{5}x^5}.$$

Multiplying (2) by the I.F. and integrating, we have:

$$ux e^{\frac{2}{5}x^5} = -\int x^4 . e^{\frac{2}{5}x^5}\, dx + c' = -\frac{1}{2} e^{\frac{2}{5}x^5} + c',$$

or $$u = \frac{c' - \frac{1}{2} e^{\frac{2}{5}x^5}}{x e^{\frac{2}{5}x^5}}.$$

Hence the general solution of (1) is

$$y = x + \frac{x e^{\frac{2}{5}x^5}}{c' - \frac{1}{2} e^{\frac{2}{5}x^5}} = \frac{x\left(c' + \frac{1}{2} e^{\frac{2}{5}x^5}\right)}{c' - \frac{1}{2} e^{\frac{2}{5}x^5}}.$$

Or $$\frac{y}{x} = \frac{c' + \frac{1}{2} e^{\frac{2}{5}x^5}}{c' - \frac{1}{2} e^{\frac{2}{5}x^5}}.$$

Or $\frac{y-x}{y+x} = \frac{e^{\frac{2}{5}x^5}}{2c'} = ce^{\frac{2}{5}x^5}$ is the required solution.

SUMMARY AND DISCUSSION

In this chapter, we discussed some elementary aspects of differential equations. This is a very important topic that can be applied to many real-life problems.

The study of differential equations is a wide sphere in pure and applied mathematics, physics, and engineering. All of these disciplines are concerned with the state of differential equations of various types. Pure sum focuses on the life and singularity of solutions, while applied tally emphasizes the rigorous occasion of the methods for approximating solutions. Differential equations play an important role in modeling virtually every physical, technical, or biological process, from celestial motion, to bridge design, to interactions between neurons. Differential equations, such as those used to solve real-life problems, may not necessarily be directly solvable, i.e. do not have closed pattern solutions. Instead, solutions can be approximated using numerical methods.

Applications. Many fundamental theories of physics and chemistry can be formulated as differential equations. In biology and economics, differential equations are used to mold the behavior of complex systems. However, diverse problems, sometimes originating in quite distinct scientific fields, may give rise to identical differential equations. Whenever this happens, the equations can be viewed as a unifying factor for diverse phenomena. As an example, consider the diffusion of light and sound in the atmosphere, and of waves on the surface of a pond. All of these may be described by the same second-order partial differential equation, the breaker equation, which allows us to think of light and sound as forms of waves. The conduction of heat, developed by Joseph Fourier, is governed by another second-order partial differential equation, the heat equation. It turns out that many scattering processes, while seemingly different, are described by the same equation; the Black-Scholes equation in finance is, for instance, related to the heat equation.

The Euler-Lagrange equation in classical mechanics, Hamilton's equations in classical mechanics, Radioactive decay in nuclear physics, Newton's statute of cooling in thermodynamics, the breaker equation, the heat equation in thermodynamics, Laplace's equation, Poisson's equation, the geodesic equation, the Navier-Stokes equations in fluid dynamics, the Diffusion equation in stochastic processes, the Convection-diffusion equation in fluid dynamics, the Cauchy-Riemann equations in complex analysis, the Poisson-Boltzmann equation in molecular dynamics, the shallow water equations and the Universal differential equation.

Differential Equation. An equation involving one dependent variable and its derivatives with respect to one or more independent variables is called a differential equation. For example,

- $\dfrac{dy}{dx} + y\cos x = \sin x.$

Ordinary Differential Equation (O. D. E). A differential equation that contains ordinary derivatives of the dependent variable with respect to a single independent variable occur is known as an ordinary differential equation (O. D. E).

Partial Differential Equation. A differential equation involving partial derivatives of the dependent variable with respect to more than one independent variable is called a partial differential equation.

$$x\frac{\partial z}{\partial x}+y\frac{\partial z}{\partial y}=nx$$

Order of Differential Equation. The order of a differential equation is the order of the highest derivative that occurs in the equation.

Degree of a Differential Equation. The degree of a differential equation is the greatest exponent of the highest order derivative that appears in the equation. (The dependent variable and its derivatives should be expressed in a form free from radicals and fractions).

$$\frac{dy}{dx}+y\cos x=\sin x$$

(order 1, degree 1)

$$\frac{d^2y}{dx^2}+xy\left(\frac{dy}{dx}\right)^2=0$$

(order 2, degree 1)

$$\left[1+\left(\frac{dy}{dx}\right)^2\right]^{\frac{3}{2}}=\frac{d^2y}{dx^2}$$

(order 2, degree 2)

Linear Differential Equation. An ordinary differential equation $F\left(x, y, \frac{dy}{dx}, \frac{d^2y}{dx^2}, \ldots, \frac{d^ny}{dx^n}\right)=0$ is said to be linear if F is a linear function in the variables $x, \frac{dy}{dx}, \frac{d^2y}{dx^2}, \ldots, \frac{d^ny}{dx^n}$

It should be carefully noted that in a linear ordinary differential equation:

- The dependent variable y and its derivative are all of degree one.
- No product of y or any of its derivative appear.
- No transcendental function of y and / or its derivative appear.

Nonlinear Differential Equation. A differential equation that is not linear is called a nonlinear differential equation. Differential equations occur in the mathematical formulation of many problems in science and engineering. Some such problems include:

- Determining the motion of projectiles, rockets, satellites and planets.
- Calculating the charge or current in an electric circuit.
- The study of chemical reactions.
- The determination of curves with given geometrical properties.

Solution of a Differential Equation

A solution (or integral) of a differential equation is a relationship between the variables not containing derivatives, such that this relationship and the derivatives obtained from it satisfy the given differential equation identically.

Initial Conditions

It is often required to find the solution of a differential equation subject to certain conditions. If the conditions relate to one value of the independent variable, such as $y = y_0$ at $x = x_0$ (written as $y(x_0) = y_0$) and $\frac{dy}{dx} = y'(x_0)$ at $x = x_0$, where x_0 belongs to some interval $[\alpha, \beta]$, then they are denoted as the initial conditions (or one- point boundary conditions), and x_0 is called the initial point.

Boundary Conditions

The problem of finding the solution of a differential equation such that all the associated constraints relate to two different values of the independent variable is called a two-point boundary value problem (or simply a boundary value problem). The associated supplementary boundary conditions are called two-point boundary conditions.

Separable Equations

Definition. A differential equation of the type:

$$F(x)G(x)dx + f(x)g(y)dy = 0 \tag{1}$$

is called an equation with separable variables or simply a separable equation. Equation (1) may be written as:

$$\frac{F(x)}{f(x)}dx+\frac{g(y)}{G(y)}dy=0,$$

which can be easily integrated.

Homogenous Equations

Definition. A function $f(x,y)$ is called homogeneous of degree n if $f(tx,ty)=t^n f(x,y)$, where t is a nonzero real number. Thus $\sqrt{xy}, \frac{x^{10}+y^{10}}{x^2+y^2}$ and $\sin\left(\frac{x}{y}\right)$ are homogeneous functions of degree 1,8 and 0 respectively.

Exact Equation. The expression $M(x,y)dx+N(x,y)dy$ is called an exact differential equation if there exists a continuously differentiable function $f(x,y)$ of two real variables x and y such that the expression equals the total differential df. We know from calculus that $df=\frac{\partial f}{\partial x}dx+\frac{\partial f}{\partial y}dy.$ Thus, if it is exact then $M(x,y)=\frac{\partial f}{\partial x}=f_x, N(x,y)=\frac{\partial f}{\partial y}=f_y.$

Integrating Factors. Consider the non-exact differential equation $M(x,y)dx+N(x,y)dy=0.$ Assume that multiplication with function $\mu(x,y)$ results in the exact equation $\mu(x,y)M(x,y)dx+\mu(x,y)N(x,y)dy=0$, then $\mu(x,y)$ is called an integrating factor (I.F) of the differential equation. The number of integrating factors of an equation may be infinite.

We list below (without proofs) some rules that can be used to find the integrating factors of equations of special types.

Rule 1. If $M(x,y)dx+N(x,y)dy=0$ (1)

is not exact and $\frac{M_y - N_x}{N} = P,$ where P is a function of x only, then (1) has an integrating factor $\mu(x)$, which also depends on x. $\mu(x)$ is the solution of the differential equation:

$$\frac{d\mu}{dx}=P\mu,$$

i.e., $\mu(x)=\exp\int P\,dx.$

Note that $M_y = \frac{\partial M}{\partial y}, N_x = \frac{\partial N}{\partial x}.$

Rule II. If $\frac{N_x - M_y}{M} = Q,$ where Q is a function of y only, then the differential equation:

$Mdx + Ndy = 0$ has an integrating factor

$\mu(y) = \exp\int Q\,dy.$

Rule III. If $M\,dx + Ndy = 0$ (1)

is homogeneous and $xM + yN \neq 0,$ then $\frac{1}{xM+yN}$ is an I.F of (1).

Rule IV. If $M\,dx + Ndy = 0$ is of the form $yf(x,y)dx + xg(x,y)dy = 0,$ and $xM - yN \neq 0,$ then $\frac{1}{xM-yN}$ is an I.F of (1).

The following differential formulas are useful in the calculation of certain exact equations:

$$d\left(\frac{y}{x}\right)=\frac{x\,dy-y\,dx}{x^2}.$$

$$d\left(\frac{x}{y}\right)=\frac{y\,dx-x\,dy}{y^2}.$$

$$d(xy) = x\,dy + y\,dx.$$

$$d\left(x^2 + y^2\right) = 2\left(x\,dx + y\,dy\right).$$

$$d\left(\ln\frac{x}{y}\right) = \frac{y\,dx - x\,dy}{xy}.$$

$$d\left(arc\tan\frac{x}{y}\right) = \frac{y\,dx - x\,dy}{x^2 + y^2}.$$

Linear Equations

A first order ordinary differential equation (ODE) is linear in the dependent variable y and the independent variable x if it is of, or can be written in, the form:

$$\frac{dy}{dx} + P(x)y = Q(x), \tag{1}$$

where P and Q are functions of x.

The Bernoulli Equation

Definition. An equation of the form $\frac{dy}{dx} + P(x)y = Q(x)y^n$ (1)
is known as the Bernoulli differential equation. This equation is linear if $n = 0$ or 1. If n is not zero or 1, then (1) is reducible to a linear equation.

Clairaut's Equation

Definition. The equation $y = xp + f(p),$ (1)
is known as Clairaut's equation.

Envelope

Let $f(x, y, c) = 0$ be a one-parameter family of curves. Suppose all members of the family of curves are drawn for various values of the parameter *c,* arranged in order of magnitude. The two curves that correspond to two consecutive values of c will be designated as neighboring curves. The locus of the ultimate points of intersection of the neighboring curves is denoted as the envelope of the family $f(x, y, c) = 0.$

Singular Solution: A solution of a differential equation $f(x, y, p) = 0$ is called a singular solution (S.S) if:

- It is not derived from the general solution by giving any particular value to the arbitrary constants.
- At each of its points, it is tangent to some member of the one parameter family of curves represented by the general solution.

The Ricatti Equation

Definition. We have already studied first order linear differential equations of the form:

$$y' + P(x)y = R(x) \tag{1}$$

If we add the term $q(x)y^2$ to the left-hand side of (1), we obtain a nonlinear differential equation

$$y' + P(x)y + Q(x)y^2 = R(x). \tag{2}$$

Equation (2) is known as the Ricatti equation.

CHAPTER 2

HIGHER ORDER DIFFERENTIAL EQUATIONS

INTRODUCTION

In the previous chapter, we studied methods of solving special types of linear and nonlinear differential equations of the first order. In this chapter, we shall consider systematic methods for the solution of certain classes of differential equations of orders greater than 1.

LINEAR DIFFERENTIAL EQUATIONS

Definition. A linear differential equation of order n in the dependent variable y and the independent variable x is of the form:

$$a_0(x)\frac{d^n y}{dx^n} + a_1(x)\frac{d^{n-1}y}{dx^{n-1}} + \ldots + a_{n-1}\frac{dy}{dx} + a_n(x)y = F(x), \tag{1}$$

where $a_0(x), a_1(x), \ldots, a_{n-1}(x), a_n(x)$ and $F(x)$ are functions of the independent variable x only, and $a_0(x)$ is not identically zero. Using prime notation, (1) is also written as:

$$a_0 y^n + a_1 y^{n-1} + \ldots + a_{n-1}y' + a_n y = F(x) \tag{2}$$

where $a_0, a_1, \ldots, a_{n-1}, a_n$ are real constants.

Equation (1) is with variable coefficients while (2) is with constant coefficients. In order to solve (2), we shall first consider the equation:

$$a_0y^n + a_1y^{n-1} + \ldots + a_{n-1}y' + a_ny = 0. \tag{3}$$

The coefficient of y^n may be made 1 by dividing (3) by a_0. The differential equation (3) is called a homogeneous linear differential equation of order *n*. The use of the word homogeneous here is quite different from that already mentioned in the previous chapter. If $F(x)$ is not identically zero then (2) is non - homogeneous and (3) is called the associated homogeneous equation of (2).

Definition. If $y_1(x), y_2(x), \ldots, y_m(x)$ are *m* functions of an independent variable x and $c_1, c_2, \ldots, c_m$ are constants, then the expression $c_1y_1(x) + c_2y_2(x) + \ldots + c_my_m(x)$ is called a linear combination of $y_1(x), y_2(x), \ldots, y_m(x)$. We usually write $y_1, y_2, \ldots, y_m$ instead of $y_1(x), y_2(x), \ldots, y_m(x)$ when it is clear from the context that $y_1, y_2, \ldots, y_m$ are functions of x.

As in vector space, the *m* nonzero functions $y_1, y_2, \ldots, y_m$ are linearly dependent if and only if, there exist constants $c_1, c_2, \ldots, c_m$ with at least one of which as nonzero, such that:

$$c_1y_1 + c_2y_2 + \ldots + c_my_m = 0.$$

The functions $y_1, y_2, \ldots, y_m$ are called linear independent if and only if, they are not linearly dependent, i.e., if and only if $c_1y_1 + c_2y_2 + \ldots + c_my_m = 0$ implies that $c_1 = c_2 = \ldots = c_m = 0.$

Now, before investigating a solution of (2), we state the following facts:

i. Every homogeneous linear n^{th}-order differential equation $a_0y^n + a_1y^{n-1} + \ldots + a_{n-1}y' + a_ny = 0$ has *n* linearly independent solutions $y_1, y_2, \ldots, y_n$.

ii. If $y_1, y_2, \ldots, y\_m$ are *n* linearly independent solutions of (3), then any linear combination $y_c = c_1y_1 + c_2y_2 + \ldots + c_ny_n$ of

$y_1, y_2, \ldots, y_n$ is the general solution of (3), $c_1, c_2, \ldots, c_m$ being arbitrary constants.

iii. Let y_p be any particular solution of (2). i.e., y_p does not contain any constants, then $y_c + y_p$ is the general solution of (2).

(ii) and (iii) can be easily checked by actual substitutions into (3) and (2).

Thus, to find the general solution of (2), we have to find a linearly independent set of *n* solutions of (3). Thus, we determine y_c and a particular solution y_p of (2) and then obtain

$$y = y_c + y_p \tag{4}$$

as the general solution of (2). In the general solution (4), y_c is denoted as the Complementary Function (C.F) and y_p the Particular Integral (P.I) of (2).

These statements are also true for equation (1) with the variable coefficients.

HOMOGENEOUS LINEAR EQUATION

Consider the equation $a_0 \dfrac{d^n y}{dx^n} + a_1 \dfrac{d^{n-1} y}{dx^{n-1}} + \ldots + a_{n-1} \dfrac{dy}{dx} + a_n\, y, = 0,$ (1)

where $a_0, a_1, \ldots, a_{n-1}$ are real constants. To find a solution of (1), we shall try a sensible guess. The differential equation (1) requires a function y with the property that if y and its successive derivatives are each multiplied by constants $a_j, j = n, n-1, \ldots, 1, 0,$ and the resulting products are then added, the sum should equal zero. This can only happen if a function is such that its various derivatives are constant multiples of itself. The exponential function $y = e^{mx}$, *m* being a constant, has such properties. Thus, we have:

$$\frac{dy}{dx}=me^{mx},\frac{d^2y}{dx^2}=m^2e^{mx},\ldots,\frac{d^ny}{dx^n}=m^ne^{mx}.$$

Substituting into (1) we have

$$a_0m^ne^{mx}+a_1m^{n-1}e^{mx}+\ldots+a_{n-1}me^{mx}+a_ne^{mx}=0.$$

Or $e^{mx}\left(a_0m^n+a_1m^{n-1}+\ldots+a_{n-1}m+a_n\right)=0$

Since $e^{mx}\neq 0,$ we have:

$$a_0m^n+a_1m^{m-1}+\ldots+a_{n-1}m+a_n=0. \quad (2)$$

Thus, $y=e^{mx}$ is a solution of (1) if and only if m is a solution of (2). Equation (2) is called the characteristic (or auxiliary) equation of the given differential equation (1). Observe that (2) can be obtained from (1) by merely replacing the kth derivative in (1) by:

$m^k\left(k=n,n-1,\ldots,3,2,1\right).$ Three cases arise according as the roots of (2):

- Real and distinct
- Real and repeated
- Complex

Case I. Distinct Real Roots

Let $m_1,m_2,\ldots,m_n$ be *n* distinct real roots of (2). Then $e^{m_1x},e^{m_2x},\ldots,e^{m_nx}$ are *n* distinct solutions of (1). These n solutions are linearly independent. Hence the general solution of (2) is

$$y=c_1e^{m_1x}+c_2e^{m_2x}+\ldots+c_ne^{m_nx},$$

where $c_1,c_2,\ldots,c_n$ are arbitrary constants.

Case II. Repeated Real Roots

In equation (1), writing $D\equiv\frac{d}{dx},$ we have

$\left(a_0D^n+a_1D^{n-1}+\ldots+a_{n-1}D+a_n\right)y=0$ or $\left[f\left(D\right)\right]y=0,$

where $f(D)=a_0D^n+a_1D^{n-1}+\ldots+a_{n-1}D+a_n$.

Note that writing D for m in the characteristic equation (2) is the same as $f(D)=0.$ If $m_1,m_2,\ldots,m_n$ are the roots of $f(D)=0,$ then (1) may be written as:

$$(D-m_1)(D-m_2)\ldots(D-m_n)y=0.$$

Let the root m_1 be repeated twice, say $m_2=m_1$.

The part of the general solution of (1) corresponding to the twice repeated root m_1 of (2) is the solution of $(D-m_1)^2y=0,$

i.e., of $(D-m_1)(D-m_1)y=0$ (3)

Let $(D-m_1)y=V.$

Then (3) becomes

$$(D-m_1)V=0.$$

Or $\frac{dV}{dx}-m_1V=0.$ (4)

Separating the variables, we have

$$\frac{dV}{V}=m_1dx.$$

Therefore, $\ln V=m_1x+k,$ where k is a constant, or $V=c_2e^{m_1x}$, where c_2 is a constant.

Replacing V in (4), we obtain

$$(D-m_1)y=c_2e^{m_1x}$$

$$\frac{dy}{dx}-m_1y=c_2e^{m_1x}, \quad (5)$$

which is a linear equation of order one. Its I.F. is $\exp\left(\int(-m_1)dx\right)=e^{-m_1x}$.

Multiplying (5) by e^{-m_1x}, we get:

$$\frac{d}{dx}\left(ye^{-m_1x}\right) = c_2$$

$$ye^{-m_1x} = c_2x + c_{1,}$$

where $y = (c_1 + c_2x)e^{m_1x}$ is the part of the general solution corresponding to the twice repeated m_1. The general solution of (1) is:

$$y = (c_1 + c_2x)e^{m_1x} + c_3e^{m_3x} + \ldots + c_ne^{m_n x}.$$

In the same manner, if the characteristic equation (2) has the triple repeated root m_1, the corresponding part of the general solution of (1) is the solution of:

$$(D - m_1)^3 y = 0$$

Proceeding as before, we can easily find

$$y = (c_1 + c_2x + c_3x^2)e^{m_1x}$$

as the part of the general solution corresponding to this triple repeated root m_1. If the characteristic equation (2) has the real root m_1 occurring k times, then the part of the general solution of (1) corresponding to the k-fold repeated root m_1 is:

$$y = (c_1 + c_2x + c_3x^2 + \ldots + c_kx^{k-1})e^{m_1x}.$$

Case III. Complex Roots

Suppose the characteristic equation has the complex number $a + ib$ as a non-repeated root. Since the coefficients of (1) are real, the conjugate complex number $a - ib$ is also a non-repeated root. The corresponding part of the general solution is:

$y = k_1e^{(a+ib)x} + k_2e^{(a-ib)x}$, where k_1, k_2 are arbitrary constants.

$$y = e^{ax}\left[k_1e^{ibx} + k_2e^{-ibx}\right]$$

$$= e^{ax}\left[k_1(\cos bx + i\sin bx) + k_2(\cos bx - i\sin bx)\right]$$

$$= e^{ax}\left[(k_1 + k_2)\cos bx + i(k_1 - k_2)\sin bx\right]$$

$$= e^{ax}\left[c_1\sin bx + c_2\cos bx\right]$$

Where $c_1 = i(k_1 - k_2), c_2 = k_1 + k_2$ are two arbitrary constants?

If $a+ib$ and $a-ib$ are conjugate complex roots, each repeated k times, then the corresponding part of the general solution of (1) may be written as:

$$y = \left[\left(c_1 + c_2x + c_3x^2 + \ldots + c_kx^{k-1}\right)\sin bx + \left(c_{k+1} + c_{k+2}x + \ldots + c_{2k}x^{k-1}\right)\cos bx\right].$$

In the examples given below we use these concepts to solve homogeneous linear differential equations.

Example: Solve $\left(D^2 + 4D + 3\right)y = 0$.

Solution: The characteristic equation is $D^2 + 4D + 3 = 0,$ with roots $D = -1, -3.$

Hence the general solution of the given equation is $y = c_1e^{-x} + c_2e^{-3x}$.

Example: Solve $\left(D^3 - 5D^2 + 7D - 3\right)y = 0.$

Solution. The characteristic equation is $D^3 - 5D^2 + 7D - 3 = 0$ (1)

By inspection, $D = 1$ is a solution of this equation? The other two roots can be found by factoring; $(D-1)\left(D^2 - 4D + 3\right) = 0$

Hence all roots of (1) are $D = 1, 1, 3.$

The general solution is $y = (c_1 + c_2x)e^x + c_3e^{3x}.$

Example. Solve $\left(D^3 - D^2 + D - 1\right)y = 0.$

Solution. The characteristic equation is $D^3 - D^2 + D - 1 = 0.$

By inspection, $D = 1$ is a root of this equation?

Now, $D^3 - D^2 + D - 1 = (D-1)(D^2+1) = 0.$

Hence the other two roots are $D = \pm i = a + ib,\ a = 0, b = 1.$

The general solution is

$$y = c_1 e^x + e^{0x}(c_2 \sin x + c_3 \cos x)$$

$$y = c_1 e^x + c_2 \sin x + c_3 \cos x.$$

Example: Solve $(D^2 + D - 12)y = 0,\ y(2) = 2, y'(2) = 0.$

Solution. The characteristic equation is $D^2 + D - 12 = 0$ with roots $D = 3, -4.$

Hence the general solution of the given equation is

$$y = c_1 e^{3x} + c_2 e^{-4x} \tag{1}$$

We now determine the co-efficient (constants) c_1, c_2 in (1) from the initial conditions as follows. Since the initial conditions are given at $x = 2,$ we rewrite the general solution in the form

$$y = k_1 e^{3(x-2)} + k_2 e^{-4(x-2)} \tag{2}$$

Where $k_1 = c_1 e^6, k_2 = c_2 e^{-8}.$

Applying initial condition, (i.e. replacing c_1, c_2 by the new constants $k_1 e^6, k_2 e^{-8}$ respectively in (1), we get

$$y(2) = 2 = k_1 + k_2. \tag{3}$$

Differentiating (2) $w.r.t.x,$ we have

$$y' = \frac{dy}{dx} = 3k_1 e^{3(x-2)} - 4k_2 e^{-4(x-2)}$$

Therefore, $y'(2) = 0 = 3k_1 - 4k_2$ (4)

Solving (3) and (4), we obtain $k_1 = \frac{8}{7}, k_2 = \frac{6}{7}.$

Hence the solution (2) of the differential equation satisfying the given conditions is

$$y = \frac{8}{7}e^{3(x-2)} + \frac{6}{7}e^{-4(x-2)}.$$

Example. Solve $(D^2 + 4D + 5) = 0, y(0) = 1, y'(0) = 0.$

Solution. The characteristic equation is $D^2 + 4D + 5 = 0,$

which has roots $D = -2 \pm i.$

The general solution is $y = e^{-2x}(c_1 \sin x + c_2 \cos x).$ (1)

Applying the given conditions, we have from (1),

$y(0) = 1 = c_2.$

Differentiating (1) $w.r.t.x,$ we get:

$$y' = -2e^{-2x}(c_1 \sin x + c_2 \cos x) + e^{-2x}(c_1 \cos x - c_2 \sin x).$$

Therefore, $y'(0) = 0 = -2c_2 + c_1,$ giving us $c_1 = 2.$

Substituting the values of c_1 and c_2 into (1), the required solution is:

$$y = e^{-2x}(2\sin x + \cos x).$$

Example. Solve $(D^3 - 3D^2 + 4)y = 0, y(0) = 1, y'(0) = -8, y''(0) = -4.$

Solution. The characteristic equation is $D^3 - 3D^2 + 4 = 0.$

$(D+1)(D^2 - 4D + 4) = 0.$

Therefore, $D = -1, 2, 2.$

The general solution is $y = c_1 e^{-x} + e^{2x}(c_2 + c_3 x).$ (1)

Now, from (1), $y' = -c_1 e^{-x} + 2e^{2x}(c_2 + c_3 x) + c_3 e^{2x}$,

$y'' = c_1 e^{-x} + 4e^{2x}(c_2 + c_3 x) + 4c_3 e^{2x}.$

Applying the initial conditions, we get from the above three equations:

$$y(0)=1=c_1+c_2 \tag{2}$$

$$y'(0)=-8=-c_1+2c_2+c_3 \tag{3}$$

$$y''(0)=-4=c_1+4c_2+4c_3. \tag{4}$$

Multiplying (3) by -4 and adding to (4), we have:

$$28=5c_1-4c_2 \tag{5}$$

Multiplying (2) by 4 and adding to (5), we obtain

$$9c_1=32\, or\, c_1=\frac{32}{9}.$$

Therefore, $c_2=1-\frac{32}{9}=-\frac{23}{9},\ c_3=\frac{6}{9}.$

The required solution is

$$y=\frac{32}{9}e^{-x}+e^{2x}\left(-\frac{23}{9}+\frac{6}{9}x\right)=\frac{1}{9}\left(32e^{-x}-23e^{2x}+6xe^{2x}\right).$$

DIFFERENTIAL OPERATORS

Let $T:V\rightarrow V,$ where V is a vector space over a field $F,$ be a linear transformation (or an operator) from $V\, to\, V.$

A linear transformation S,T,U defined on V has the following properties:

$S+T=T+S$ where $(S+T)v=S(v)+T(v),v\in V,$

$$(S+T)+U=S+(T+U)$$

$$S(T+U)=ST+SU$$

$$(ST)U=S(TU)$$

In general, $ST\neq TS$.

Here $ST:V\rightarrow V$ is a linear transformation defined by:

$(ST)(v) = S(Tv)$ for all $v \in V$.

Let X be the vector space of all real (or complex) valued real (or complex) functions possessing a kth order derivative for every $k = 1, 2, 3, \ldots$ For each $g \in X,\ y = g(x) \in R(or \in C). D^k,$ which associates with each $y = g(x)$, its kth derivative, is a linear operator on X. We write $D^k y$ as $\frac{d^k y}{d^k x}$. Since the sum and product of the linear operator and scalar multiple of a linear operator are also linear operators, the linear combination:

$$f(D) = a_0 D^n + a_1 D^{n-1} + \ldots + a_{n-1} D + a_n I \tag{1}$$

of $D^n, D^{n-1}, \ldots, D, I$ is also a linear operator. Here, I is the identity linear operator defined by $I(y) = y$ for all $y = g(x)$ and $a_0, a_1, \ldots, a_n$ are scalars. The image of a $y = g(x)$ under $f(D)$ is written as

$$f(D)y = a_0 D^n y + a_1 D^{n-1} y + \ldots + a_{n-1} Dy + a_n y$$

$$= a_0 \frac{d^n y}{dx^n} + a_1 \frac{d^{n-1} y}{dx^{n-1}} + \ldots + a_{n-1} \frac{dy}{dx} + a_n y.$$

The equation $f(D)y = 0.$ is then a linear differential equation of order *n*.

The operators $D, D^2, \ldots, D^n$ and $f(D),$ given by (1), are called differential operators. We have written D for $\frac{d}{dx}, D^2$ for $\frac{d^2}{dx^2}$ and so on.

Thus $Dy = \frac{dy}{dx}, D^2 y = \frac{d^2 y}{dx^2} and\ D^n y = \frac{d^n y}{dx^n}$.

The expression $A = f(D) = a_0 D^n + a_1 D^{n-1} + \ldots + a_{n-1} D + a_n I$ is called a differential operator of order *n*. This may be thought of as an operator which, when applied to any function y, results in:

$$Ay = f(D)y = a_0 \frac{d^n y}{dx^n} + a_1 \frac{d^n - 1y}{dx^{n-1}} + \ldots + \frac{a_{n-1} dy}{dx} + a_n Iy.$$

Nonhomogeneous Linear Equations

In this section, we discuss the solution of the nonhomogeneous linear equation

$a_0 y^n + a_1 y^{n-1} + \ldots + a_{n-1} y' + a_n y = F(x)$.

SOLUTION OF THE EQUATION

$f(D)y = F(x)$

$$(1)$$

Where $f(D) = a_0 D^n + a_1 D^{n-1} + \ldots + a_{n-1} D + a_n$.

We have already discussed that the general solution of (1) consist of two parts, namely:

- The complementary function (C.F)
- The particular solution (P.I).

The C.F is the solution of the homogeneous equation $f(D)y = 0$. Above we described different cases of its solution.

Now to find the $P.I$ of (1), we can write:

$$y = \frac{1}{f(D)} F(x)$$

and try to evaluate $y = \frac{1}{f(D)} F(x)$.

A real number m is said to be a zero of $f(D)$ if $f(m) = 0$.

Suppose $f(D)$ has *n* distinct zeros $m_1, m_2, \ldots, m_n$.

Then $\frac{1}{f(D)} F(x) = \frac{1}{(D - m_1)(D - m_2)\ldots(D - m_n)} F(x)$.

Suppose that $\frac{1}{D - m_n} F(x) = y^*$.

Then $(D - m_n) y^* = F(x)$,

or $\frac{dy}{dx} - m_n y^* = F(x),$

which is a linear equation of the first order. Its I.F. is given as:

$\exp\left(\int(-m_n)\right)dx = \exp(-m_n x)$.

Hence $\frac{d}{dx}\left(ye^{-m_n x}\right) = F(x)e^{-m_n x}$ and

$y^* = e^{m_n x}\int F(x)e^{-m_n x}dx.$

The constant of integration is omitted, since we are looking for a particular solution, Therefore:

$$\frac{1}{f(D)}F(x) = \frac{1}{(D-m_1)\ldots(D-m_{n-1})}e^{m_n x}\int e^{-m_n x}F(x)dx.$$

Next, we evaluate $\frac{1}{D-m_{n-1}}e^{m_n x}\int e^{-m_n x}F(x)dx$ as before, and continue the process. This is the required P.I of the equation (1).

Alternatively, we may also write:

$$\frac{1}{f(D)}F(x) = \frac{1}{(D-m_1)(D-m_2)\ldots(D-m_n)}F(x)$$

$= \left[\frac{A_1}{D-m_1} + \frac{A_2}{D-m_2} + \ldots + \frac{A_n}{D-m_n}\right]F(x)$ after resolving $\frac{1}{f(D)}$ into partial fractions

$= A_1 e^{m_1 x}\int e^{-m_1 x}F(x)dx + A_2 e^{m_2 x}\int e^{-m_2 x}F(x)dx + \ldots + A_n e^{m_n x}\int e^{-m_n x}F(x)dx$.

This is the required P.I of the equation (1).

Example: Solve $\left(D^3 - D\right)y = e^x$.

Solution. The characteristic equation is $D^3 - D = 0,$ with roots $0, 1, -1$

$C.F = y_c = c_1 + c_2e^x + c_3e^{-x}.$

P.I = $y_p = \frac{e^x}{D(D+1)(D-1)}$

Now $\frac{1}{D-1}e^x = e^x \int e^{-x}.e^x dx = xe^x,$ by the method already described.

Thus $\frac{1}{D(D+1)(D-1)}e^x = \frac{1}{D(D+1)}xe^x$

Again,

$$\frac{1}{D+1}\left(xe^x\right) = e^{-x}\int e^x.xe^x dx = \frac{xe^x}{2} - \frac{e^x}{4}$$

Hence $\frac{1}{D(D+1)}\left(xe^x\right) = \frac{1}{D}\left(\frac{xe^x}{2} - \frac{e^x}{4}\right) = \int \frac{xe^x}{2}dx - \int \frac{e^x}{4}dx$

$$= \frac{xe^x}{2} - \frac{3}{4}e^x$$

Therefore, the general solution of the equation is

$$y = y_c + y_p = c_1 + c_2 e^x + c_3 e^{-x} + \frac{xe^x}{2} - \frac{3}{4}e^x.$$

$$= c_1 + \left(c_2 - \frac{3}{4}\right)e^x + c_3e^{-x} + \frac{xe^x}{2}$$

$$= c_1 + c_2'e^x + c_3e^{-x} + \frac{xe^x}{2}.$$

Alternative Method

Consider $f(D)^{-1}$ as the inverse of the operator $f(D)$, so that:

$$y_p = \frac{1}{D(D+1)(D-1)}e^x$$

$$=\left(-\frac{1}{D}+\frac{1}{2}\cdot\frac{1}{1+D}+\frac{1}{2}\cdot\frac{1}{D-1}\right)e^{x}.$$

$$=-\int e^{x}dx+\frac{1}{2}e^{-x}\int e^{x}.e^{x}\,dx+\frac{1}{2}e^{x}\int e^{-x}.e^{x}\,dx$$

$$=-e^{x}+\frac{1}{4}e^{x}+\frac{1}{2}xe^{x}$$

$=\frac{xe^{x}}{2}$, since e^{x} already occurs in the C.F and so the terms involving e^{x} are omitted.

RULES FOR DETERMINING THE P.I.

The method of evaluating $\frac{1}{f(D)}F(x)$ involves successive integrations and is quite long. We list below short methods to be used for determining the P.I. when $F(x)$ is a function of a particular type.

For example, evaluate $\frac{1}{f(D)}e^{ax}$, where $f(D)=a_0D^n+a_1D^{n-1}+\ldots+a_{n-1}D+a_n$.

We know that $D^k e^{ax}=a^k e^{ax}$, where k is a positive integer.

Hence $f(D)e^{ax}=\left(a_0D^n+a_1D^{n-1}+\ldots+a_{n-1}D+a_n\right)e^{ax}$

$$=a_0.D^n e^{ax}+a_1.D^{n-1}e^{ax}+\ldots+a_{n-1}.De^{ax}+a_n.e^{ax}$$

$$=a_0.a^n e^{ax}+a_1.a^{n-1}e^{ax}+\ldots+a_{n-1}.a\,e^{ax}+a_n.e^{ax}.$$

$$=f(a)e^{ax}. \qquad (1)$$

Applying $\frac{1}{f(D)}$ to both sides of (1), we obtain

$$\frac{f(D)}{f(D)}e^{ax} = \frac{1}{f(D)}f(a)e^{ax}$$

$$e^{ax} = f(a).\frac{1}{f(D)}e^{ax},$$

i.e., $\frac{1}{f(D)}e^{ax} = \frac{e^{ax}}{f(a)}, if\ f(a) \neq 0$

if $f(a)=0,$ then a is a root of $f(D)=0.$

Let a be a k-fold root of $f(D)=0,$ such that $(D-a)^k$ is a factor of $f(D).$ Then

$$f(D)=(D-a)^k\varnothing(D), where \varnothing(a)\neq 0.$$

We first check the effect of $(D-a)^k\ on\ e^{ax}p(x),$ for a polynomial $p(x)$ in x.

We have $(D-a)(e^{ax}p(x))=D(e^{ax}p(x))-a\,e^{ax}p(x)=e^{ax}Dp(x)$

$$(D-a)^2(e^{Ax}p(x))=(D-a)(e^{ax}D\,p(x))=e^{ax}D^2p(x).$$

Continuing in this way, we get

$$(D-a)^k(e^{Ax}p(x))=e^{ax}D^kp(x). \tag{3}$$

Setting $p(x)=x^k$ in (3), we are led to:

$$(D-a)^k(x^ke^{ax})=e^{ax}D^kx^k = k!e^{ax}$$

$$\varnothing(D)\,(D-a)^k n(x^ke^{ax})=\varnothing(D)k!e^{ax}$$

$=k!\varnothing(D)e^{ax} = k!\varnothing(a)e^{ax},$ by (1).

Applying $\frac{1}{\varnothing(D)\,(D-a)^k}$ to both sides:

$$x^k e^{ax} = \frac{1}{\varnothing(D)\,(D-a)^k} k!\varnothing(a)e^{ax}$$

$$\frac{1}{\varnothing(D)\,(D-a)^k} e^{ax} = \frac{x^k\, e^{ax}}{k!\varnothing(a)} . \qquad (4)$$

Principle of Superposition

Let $f(D)y = \left(a_0 D^n + a_1 D^{n-1} + \ldots + a_{n-1} D + a_n\right) y = F(x)$ (1)

be a linear differential equation of order $n.$ if $F(x) = F_1(x) + F_2(x),$ then the P.I. of (1) is the sum of the particular integrals of

$F(D)y = F_1(x)$ (2)

and $F(D)y = F_2(x)$ (3)

Proof: Let $y_1 \; and \; y_2$ be particular integrals of (2) and (3) respectively.

Then $f(D)y_1 = F_1(x)$

And $f(D)y_2 = F_2(x).$

Suppose $y = y_1 + y_2.$ Setting this value of y into (1), we obtain

$f(D)y = f(D)(y_1 + y_2)$

$= f(D)y_1 + f(D)y_2$

$= F_1(x) + F_2(x) = F(x).$

Showing that y is a solution of (1).

Example: Solve $\left(D^3+1\right)y = 1 + e^{-x} + e^{2x}.$

Solution. The characteristic equation is

$D^3 + 1 = 0$ with roots $-1\,, \frac{1 \pm i\sqrt{3}}{2}$.

Therefore, the $C.F$ is

$$y_c = c_1 e^{-x} + e^{\frac{x}{2}}\left(c_2 \sin\frac{\sqrt{3}}{2}x + c_3 \cos\frac{\sqrt{3}}{2}x\right).$$

The P.I. is given by $y_p = \frac{1}{D^3+1}\left(1+e^{-x}+e^{2x}\right)$

$$= \frac{1}{D^3+1}e^{0x} + \frac{1}{D^3+1}e^{2x} + \frac{1}{(D+1)(D^2-D+1)}e^{-x}$$

$$= 1 + \frac{e^{2x}}{9} + \frac{1}{3(D+1)}e^{-x},$$ using $\frac{1}{f(D)}e^{ax} = \frac{1}{f(a)}e^{ax},$

for $\frac{1}{D^2-D+1}e^{-x}, with\, D = -1\, when\, f(-1) \neq 0$

$$= 1 + \frac{e^{2x}}{9} + \frac{1}{3}xe^{-x}.$$

Hence the general solution of the equation is

$$y = y_c + y_p$$

$$= c_1 e^{-x} + e^{\frac{x}{2}}\left(c_2 \sin\frac{\sqrt{3}}{2}x + c_3 \cos\frac{\sqrt{3}}{2}x\right) + 1 + \frac{e^{2x}}{9} + \frac{1}{3}x.e^{-x}.$$

To Find the P.I when $F(x) = \sin ax$ or $\cos ax$.

Here, we have to evaluate $\frac{1}{f(D)}\sin ax$ and $\frac{1}{f(D)}\cos ax.$

From Euler's theorem, we have

$$e^{iax} = \cos ax + i \sin ax.$$

Thus $\frac{1}{f(D)}\sin ax$ and $\frac{1}{f(D)}\cos ax$ are the imaginary and real parts of $\frac{1}{f(D)}e^{iax}$, respectively. If $f(D)$ contains only even powers of D, say $f(D)=f(D^2)$, it is easy to see that

$$\frac{1}{f(D^2)}\begin{cases}\sin ax\\ \cos ax\end{cases}=\frac{1}{f(-a^2)}\begin{cases}\sin ax\\ \cos ax\end{cases}, \textit{ provided } -a^2 \textit{ is not a zero of } f(D^2).$$

Example: Solve $(D^2-5D+6)y=\sin 3x.$

Solution: The roots of the characteristic equation $D^2-5D+6=0$ are 2 and 3. Hence,

$C.F=y_c=c_1e^{2x}+c_2e^{3x}$ and

$$P.I=y_p=\frac{1}{D^2-5D+6}\sin 3x,$$

which is the imaginary part of $\frac{1}{(D-2)(D-3)}e^{3ix}$.

Now $\frac{1}{(D-2)(D-3)}e^{3ix}=\frac{e^{3ix}}{(3i-2)(3i-3)},$

$$=\frac{-1+5i}{78}e^{3ix}$$

$$=\left(-\frac{1}{78}+\frac{5}{78}i\right)(\cos 3x+i\sin 3x).$$

Its imaginary part is $-\frac{1}{78}\sin 3x+\frac{5}{78}\cos 3x=y_p$.

Hence, the required general solution is

$$y=c_1e^{2x}+c_2e^{3x}-\frac{1}{78}\sin 3x+\frac{5}{78}\cos 3x.$$

Theorem. If a is not a zero of $f(D)$, then

$$\frac{1}{f(D-a)}e^{ax}F(x)=e^{ax}\frac{1}{f(D)}F(x).$$

Replacing of $D\,by\,D+a$ is known as exponential shift.

Proof. We have already shown in the previous section that $(D-a)^k\left(e^{ax}p(x)\right)=e^{ax}D^k p(x)$.

Using the linearity of differential operators, we conclude that, when $f(D)$ is a polynomial (with constant coefficients), then

$$f(D-a)\left(e^{ax}p(x)\right)=e^{ax}f(D)p(x). \qquad (1)$$

Suppose $f(D)p(x)=F(x)$.

Then $p(x)=\frac{1}{f(D)}F(x)$. (2)

From (1) and (2), we have:

$$f(D-a)e^{ax}\frac{1}{f(D)}F(x)=e^{ax}F(x).$$

Operating on both sides by $\frac{1}{f(D-a)}$, we get

$$e^{ax}\frac{1}{f(D)}F(x)=\frac{1}{f(D-a)}e^{ax}F(x),$$

or $\frac{1}{f(D-a)}e^{ax}F(x)=e^{ax}\frac{1}{f(D)}F(x),$ provided that a is not a zero of $f(D)$.

Example: Solve $\left(D^3+D^2-4D-4\right)y=e^{2x}\cos 3x.$

Solution. The characteristics equation is $D^3+D^2-4D-4=0.$

By inspection $D=2$ is a root of this equation. The other roots are $D=-2,-1$ and hence

$$C.F.=y_c=c_1e^{-x}+c_2e^{-2x}+c_3e^{2x},$$

$$P.I=y_p=\frac{1}{(D-2)(D+2)(D+1)}e^{2x}\cos 3x$$

$$=e^{2x}\frac{1}{D(D+4)(D+3)}\cos 3x,$$ by the exponential shift

$$=e^{2x}\frac{1}{(D^3+7D^2+12D)}\cos 3x$$

$$=e^{2x}\frac{1}{-9D-63+12D}\cos 3x,$$

putting $D^2=-3^2$

$$=e^{2x}\frac{1}{3(D-21)}\cos 3x$$

$$=e^{2x}\frac{(D+21)}{3(D^2-441)}\cos 3x$$

$$=e^{2x}.\frac{-1}{3\times 450}[-3\sin 3x+21\cos 3x]$$

$$=\frac{e^{2x}}{450}(\sin 3x-7\cos 3x).$$

ALTERNATIVE METHOD

Here, for $\frac{1}{D(D+4)(D+3)}\cos 3x,$ we proceed as follows:

$$\frac{\cos 3x}{(D^3+7D^2+12D)}=Re\frac{1}{(D^3+7D^2+12D)}e^{3ix},$$

$$= Re\frac{1}{(-27i-63+36i)}$$

$$= Re\frac{e^{3ix}}{9(i-7)}$$

$$= Re\frac{(i+7)e^{3ix}}{-450}$$

$$= Re\frac{(i+7)(cos 3x+i \sin 3x)}{-450}$$

$$= -\frac{7}{450}\cos 3x+\frac{1}{450}\sin 3x.$$

Therefore, $y_p = \frac{e^{2x}}{450}(\sin 3x-7\cos 3x)$ as before.

The general solution is $y = y_c + y_p$

$$= c_1e^{-x} + c_2e^{-2x} + c_3e^{2x} - \frac{7}{450}\cos 3x + \frac{1}{450}\sin 3x.$$

$$= c_1e^{-x} + c_2e^{-2x} + c_3e^{2x} + \frac{e^{2x}}{450}(\sin 3x - 7\cos 3x).$$

SERIES EXPANSION OF POLYNOMIALS OR EXPRESSIONS INVOLVING OPERATORS.

This method is useful when $F(x)$ is a polynomial in x.

Let $f(D)y=F(x)$ be such that $F(x)$ is a polynomial in x. To evaluate the particular integral

$$y_p = \frac{1}{f(D)}F(x),$$

it is often useful to expand $\frac{1}{f(D)}$ in a series in D using the Binomial

Theorem for negative exponents such that:

$$y_p = \left(1 + a_1 D + a_2 D^2 + a_3 D^3 + \ldots\right) F(x).$$

The derivatives on the right will vanish after a certain stage as

$D^n x^r = 0 \; if \; n > r.$

The following binomial expansions will be useful for this:

$$\frac{1}{1-x} = 1 + x + x^2 + x^3 + \ldots.$$

$$\frac{1}{1+x} = 1 - x + x^2 - x^3 + \ldots.$$

Example: Solve $\left(D^3 - 2D + 1\right) y = 2x^3 - 3x^2 + 4x + 5$.

Solution. The characteristic equation is $D^3 - 2D + 1 = 0$.

$$\left(D - 1\right)\left(D^2 + D - 1\right) = 0$$

$$D = 1, \frac{-1+\sqrt{5}}{2}, \frac{-1-\sqrt{5}}{2}$$

$$y_c = c_1 e^x + c_2 e^{\frac{-1+\sqrt{5}}{2}x} + c_3 e^{\frac{-1-\sqrt{5}}{2}x}.$$

Next, $y_p = \dfrac{1}{1 - 2D + D^3}\left(2x^3 - 3x^2 + 4x + 5\right)$.

Now $\dfrac{1}{1 - 2D + D^3} = \left[1 - \left(2D - D^3\right)\right]^{-1}$

$$= 1 + \left(2D - D^3\right) + \left(2D - D^3\right)^2 + \left(2D - D^3\right)^3 + \ldots$$

$$= 1 + 2D - D^3 + 4D^2 + 8D^3.$$

Neglecting D^4 and higher powers of D of the polynomial $p(x)$ in x:

$$y_p = \left(1 + 2D + 4D^2 + 7D^3\right)\left(2x^3 - 3x^2 + 4x + 5\right)$$

$$= 2x^3 - 3x^2 + 4x + 5 + 2\left(6x^2 - 6x + 4\right) + 4\left(12x - 6\right) + 7\left(12\right)$$

$=2x^3+9x^2+40x+73.$

The general solution is $y=y_c+y_p$

$$=c_1e^x+c_2e^{\frac{-1+\sqrt{5}}{2}x}+c_3e^{\frac{-1-\sqrt{5}}{2}x}+2x^3+9x^2+40x+73.$$

THE METHOD OF UNDETERMINED COEFFICIENTS (U. C. METHOD)

We have studied some special cases in which the particular integral can be evaluated by the inverse operator. Now we consider the method of undetermined coefficients, which can prove simpler in finding the particular integral of the equation $f(D)y=F(x)$ when $F(x)$ is:

- An exponential function: e^{ax}.
- A polynomial: $\left(b_0x^n+b_1x^{n-1}+\ldots+b_n\right)$.
- A Sinusoidal function: $\sin ax \text{ or } \cos bx$.
- The more general case in which $F(x)$ is the sum of a product of terms of the above types, such as $F(x)=e^{ax}\left(b_0x^n+b_1x^{n-1}+\ldots+b_n\right)\begin{Bmatrix}\sin ax\\ \cos bx\end{Bmatrix}$.

The P.I., i.e., y_p will be constructed according to the following table:

$F(x)$ *is of the form*	*Take* y_p *as*
a	$A.x^k$
ax^n (n *is a +ve integer*)	$x^k\left(A_0x^n+A_1x^{n-1}+\ldots+A_n\right)$
ax^ne^{rx} (n *is a +ve integer*)	$x^k\left(A_0x^n+A_1x^{n-1}+\ldots+A_{n-1}x+A_n\right)e^{rx}$

$Cx^n \cos ax$	$x^k\left(A\cos ax + B\sin ax\right)$
$Cx^n \sin ax$	
$Cx^n e^{rx} \cos ax$	$x^k\ [\left(A_0x^n + A_1x^{n-1} + \ldots + A_{n-1}x + A_n\right)e^{rx}\cos ax$ $+$
$Cx^n e^{rx} \sin ax$	$\left(B_0x^n + B_1x^{n-1} + \ldots + B_{n-1}x + B_n\right)e^{rx}\sin ax$

For x^k, k is the smallest nonnegative integer which will ensure that no terms in y_p is already in the C.F.

If $F(x)$ is sum of several terms, write y_p for each term individually and then add up all of them. The y_p and its derivatives will be substituted into the equation $f(D)y = F(x)$ and coefficients of like terms on the left hand and right hand sides will be equated to determine the $U.C. A_0, A_1, \ldots, A_n, B_0, B_1, \ldots, B_n$. The method is illustrated by examples as follows.

Example: Solve $y'' - 3y' + 2y = 2x^3 - 9x^2 + 6x$. (1)

Solution: C.F. is easily found as $y_c = c_1e^x + c_2e^{2x}$.

For a particular solution, we assume

$y_p = x^k\left(Ax^3 + Bx^2 + Cx + D\right)$.

Since no term of the C.F. is present in y_p, we take $k = 0$ and thus

$y_p = Ax^3 + Bx^2 + Cx + D$

$y_p' = 3Ax^2 + 2Bx + C$

$y_p'' = 6Ax + 2B$.

Substituting for y_p, y_p' and y_p'' into (1), we have

$$(6Ax+2B)-3(3Ax^2+2Bx+C)+2(Ax^3+Bx^2+Cx+D)=2x^3-9x^2+6x$$

Or $2Ax^3+x^2(2B-9A)+x(6A-6B+2C)+2B-3C+2D=2x^3-9x^2+6x.$

Equating coefficients of like terms, we obtain

Coeff. of $x^3: 2A=2 \quad or\ A=1.$

Coeff. of $x^2: 2B-9A=-9 \quad or\ 2B=9A-9\ giving\ B=0$.

Coeff. of $x: 6A-6B+2c=6 \quad or\ C=0.$

Coeff. of $x^0: 2B-3C+2D=0 \quad or\ D=0.$

Thus, $y_p=x^3$.

The required general solution is $y=c_1e^x+c_2e^{2x}+x^3.$

Example: Solve $y''-3y'+2y=x^2e^x.$ (1)

Solution. The C.F. is $y_c=c_1e^x+c_2e^{2x}.$

To construct a particular solution, we use (3) in the table above:

$y_p=x^k\left(Ax^2+Bx+C\right)e^x.$

Since e^x is already in the $C.F.,$ we take $k=1$ such that no terms of the C.F. are in $y_p.$ Hence the modified P.I. is $y_p=\left(Ax^3+Bx^2+Cx\right)e^x$:

$y_p'=\left(Ax^3+Bx^2+Cx\right)e^x+\left(3Ax^2+2Bx+C\right)e^x$

$y_p''=\left(Ax^3+Bx^2+Cx\right)e^x+2\left(3Ax^2+2Bx+C\right)e^x+\left(6Ax+2B\right)e^x.$

Substituting for $y_p, y_p'\ and\ y_p''$ into (1), we have:

$$\left(Ax^3+Bx^2+Cx\right)e^x+2\left(3Ax^2+2Bx+C\right)e^x+\left(6Ax+2B\right)e^x+\left(6Ax+2B\right)e^x$$

$$-3\left(Ax^3+Bx^2+Cx\right)e^x-3\left(3Ax^2+2Bx+C\right)e^x+2\left(Ax^3+Bx^2+Cx\right)e^x=x^3e^x$$

$$\left[-3Ax^2+(6A-2B)x+C\right]e^x=x^3e^x.$$

Equating coefficients of like terms, we obtain:

$$Co-eff.of\ x^3 :-3A=1\ or\ A=-\frac{1}{3}$$

$$Co-eff.of\ x :6A-2B=0\ or\ B=-1$$

$$Co-eff.of\ x^0 :C=0$$

Hence, $y_p = -\frac{1}{3}x^3e^x - x^2e^x.$

The required solution is

$$y = c_1e^x + c_2e^{2x} - \frac{1}{3}x^3e^x - x^2e^x.$$

Example: Solve $y'' + 4y = xe^x + x\sin 2x$. (1)

Solution. The C.F., as detailed above, is $y_c = c_1\sin 2x + c_2\cos 2x.$

For the P.I of (1), we find the P.I. of $y'' + 4y = xe^x$ (2)

and $y'' + 4y = x\sin 2x,$ (3)

separately. Their sum will be the P.I of (1) (by the principle of superposition). For a particular solution of (2), we have:

$$y_p = x^k(Ax+B)e^x.$$

Since no term of the C.F. is in the y_p, we must take $k=0$. Therefore,

$$y_p = (Ax+B)e^x$$

$$y_p' = (Ax+B)e^x + Ae^x$$

$$y_p'' = (Ax+B)e^x + 2Ae^x.$$

Substituting for $y_p, y_p'\ and\ y_p''$ into (2), we have

$$(Ax+B)e^x + 2Ae^x + 4(Ax+B)e^x = xe^x,$$

or $5Axe^x + (2A+5B)e^x = xe^x.$

Equating coefficients of like terms, we obtain

Coeff. of $xe^x : 5A = 1 \, or \, A = \frac{1}{5}$.

Coeff. of $e^x : 2A + 5B = 0 \, or \, B = -\frac{2}{25}$.

Thus, the P.I of (2) is $y_p = \frac{1}{5}xe^x - \frac{2}{25}e^x$.

$$y_p = x^k\left[(Cx + D)\sin 2x + (Ex + F)\cos 2x\right].$$

Since $\sin 2x \, and \, \cos 2x$ are already in the C.F., we must take $k = 1$, so that the P.I. is

$$y_p = \left(Cx^2 + Dx\right)\sin 2x + \left(Ex^2 + Fx\right)\cos 2x$$

$$y_p' = \left(Cx^2 + Dx\right)(2\cos 2x) + (2Cx + D)\sin 2x$$

$$+\left(Ex^2 + Fx\right)(-2\sin 2x) + \left(2E(x) + F\right)\cos 2x$$

$$y_p'' = \left(Cx^2 + Dx\right)(-4\sin 2x) + (2Cx + D)(2\cos 2x) + (2Cx + D)(2\cos 2x)$$

$$+(2Cx + D)(2\cos 2x) + 2C\sin 2x + \left(Ex^2 + Fx\right)(-4\cos 2x)$$

$$+(2Ex + F)(-2\sin 2x) + (2Ex + F)(-2\sin 2x) + 2E\cos 2x.$$

Substituting into (3), we have

$$\left(Cx^2 + Dx\right)(-4\sin 2x) + 4(2Cx + D)\cos 2x + 2C\sin 2x + (Ex + Fx)(-4\cos 2x)$$

$$-4(2Ex + F)\sin 2x + 2E\cos 2x + 4\left(Cx^2 + Dx\right)\sin 2x + 4\left(Ex^2 + Fx\right)\cos 2x = x\sin 2x.$$

Equating the coefficients of like terms, we get

Coeff. of $x\sin 2x$: $-8E = 1 \quad or \, E = -\frac{1}{8}$.

Coeff. of $x\cos 2x : 8C = 0 \, or \, C = 0$.

Coeff. of $\sin 2x : 2C - 4F = 0 \, or \, F = 0$.

Coeff. of $\cos 2x : 2E + 4D = 0 \; or \; D = \frac{1}{16}.$

Hence, the P.I of (3) is:

$$y_p = \frac{1}{16}x\sin 2x - \frac{1}{8}x^2\cos 2x. \tag{5}$$

The sum of (4) and (5) is the required P.I of (I), i.e.,

$y_p = \frac{1}{5}xe^x - \frac{2}{25}e^x + \frac{1}{16}x\sin 2x - \frac{1}{8}x^2\cos 2x$ is the particular solution of (1).

Hence the general solution of (1) is:

$$y = c_1\sin 2x + c_2\cos 2x + \frac{1}{5}xe^x - \frac{2}{25}e^x + \frac{1}{16}x\sin 2x - \frac{1}{8}x^2\cos 2x.$$

THE CAUCHY-EULER EQUATION

An equation of the form

$$a_0x^n\frac{d^ny}{dx^n} + a_1x^{n-1}\frac{d^{n-1}y}{dx^{n-1}} + \ldots + a_{n-1}x\frac{dy}{dx} + a_ny = F(x)$$

is called Cauchy-Euler (or equidimensional) equation, where $a_0, a_1, \ldots, a_{n-1}, a_n$ are real constants. The equation can be reduced to a linear differential equation with constant coefficients by the transformation $x = e^t \; or \; t = \ln x$.

Thus, $\frac{dy}{dx} = \frac{dy}{dt}.\frac{dt}{dx} = \frac{1}{x}\frac{dy}{dt}$, and

$$\frac{d^2y}{dx^2} = \frac{d}{dx}\left(\frac{1}{x}\frac{dy}{dt}\right) = \frac{1}{x}\frac{d}{dx}\left(\frac{dy}{dt}\right) + \frac{dy}{dt}.\frac{d}{dx}\left(\frac{1}{x}\right)$$

$$= \frac{1}{x}\frac{d}{dt}\left(\frac{dy}{dt}\right).\frac{dt}{dx} + \frac{dy}{dt}\left(-\frac{1}{x^2}\right) = \frac{1}{x^2}\left(\frac{d^2y}{dt^2} - \frac{dy}{dt}\right).$$

$$\frac{d^3y}{dx^3} = \frac{d}{dx}\left[\frac{1}{x^2}\frac{d^2y}{dt^2} - \frac{1}{x^2}\frac{dy}{dt}\right]$$

$$= \frac{1}{x^2}\frac{d}{dt}\left(\frac{d^2y}{dt^2}\right).\frac{dt}{dx} - \frac{2}{x^3}\frac{d^2y}{dt^2} - \frac{1}{x^2}\frac{d}{dt}\left(\frac{dy}{dt}\right).\frac{dt}{dx} + \frac{2}{x^3}\frac{dy}{dt}$$

$$= \frac{1}{x^3}\frac{d^3y}{dt^3} - \frac{3}{x^3}\frac{d^2y}{dt^2} + \frac{2}{x^3}\frac{dy}{dt}$$

$$= \frac{1}{x^3}\left(\frac{d^3y}{dt^3} - 3\frac{d^2y}{dt^2} + 2\frac{dy}{dt}\right).$$

If we write $D = \frac{d}{dx}$ *and* $\Delta = \frac{d}{dt}$:

$$xD = \Delta$$

$$x^2D^2 = \Delta^2 - \Delta = \Delta(\Delta - 1)$$

$$x^3D^3 = \Delta^3 - 3\Delta^2 + 2\Delta = \Delta(\Delta-1)(\Delta-2)$$

$$\vdots \qquad \vdots \qquad \vdots$$

$$x^nD^n = \Delta(\Delta-1)(\Delta-2)\ldots(\Delta-n+1).$$

Substituting these values of $xD, x^2D^2, \cdots$ into (1), we obtain an equation of nth order with constant coefficients having t as the independendent variable. This equation can be solved by the previous methods.

Example: Solve $x^2\frac{d^2y}{dx^2} - 2x\frac{dy}{dx} + 2y = x^3.$ (1)

Solution: Let $x = e^t$. Then we have $t = \ln x$,

$$xD = \Delta$$

$$x^2D^2 = \Delta(\Delta - 1).$$

Substituting into (1), we get:

$$\left[\Delta(\Delta-1) - 2\Delta + 2\right]y = e^{3t}$$

$$\left[\Delta^2 - 3\Delta + 2\right]y = e^{3t}. \qquad (2)$$

The characteristic equation has roots 2 and 1. Therefore, the C.F. of (2) is:

$$y_c = c_1 e^t + c_2 e^{2t}$$

and P.I. of equation (2) is $y_p = \frac{1}{(\Delta - 1)(\Delta - 2)} e^{3t} = \frac{1}{2} e^{3t}.$

The general solution of (2) is $y = c_1 e^t + c_2 e^{2t} + \frac{1}{2} e^{3t}.$

Replacing t by $\ln x \left(or\, e^t by\, x \right),$ we have

$y = c_1 x + c_2 x^2 + \frac{1}{2} x^3$ as the general solution of (1).

Example: Solve $x^4 \frac{d^4 y}{dx^4} + 6x^3 \frac{d^3 y}{dx^3} + 9x^2 \frac{d^2 y}{dx^2} + 3x \frac{dy}{dx} + y = (1 + \ln x)^2.$ (1)

Solution: We let $x = e^t$ so that $t = \ln x$

$$xD = \Delta$$

$$x^2 D^2 = \Delta(\Delta - 1) = \Delta^2 - \Delta.$$

$$x^3 D^3 = \Delta(\Delta - 1)(\Delta - 2) = \Delta^3 - 3\Delta^2 + 2\Delta$$

$$x^4 D^4 = \Delta(\Delta - 1)(\Delta - 2)(\Delta - 3) = \Delta^4 - 6\Delta^3 + 11\Delta^2 - 6\Delta.$$

Substituting into (1), we get:

$$\left[\Delta^4 - 6\Delta^3 + 11\Delta^2 - 6\Delta + 6\left(\Delta^3 - 3\Delta^2 + 2\Delta\right) + 9\left(\Delta^2 - \Delta\right) + 3\Delta + 1\right] y = (1 + t)^2$$

$$\left(\Delta^4 + 2\Delta^2 + 1\right) y = 1 + 2t + t^2 \quad (2)$$

The roots of the characteristic equation of (2) arc $\pm i, \pm i.$

Therefore, the C.F. of (2) is $y_c = \left(c_1 + c_2 t\right)\sin t + \left(c_3 + c_4 t\right)\cos t.$

To find a particular integral of (2), we use the method of $U.C.$ Let the particular integral of (2) be

$$y_p = At^2 + Bt + C$$

$y_p^{'} = 2At + B$

$y_p^{''} = 2A.$

Substituting into (2), we obtain

$4A + At^2 + Bt + C = 1 + 2t + t^2$

$At^2 + Bt + (4A + C) = t^2 + 2t + 1.$

Equating coefficients of like terms, we have

$A = 1, B = 2, 4A + C = 1 \, or \, C = -3.$

Therefore, a particular integral of (2) is

$y_p = t^2 + 2t - 3.$

Hence the general solution of (2) is:

$$y = (c_1 + c_2 t)\sin t + (c^3 + c_4 t)\cos t + t^2 + 2t - 3. \qquad (3)$$

Replacing t by $\ln x$ in (3), we get

$$y = (c_1 + c_2 \ln x)\sin(\ln x) + (c_3 + c_4 \ln x)\cos(\ln x) + (\ln x)^2 + 2\ln x - 3$$

as the general solution of (1).

Reduction of Order

If one solution of the second order linear equation

$$\frac{d^2 y}{dx^2} + P\frac{dy}{dx} + Qy = 0 \qquad (1)$$

(where P, Q are not necessarily constants and may be functions of x) is known, then we can use it to find the general solution of

$$\frac{d^2 y}{dx^2} + P\frac{dy}{dx} + Qy = F(x) \qquad (2)$$

The procedure to be used is known as the method of reduction of order.

Suppose it is known that $y = y_1$ is a solution of (1). We assume that

$$y = vy_1 \tag{3}$$

is a solution of (2), where v is a function of x to be determined. From (3), we get:

$$\frac{dy}{dx} = v\frac{dy_1}{dx} + y_1\frac{dv}{dx},$$

$$\frac{d^2y}{dx^2} = v\frac{d^2y_1}{dx^2} + 2\frac{dv}{dx}\frac{dy_1}{dx} + y_1\frac{d^2v}{dx^2}.$$

Substituting into (2), we have

$$v\frac{d^2y_1}{dx^2} + 2\frac{dv}{dx}\frac{dy_1}{dx} + y_1\frac{d^2v}{dx^2} + Pv\frac{dy_1}{dx} + Py_1\frac{dv}{dx} + Qvy_1 = F(x)$$

$$y_1\frac{d^2v}{dx^2} + \left(2\frac{dy_1}{dx} + Py_1\right)\frac{dv}{dx} + \left(\frac{d^2y_1}{dx^2} + P\frac{dy_1}{dx} + Qy_1\right)v = F(x). \tag{4}$$

Since $y = y_1$ is a solution of (1)

$$\frac{d^2y_1}{dx^2} + P\frac{dy_1}{dx} + Qy_1 = 0,$$

and, therefore, equation (4) reduces to

$$y_1\frac{d^2v}{dx^2} + \left(2\frac{dy_1}{dx} + Py_1\right)\frac{dv}{dx} = F(x). \tag{5}$$

Setting $\frac{dv}{dx} = u$ in (5), we obtain

$$y_1\frac{du}{dx} + \left(2\frac{dy_1}{dx} + Py_1\right)u = F(x).$$

Which is a linear equation of the first order in u, and can be solved for u. From $\frac{dv}{dx} = u,$ we determine v and hence the solution $y = vy_1$. The method is illustrated by means of an example.

Example: Solve $\frac{d^2y}{dx^2}+y=\csc x.$ (1)

Solution. The C.F. of (1) is $y_c = c_1 \sin x + c_2 \cos x.$

We may take any special value of y_c as y_1. Let us take y_1 to be the value of y_c when $c_1 = 1$ and $c_2 = 0$. Then assume that

$y = v \sin x$. (2)

is a solution of (1). From (2), we get

$$\frac{dy}{dx} = \sin x \frac{dv}{dx} + v\cos x$$

$$\frac{d^2y}{dx^2} = \sin x \frac{d^2v}{dx^2} + 2\cos x \frac{dv}{dx} - v\sin x.$$

Substituting into (1), we obtain

$$\sin x \frac{d^2v}{dx^2} + 2\cos x \frac{dv}{dx} - v\sin x + v\sin x = \csc x$$

$$\frac{d^2v}{dx^2} + 2\cot x \frac{dv}{dx} = \csc^2 x.$$

Setting $u = \frac{dv}{dx}$, the above equation becomes

$$\frac{du}{dx} + 2\cot x u = \csc^2 x$$. (3)

This is a first order linear equation. An integrating factor of (3) is:

$$exp\left[\int 2\cot x dx\right] = \exp\left[\ln \sin^2 x\right] = \sin^2 x.$$

Multiplying (3) by $\sin^2 x$ and integrating, we have

$u \sin^2 x = x,$

neglecting the constant of integration as we seek only a particular solution:

$u x \csc^2 x$

$$\frac{dv}{dx} = x\csc^2 x$$

$$v = \int x\csc^2 x\,dx = -x\cot x + \ln|\sin x|$$ (on integrating by parts).

Hence, a particular solution of (1) is:

$$y_p = v\sin x = -x\cos x + \sin x\ (\ln|\sin x|),$$

and the general solution of (1) is

$$y = c_1\sin x + c_2\cos x - x\cos x + \sin x(\ln|\sin x|).$$

Example: Solve $(x+2).\dfrac{d^2y}{dx^2} - (2x+5).\dfrac{dy}{dx} + 2y = (x+1)e^x.$ (1)

Solution: We note that $y = e^{2x}$ makes the left-hand side of (1) zero. Therefore, we let

$$y = ve^{2x}$$

$$\frac{dy}{dx} = \left(\frac{dv}{dx} + 2v\right)e^{2x}$$

$$\frac{d^2y}{dx^2} = \left(\frac{d^2v}{dx^2} + 4\frac{dv}{dx} + 4v\right)e^{2x}.$$

Substituting into (1), we obtain

$$(x+2)\left(\frac{d^2v}{dx^2} + 4\frac{dv}{dx} + 4v\right)e^{2x} - (2x+5)\left(\frac{dv}{dx} + 2v\right)e^{2x} + 2ve^{2x} = (x+1)e^x.$$

$$(x+2)\frac{d^2v}{dx^2}e^{2x} + \left[4(x+2) - (2x+5)\right]\frac{dv}{dx}e^{2x} + \left[4(x+2) - 2(2x+5) + 2\right]ve^{2x} = (x+1)e^x$$

$$(x+2)\frac{d^2v}{dx^2} + (2x+3)\frac{dv}{dx} = (x+1)e^{-x}.$$

Writing $u = \dfrac{dv}{dx},$ the above equation becomes:

$$(x+2)\frac{du}{dx} + (2x+3)u = (x+1)e^{-x}$$

$$\frac{du}{dx}+\frac{3x+3}{x+2}u=\frac{x+1}{x+2}e^{-x}. \tag{2}$$

This is a first order linear equation. An integrating factor of (2) is:

$$exp\left[\int\frac{2x+3}{x+2}dx\right]=\exp\left[\int\left(2-\frac{1}{x+2}\right)dx\right]$$

$$=\exp\left[2x-\ln(x+2)\right]=\frac{e^{2x}}{x+2}.$$

Multiplying (2) by $\frac{e^{2x}}{x+2}$, we have

$$\frac{d}{dx}\left(u\frac{e^{2x}}{x+2}\right)=\frac{x+1}{(x+2)^2}.e^{x}$$

$$u\frac{e^{2x}}{x+2}=\int\frac{x+1}{(x+2)^2}.e^{x}\,dx+c_1$$

$$=\int\frac{e^x}{x+2}dx-\int\frac{e^x}{(x+2)^2}dx+c_1$$

$$=\frac{e^x}{x+2}+c_1$$

$$u=e^{-x}+c_1(x+2)e^{-2x}.$$

Therefore, $\frac{dv}{dx}=e^{-x}+c_1(x+2)e^{-2x}.$

Integrating, we get

$$v=-e^{-x}-\frac{1}{4}c_1(2x+5)e^{-2x}+c_2.$$

Hence $y=ve^{2x}=-e^{x}-\frac{1}{4}c_1(2x+5)+c_2e^{2x}$ is the general solution of (1).

Note. It is easy to see that $y'' + Py' + Qy = 0$ is satisfied by $y = e^{\pm x}$, if $1 \pm P + Q = 0$, and by $y = x$ if $P + Qx = 0$. These can be used to find a solution $y'' + Py' + Q = 0$ by inspection.

THE WRONSKIAN

Definition. If y_1, y_2 are two differentiable functions of x on $I = [a, b]$, then their Wronskian, denoted by $W = W[y_1, y_2]$, is obtained by:

$$W = W[y_1, y_2] = y_1 y_2' - y_1' y_2 = \begin{vmatrix} y_1 & y_2 \\ y_1' & y_2' \end{vmatrix}.$$

Similarly, the Wronskian of three differentiable functions y_1, y_2, y_3 on $I = [a, b]$ is defined by:

$$W = W[y_1, y_2, y_3] = \begin{vmatrix} y_1 & y_2 & y_3 \\ y_1' & y_2' & y_3' \\ y_1'' & y_2'' & y_3'' \end{vmatrix}.$$

The definition can be extended in a similar manner for the Wronskian of *n* differentiable functions on $I = [a, b]$.

Theorem. Let y_1, y_2 be two solutions of the homogeneous linear equation

$$\frac{d^2 y}{dx^2} + P\frac{dy}{dx} + Qy = 0, \qquad (1)$$

where P, Q are functions of x and are continuous on $I = [a, b]$. Their Wronskian $W = W[y_1, y_2]$ is then either identically zero or is never zero on I.

Proof. We have $W = y_1 y_2' - y_1' y_2$

Differentiating $w.r.t.x,$ we get:

$$W' = y_1 y_2'' + y_1' y_2' - y_1' y_2' - y_1'' y_2$$

$$= y_1 y_2'' - y_1'' y_2.$$

Since y_1, y_2 are solutions of (1), we have:

$$y_1'' + Py_1' + Qy_1 = 0 \quad (2)$$

and $y_2'' + Py_2' + Qy_2 = 0.$ (3)

Multiply (2) by y_2 and (3) by y_1 to have the equivalent system

$$y_1'' y_2 + P y_1' y_2 + Qy_1 y_2 = 0 \quad (4)$$

$$y_1 y_2'' + P y_2' y_1 + Qy_1 y_2 = 0 \quad (5)$$

Subtracting (4) from (5), we obtain:

$$\left(y_1 y_2'' - y_2 y_1''\right) + P\left(y_1 y_2' - y_1' y_2\right) = 0.$$

$$\frac{dW}{dx} + PW = 0.$$

The general solution of the above equation is readily found as $W = c\exp(-\int P\,dx\,).$

Since the exponential function is never zero, W is identically zero if $c = 0$ or is never zero when $c \neq 0.$

Theorem. If y_1, y_2 are two solutions of $\frac{d^2y}{dx^2} + P\frac{dy}{dx} + Qy = 0,$ their Wronskian $W = W\left[y_1, y_2\right] = 0$ if and only if y_1, y_2 are linearly dependent.

Proof. If one of the two solutions is identically zero, the theorem is obviously true. Assume that y_1, y_2 are both nonzero and let y_1, y_2 be linearly dependent. Then $y_1 = cy_2,$ where *c* is a constant,

i.e., $\frac{y_1}{y_2} = c$,

or $\frac{d}{dx}\left(\frac{y_1}{y_2}\right)=\frac{y_2y_1'-y_1y_2'}{y_2^2}=\frac{dc}{dx}=0$,

i.e., $y_1y_2'-y_1'y_2=0$

$W[y_1,y_2]=0.$

Conversely, if $W[y_1,y_2]=0$, then $\frac{d}{dx}\left(\frac{y_1}{y_2}\right)=\frac{W}{y_2^2}=0$ and $\frac{y_1}{y_2}=c$, where c is a constant.

Thus $y_1=cy_2$ and so y_1,y_2 are linearly dependent.

Corollary. $W[y_1,y_2]\neq 0$ if and only if y_1,y_2 are linearly independent.

The above theorems can only be applied to check the linear dependence or linear independence of differentiable functions that are solutions of a linear homogeneous O.D.E. Arbitrary differentiable functions exist with their Wronskian as zero, however, they are linearly independent.

Consider the functions $y_1=x^3, y_2=|x^3|$ on $[-\infty,\infty]$. Here $W(y_1,y_2)=0$, but y_1 and y_2 are linearly independent.

VARIATION OF PARAMETERS.

We found the solution of the equation $\frac{d^2y}{dx^2}+P\frac{dy}{dx}+Qy=F(x)$, (1)

where P,Q are functions of x, in the previous section, by the reduction of order of (1). The solution of (1) can be determined by a procedure known as the method of variation of parameters. This method can be applied even to equations of higher orders.

Suppose that linearly independent solutions of

$$\frac{d^2y}{dx^2}+P\frac{dy}{dx}+Q=0 \qquad (2)$$

are given by $y = y_1(x)$ and $y = y_2(x)$. Then the complementary function of (1) is

$$y_c = c_1 y_1 + c_2 y_2,$$

where c_1 and c_2 are arbitrary constants. We replace the arbitrary constants c_1 and c_2 by unknown functions $u_1(x)$ and $u_2(x)$ and require that

$$y_p = u_1 y_1 + u_2 y_2 \qquad (3)$$

be a particular solution of (1). In order to determine the two functions u_1 and u_2, we need two conditions. One condition is that (3) must satisfy (1). A second condition can be imposed arbitrarily.

Differentiating (3) $w.r.t.x$, we have:

$$y_p' = \left(u_1' y_1 + u_2' y_2\right) + u_1 y_1' + u_2 y_2'.$$

If we differentiate the above equation, y_p'' will contain u_1'' *and* u_2''. To avoid second derivatives of u_1 and u_2. we let:

$$u_1' y_1 + u_2' y_2 = 0. \qquad (4)$$

With this condition, we have $y_p' = u_1 y_1' + u_2 y_2'$, such that $y_p' = u_1 y_1' + u_2 y_2'$ and $y_p'' = u_1' y_1' + u_2' y_2' + u_1 y_1'' + u_2 y_2''$.

Substituting for y_p, y_p', y_p'' into equation (1), we get:

$$\left(u_1' y_1' + u_2' y_2' + u_1 y_1'' + u_2 y_2''\right) + P\left(u_1 y_1' + u_2 y_2'\right) + Q\left(u_1 y_1 + u_2 y_2\right) = F(x)$$

or $u_1\left(y_1'' + P y_1' + Q y_1\right) + u_2\left(y_2'' + P y_2' + Q y_2\right) + u_1' y_2' = F(x).$

Expressions within the parenthesis are zero since y_1 and y_2 are solutions of (2). Hence

$$u_1' y_1' + u_2' y_2' = F(x) \qquad (5)$$

Taking (4) and (5) together, we have two equations in the two unknowns u_1' and u_2';

$u_1'y_1 + u_2'y = 0$ and $u_1'y_1' + u_2'y_2' = F(x)$.

Solving these, we have:

$$\left.\begin{aligned} u_1' &= \frac{-y_2F(x)}{y_1y_2' - y_1'y_2} \\ u_2' &= \frac{y_1F(x)}{y_1y_2' - y_1'y_2} \end{aligned}\right\} \qquad (6)$$

In (6), the Wronskian of y_1, y_2 namely $W(y_1, y_2) = y_1y_2' - y_1'y_2 \neq 0$, since y_1, y_2 are linearly independent solutions of (2). Integrating (6), we find u_1 and u_2 as

$$u_1 = \int \frac{-y_2F(x)}{y_1y_2' - y_1'y_2}dx = \int \frac{-y_2F(x)}{W}dx \qquad (7)$$

$$u_2 = \int \frac{y_1F(x)}{y_1y_2' - y_1'y_2}dx = \int \frac{y_1F(x)}{W}dx, \qquad (8)$$

where W is the Wronskian of y_1, y_2.

Thus y_p is completely determined. In numerical problems, instead of performing the complete process, the formulas in (7) will be directly applied to evaluate u_1 and u_2.

Example. Find the general solution of

$$\frac{d^2y}{dx^2} - 2\frac{dy}{dx} + y = xe^x \ln x. \qquad (1)$$

Solution. The C.F. of (1) is

$y_c = c_1e^x + c_2xe^x$.

Let $y_p = u_1e^x + c_2xe^x$. Here $y_1 = e^x, y_2 = xe^x, F(x) = xe^x \ln x$

$W = W(y_1, y_2) = y_1y_2' - y_1'y_2 = e^x(e^x + xe^x) - xe^x.e^x = e^{2x}$.

From the formulas of (7) and (8), we have,

$$u_1 = \int \frac{-y_2 F(x)}{W} dx = \int \frac{-xe^x . xe^x \ln x}{e^{2x}} dx$$

and $$u_2 = \int \frac{y_1 F(x)}{W} dx = \int \frac{e^x . xe^x \ln x}{e^{2x}} dx$$,

i.e., $$u_1 = -\int x^2 \ln x dx = -(\ln x).\frac{x^3}{3} + \int \frac{x^3}{3} dx = \frac{x^3}{9} - \frac{x^3}{3} \ln x$$

$$u_2 = \int x \ln x \, dx = \frac{x^2}{2} \ln x - \int \frac{x}{2} dx = -\frac{x^2}{4} + \frac{x^2}{2} \ln x.$$

$$y_p = \left(\frac{x^3}{9} - \frac{x^3}{3} \ln x \right) e^x + \left(-\frac{x^2}{4} + \frac{x^2}{2} \ln x \right) x e^x$$

$$= e^x \left(\frac{x^3}{9} - \frac{x^3}{3} \ln x - \frac{x^3}{4} + \frac{x^3}{2} \ln x \right)$$

$$= e^x \left(-\frac{5x^3}{36} + \frac{x^3}{6} \ln x \right).$$

The general solution of (1) is

$$y = y_c + y_p = c_1 e^x + c_2 x e^x = e^x \left(-\frac{5x^3}{36} + \frac{x^3}{6} \ln x \right).$$

Example. Find the general solution of

$$x^2 \frac{d^2 y}{dx^2} - x(x+2) \frac{dy}{dx} + (x+2) y = x^3, \quad (1)$$

given that $y = xe^x$ is a solution of the associated homogeneous equation

$$x^2 \frac{d^2 y}{dx^2} - x(x+2) \frac{dy}{dx} + (x+2) y = 0. \quad (2)$$

Solution. The given equation in the standard form is

$$\frac{d^2 y}{dx^2} - \frac{x+2}{x} \frac{dy}{dx} + \frac{x+2}{x^2} y = x.$$

Since $P+Qx=-\frac{x+2}{x}+x\frac{x+2}{x^2}=0, y=x$ is also a solution of (2). The two solutions, $y=x$ and $y=xe^x$, are linearly independent. The complementary function of (1) is

$y_c=c_1x+c_2xe^x.$

We assume that $y_p=u_1x+u_2xe^x$ (3)

is a particular solution of (1).

Here $y_1=x, y_2=xe^x, F(x)=x$ and

$W=W(y_1,y_2)=y_1y_2'-y_1'y_2=x(xe^x+e^x)-xe^x=x^2e^x.$

From the formulas of (7) and (8), we have:

$u_1=\int\frac{-y_2F(x)}{W}dx$ and $u_2=\int\frac{y_1F(x)}{W}dx$

i.e., $u_1=-\int\frac{x^2e^x}{x^2e^x}dx=-x$ and $u_2=\int\frac{x^2}{x^2e^x}dx=-e^{-x}$.

Substituting for u_1,u_2 into (3), a particular solution of (1) is

$y=y_c+y_p=c_1x+c_2xe^x-x^2-x.$

Example. Explain how the method of variation of parameters can be applied to find a particular solution of a nonhomogeneous linear third order differential equation whose complementary function is known. Apply the method to find a particular solution of

$$\frac{d^3y}{dx^3}+\frac{dy}{dx}=\cos x.$$

Solution. Suppose the C.F. of a linear third order differential equation

$$\frac{d^3y}{dx^3}+P\frac{d^2y}{dx^2}+Q\frac{dy}{dx}+Ry=F(x) \quad (1)$$

is known to be $y_c=c_1y_1+c_2y_2+c_3y_3$; with y_1,y_2,y_3 being linearly independent solutions of the associated homogeneous equation of (1).

We assume that $y_p = u_1 y_1 + u_2 y_2 = u_3 y_3$, where u_1, u_2, u_3 are functions of x.

$$y_p' = u_1 y_1' + u_3 y_3' + u_1' y_1 + u_2' y_2 + u_3' y_3.$$

We set $u_1' y_1 + u_2' y_2 + u_3' y_3 = 0.$

Then $y_p' = u_1 y_1' + u_2 y_2' + u_3 y_3'$

$$y_p'' = u_1 y_1'' + u_2 y_2'' + u_3 y_3''$$

$$y_p''' = u_1 y_1''' + u_2 y_2''' + u_3 y_3''' + u_1' y_1'' + u_2' y_2'' + u_3' y_3''.$$

Substituting for y_p, y_p', y_p'', y_p''' into equation (1), we have:

$$\left(u_1 y_1''' + u_2 y_2''' + u_3 y_3''' + u_1' y_1'' + u_2' y_2'' + u_3' y_3''\right) + P\left(u_1 y_1'' + u_2 y_2'' + u_3 y_3''\right)$$

$$+Q\left(u_1 y_1' + u_2 y_2' + u_3 y_3'\right) + R\left(u_1 y_1 + u_2 y_2 + u_3 y_3\right) = F(x)$$

$$u_1\left(y_1''' + Py_1'' + Qy_1' + Ry_1\right) + u_2\left(y_2''' + Py_2'' + Qy_2' + Ry_2\right)$$

$$+u_2\left(y_2''' + Py_2'' + Qy_2' + Ry_2\right)$$

$$+u_3\left(y_3''' + Py_3'' + Qy_3' + Ry_3\right) + u_1' y_1'' + u_2' y_2'' + u_3' y_3'' = F(x). \tag{4}$$

From the system of equations (2), (3) and (4), we find u_1', u_2' and u_3', and upon integration we get u_1, u_2 and u_3.

We then apply this method to find the P.I. of $\dfrac{d^3y}{dx^3} + \dfrac{dy}{dx} = \csc x.$

The C.F. of the equation is obtained by solving the equation $D\left(D^2+1\right) = 0.$ Thus

$y_c = c_1 + c_2 \cos x + c_3 \sin x$ with $y_1 = 1, y_2 = \cos x$ and $y_3 = \sin x.$

For the particular integral, let $y_p = u_1 y_1 + u_2 y_2 + u_3 y_3$,

$= u_1 + u_2 \cos x + u_3 \sin x.$

Substituting for y_1, y_2, y_3 and their derivatives into equations (2), (3) and (4) above, we have the following system of linear equations in u_1', u_2', u_3':

$$u_1' + u_2' \cos x + u_3' \sin x = 0$$

$$u_1'.0 - u_2' \sin x + u_3' \cos x = 0$$

$$u_1'.0 - u_2' \cos x - u_3' \sin x = \csc x.$$

We solve this system using the Gaussian Elimination method. The augmented matrix of the system is given as

$$\left[\begin{array}{ccc|c} 1 & \cos x & \sin x & 0 \\ 0 & -\sin x & \cos x & 0 \\ 0 & -\cos x & -\sin x & \csc x \end{array}\right]$$

$$\left[\begin{array}{ccc|c} 1 & 0 & 0 & \csc x \\ 0 & -\sin x & \cos x & 0 \\ 0 & -\cos x & -\sin x & \csc x \end{array}\right] by\, R_1 + R_3$$

$$\left[\begin{array}{ccc|c} 1 & 0 & 0 & \csc x \\ 0 & -\sin x \cos x & \cos^2 x & 0 \\ 0 & -\cos x \sin x & -\sin^2 x & 1 \end{array}\right] \begin{array}{l} by (\cos x) R_2 \\ and (\sin x) R_3 \end{array}$$

$$\left[\begin{array}{ccc|c} 1 & 0 & 0 & \csc x \\ 0 & -\sin x \cos x & \cos^2 x & 0 \\ 0 & 0 & -1 & 1 \end{array}\right].$$

Therefore, $-u_3' = 1 \, or \, u_3' = -1,$ from R_3,

$(-\sin x \cos x) u_2' + (\cos^2 x) u_3' = 0$ from R_2,

or $u_2' = -\cot x$,

$u_1' = \csc x,$ from R_1.

Hence $u_1 = \int \csc x\,dx = \ln\left|\csc x - \cot x\right|$,

$u_2 = \int \frac{-\cos x}{\sin x}\,dx = -\ln\left|\sin x\right|,$

and $u_3 = \int - dx = -x.$

$y_p = u_1 + u_2 \cos x + u_3 \sin x.$

$= \ln\left|\csc x - \cot x\right| - (\cos x) ln\left|\sin x\right| - x \sin x$

Is the required particular solution.

APPLICATION OF DIFFERENTIAL EQUATIONS

Differential equations can be used in the solution of many problems in physical, biological and social sciences. In this section, we shall study such problems, in terms of first and second order differential equations. The solution of the differential equation will lead to the solution of the given problem.

The following examples illustrate the techniques.

Example. A stone weighting 8 lbs. falls from rest toward the earth from a great height. As it falls it is acted upon by air resistance that is numerically $\frac{1}{2}v(\text{in pounds})$, where v is the velocity (in feet per second). Find the velocity and distance fallen at time t.

Solution. We use Newton's second law $F = ma$ to formulate a mathematical model of this problem. The forces acting on the body are:

- F_1, the weight of the stone acting downward and hence is positive.
- F_2, the air resistance, numerically equal to $\frac{1}{2}v$, which acts upward and is therefore negative.

Newton's second law becomes $m\frac{dv}{dt} = F_1 + F_3,$ i.e., $\frac{1}{4}\frac{dv}{dt} = 9 - \frac{1}{2}v,$ taking $g = 32.$

$$\frac{dv}{dt} = 32 - 2v$$

$$\frac{dv}{dt} + 2v = 32,$$

which is a linear equation. We now solve this differential equation. Here the $I.F$ is $e^{\int 2\,dt} = e^{2t}.$

Thus, $\frac{d}{dt}\left(ve^{2t}\right) = 32e^{2t}.$

Integrating, we get

$ve^{2t} = 16e^{2t} + c$ or $v = 16 + ce^{-2t}.$

Now $v(0) = 0.$

Hence $0 = 16 + c$ or $c = -16.$

Therefore, $v = 16\left\{1 - e^{-2t}\right\}$ (1)

is the velocity after time t. (1) may be written as:

$$\frac{dx}{dt} = 16\left(1 - e^{-2t}\right). \quad (2)$$

From (2), we have $dx = 16\left(1 - e^{-2t}\right)dt,$

or $x = 16t + 8e^{-2t} + k.$

Applying the initial condition $x(0) = 0,$ we find that $k = -8.$

Hence $x = 16t + 8e^{-2t} - 8$ is the distance fallen after time t.

Example. A body of constant mass is projected upward from the earth's surface with an initial velocity v_0. Assuming there is no air resistance, but taking into consideration the variation of the earth's gravitational field with altitude, find the smallest initial velocity for which the body will not return to the earth (this is the so-called escape velocity).

Solution. The general expression for the weight $w(x)$ of a body of mass m is obtained from Newton's inverse- square law of gravitational attraction. If *R* is the radius of the earth and x is the altitude above sea level, then $w(x)=\frac{k}{(R+x)^2}, k$ being constant. At $x=0, w=mg,$ hence $k=mgR^2$ and

$$w(x)=\frac{mg\,R^2}{(R+x)^2}.$$

The only force acting on the body is its weight, which acts downward. Thus, the equation of motion is $m\frac{dv}{dt}=\frac{-mg\,R^2}{(R+x)^2}$,

or

$$v\frac{dv}{dt}=-\frac{mg\,R^2}{(R+x)^2}.$$

Separating the variables and integrating, we have

$$\frac{1}{2}v^2=\frac{gR^2}{R+x}+c.$$

However, $v=v_0 \text{ at } t=0$ i.e., at $x=0.$

Therefore, $c=\frac{1}{2}v_0^2-gR$,

and so

$$v^2=v_0^2-2gR+\frac{2\,gR^2}{R+x}. \tag{1}$$

The escape velocity is found by requiring that *v*, given by (1), remains positive for all (positive) values of x. Thus we must have:

$v_0^2 \geq n\,2g\,R.$

Hence the escape velocity is $v_0 = \sqrt{2gR} =$ 6.9 mile/sec, taking R=4000 miles.

SUMMARY AND DISCUSSION

In this chapter, we discussed higher order differential equations in detail, as well as different methods to solve higher order differential equations.

Linear Differential Equations

Definition. A linear differential equation of order n in the dependent variable y and the independent variable x is of the form:

$$a_0(x)\frac{d^n y}{dx^n} + a_1(x)\frac{d^{n-1}y}{dx^{n-1}} + \ldots + a_{n-1}\frac{dy}{dx} + a_n(x)y = F(x), \tag{1}$$

where $a_0(x), a_1(x), \ldots, a_{n-1}(x), a_n(x)$ and $F(x)$ are functions of the independent variable x only, and $a_0(x)$ is not identically zero. Using the notation of primes, (1) is also written as:

$$a_0 y^n + a_1 y^{n-1} + \ldots + a_{n-1}y' + a_n y = F(x), \tag{2}$$

where $a_0, a_1, \ldots, a_{n-1}, a_n$ are real constants.

Definition. If $y_1(x), y_2(x), \ldots, y_m(x)$ are m functions of an independent variable x and $c_1, c_2, \ldots, c_m$ are constants, then the expression $c_1 y_1(x) + c_2 y_2(x) + \ldots + c_m y_m(x)$ is called a linear combination of $y_1(x), y_2(x), \ldots, y_m(x)$.

- Every homogeneous linear n[th]-order differential equation $a_0 y^n + a_1 y^{n-1} + \ldots + a_{n-1}y' + a_n y = 0$ has n linearly independent solutions $y_1, y_2, \ldots, y_n$.
- If $y_1, y_2, \ldots, y\_m$ are n linearly independent solutions of (3), then any linear combination $y_c = c_1 y_1 + c_2 y_2 + \ldots + c_n y_n$ of

$y_1, y_2, \ldots, y_n$ is the general solution of (3), $c_1, c_2, \ldots, c_m$ being arbitrary constants.

- Let y_p be any particular solution of (2). i.e., y_p does not contain any constants, then $y_c + y_p$ is the general solution of (2).

Homogeneous Linear Equations

Consider the equation $a_0 \dfrac{d^n y}{dx^n} + a_1 \dfrac{d^{n-1} y}{dx^{n-1}} + \ldots + a_{n-1} \dfrac{dy}{dx} + a_n y, = 0,$ (1)

where $a_0, a_1, \ldots, a_{n-1}$ are real constants.

Case I. Distinct Real Roots

Let $m_1, m_2, \ldots, m_n$ be n distinct real roots of (2). Then $e^{m_1 x}, e^{m_2 x}, \ldots, e^{m_n x}$ are n distinct solutions of (1). These n solutions are linearly independent. Hence the general solution of (2) is $y = c_1 e^{m_1 x} + c_2 e^{m_2 x} + \ldots + c_n e^{m_n x}$, where $c_1, c_2, \ldots, c_n$ are arbitrary constants.

Case II. Repeated Real Roots

$y = \left(c_1 + c_2 x \right) e^{m_1 x} + c_3 e^{m_3 x} + \ldots + c_n e^{m_n x}$

In the same manner, if the characteristic equation (2) has the triple repeated root m_1, the corresponding part of the general solution of (1) is the solution of:

$(D - m_1)^3 y = 0.$

Proceeding as before, we can easily find

$y = \left(c_1 + c_2 x + c_3 x^2 \right) e^{m_1 x}.$

Case III. Complex Roots

Suppose the characteristic equation has the complex number $a+ib$ as a non-repeated root. Since coefficients of (1) are real, the conjugate complex number $a-ib$ is also a non-repeated root. The corresponding part of the general solution is:

$y=k_1e^{(a+ib)x}+k_2e^{(a-ib)x}$, where k_1 and k_2 are arbitrary constants.

$$y=e^{ax}\left[k_1e^{ibx}+k_2e^{-ibx}\right]$$

$$=e^{ax}\left[k_1(\cos bx+i\sin bx)+k_2(\cos bx-i\sin bx)\right]$$

$$=e^{ax}\left[(k_1+k_2)\cos bx+i(k_1-k_2)\sin bx\right]$$

$$=e^{ax}\left[c_1\sin bx+c_2\cos bx\right],$$

where $c_1=i(k_1-k_2), c_2=k_1+k_2$ are two arbitrary constants.

If $a+ib$ and $a-ib$ are conjugate complex roots, each repeated k times, the corresponding part of the general solution of (1) may be written as:

$$y=\left[\left(c_1+c_2x+c_3x^2+\ldots+c_kx^{k-1}\right)\sin bx+\left(c_{k+1}+c_{k+2}x+\ldots+c_{2k}x^{k-1}\right)\cos bx\right].$$

Solution of the equation

- Complementary Function (C.F)
- Particular Solution (P.I)

The C.F is the solution of the homogeneous equation $f(D)y=0$, we have described different cases of its solution.

Now to find the $P.I$ of (1), we can write

$$y=\frac{1}{f(D)}F(x)$$

and try to evaluate $y=\dfrac{1}{f(D)}F(x)$.

Finding the P.I when $F(x) = \sin ax$ or $\cos ax$.

Here, we have to evaluate $\frac{1}{f(D)}\sin ax$ and $\frac{1}{f(D)}\cos ax$.

From Euler's theorem, we have

$e^{iax} = \cos ax + i\sin ax.$

Thus $\frac{1}{f(D)}\sin ax$ and $\frac{1}{f(D)}\cos ax$ are the imaginary and real parts of $\frac{1}{f(D)}e^{iax}$, respectively.

If $f(D)$ contains only even powers of D, say $f(D) = f(D^2)$, it is easy to see that

$$\frac{1}{f(D^2)}\begin{cases}\sin ax \\ \cos ax\end{cases} = \frac{1}{f(-a^2)}\begin{cases}\sin ax \\ \cos ax\end{cases}, \quad \textit{provided} -a^2 \textit{ is not a zero of } f(D^2).$$

Series Expansion of Polynomials or Expressions Involving Operators

This method is useful when $F(x)$ is a polynomial in x.

Let $f(D)y = F(x)$ be such that $F(x)$ is a polynomial in x. To evaluate the particular integral

$$y_p = \frac{1}{f(D)}F(x),$$

it is often useful to expand $\frac{1}{f(D)}$ in a series in D by the Binomial Theorem for negative exponents, such that:

$$y_p = \left(1 + a_1D + a_2D^2 + a_3D^3 + \ldots\right)F(x).$$

The derivatives on the right will vanish after a certain stage since $D^n x^r = 0 \; if \; n > r.$

The following binomial expansions will be useful for solving these types of problems:

$$\frac{1}{1-x} = 1 + x + x^2 + x^3 + \ldots.$$

$$\frac{1}{1+x} = 1 - x + x^2 - x^3 + \ldots.$$

The Method of Undetermined Coefficients (U. C. Method)

We have seen some special cases in which the particular integral can be evaluated by the inverse operator. Now we consider the method of undetermined coefficients, which can prove simpler in finding the particular integral of the equation $f(D)y = F(x)$ when $F(x)$ is

- An exponential function: e^{ax}.
- A polynomial: $\left(b_0 x^n + b_1 x^{n-1} + \ldots + b_n\right)$.
- A Sinusoidal function: $\sin ax$ or $\cos bx$
- The more general case in which $F(x)$ is sum of a product of terms of the above types, such as $F(x) = e^{ax}\left(b_0 x^n + b_1 x^{n-1} + \ldots + b_n\right)\begin{Bmatrix} \sin ax \\ \cos bx \end{Bmatrix}$.

The Cauchy-Euler Equation

An equation of the form

$$a_0 x^n \frac{d^n y}{dx^n} + a_1 x^{n-1} \frac{d^{n-1} y}{dx^{n-1}} + \ldots + a_{n-1} x \frac{dy}{dx} + a_n y = F(x)$$

is called a Cauchy-Euler (or equidimensional) equation, with $a_0, a_1, \ldots, a_{n-1}, a_n$ real constants. The equation can be reduced to a linear differential equation with constant coefficients using the transformation $x = e^t$ or $t = \ln x$.

Reduction of Order

If one solution of the second order linear equation

$$\frac{d^2y}{dx^2}+P\frac{dy}{dx}+Qy=0 \qquad (1)$$

(where P,Q are not necessarily constants and may be functions of x) is known, then we can use it to find the general solution of

$$\frac{d^2y}{dx^2}+P\frac{dy}{dx}+Qy=F(x) \qquad (2)$$

This procedure is known as the reduction of order method.

THE WRONSKIAN

Definition. If y_1, y_2 are two differentiable functions of x on $I=[a,b]$, then their Wronskian, denoted by $W=W[y_1,y_2]$, is obtained by:

$$W=W[y_1,y_2]=y_1y_2'-y_1'y_2=\begin{vmatrix} y_1 & y_2 \\ y_1' & y_2' \end{vmatrix}.$$

Similarly, the Wronskian of three differentiable functions y_1, y_2, y_3 on $I=[a,b]$ is defined by

$$W=W[y_1,y_2,y_3]=\begin{vmatrix} y_1 & y_2 & y_3 \\ y_1' & y_2' & y_3' \\ y_1'' & y_2'' & y_3'' \end{vmatrix}.$$

The definition can be extended in a similar manner for the Wronskian of *n* differentiable functions on $I=[a,b]$.

Theorem. If y_1, y_2 are two solutions of $\frac{d^2y}{dx^2}+P\frac{dy}{dx}+Qy=0,$ Then their Wronskian $W=W[y_1,y_2]=0$ if and only if y_1, y_2 are linearly dependent.

Corollary. $W[y_1, y_2] \neq 0$ if and only if y_1, y_2 are linearly independent.

VARIATION OF PARAMETERS.

We found the solution of the equation $\frac{d^2y}{dx^2} + P\frac{dy}{dx} + Qy = F(x)$, (1)

where P, Q are functions of x, in the previous section by reduction of order of (1). The solution of (1) can be determined by a procedure known as the method of variation of parameters. This method can be applied even to equations of higher orders.

$$u_1 = \int \frac{-y_2 F(x)}{y_1 y_2' - y_1' y_2} dx = \int \frac{-y_2 F(x)}{W} dx,$$

$$u_2 = \int \frac{y_1 F(x)}{y_1 y_2' - y_1' y_2} dx = \int \frac{y_1 F(x)}{W} dx,$$

where W is the Wronskian of y_1, y_2.

CHAPTER 3

FRICTION

INTRODUCTION

When two smooth surfaces are in contact with a point *P*, the reaction of one body on the other body is along a common normal. However, if the bodies are not smooth, the reaction of one body on the other is not along the common normal but it is inclined at an angle θ to the common normal at the point of contact *P*. Let R has component R along the common normal at P and F along the tangent plane at P Therefore,

$R = R' \cos\theta$ and $F = R' \sin\theta$,

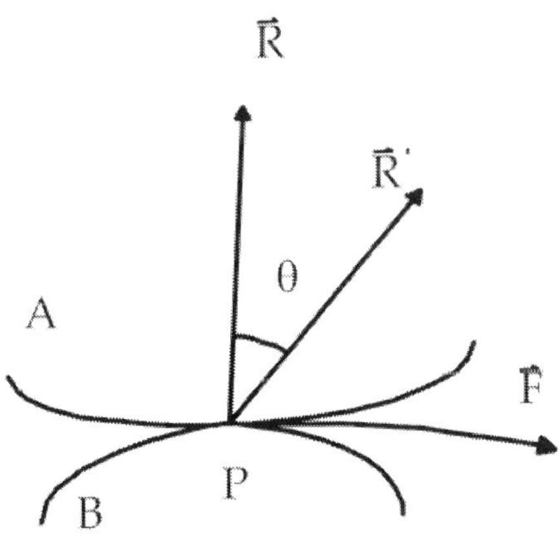

where *R* is called the normal reaction or normal pressure, and *F* is called the force of friction, or simply friction. The force of friction prevents the sliding of one body on the surface of the other. R' is the resultant of the normal reaction R and force of friction F and is also called resultant friction.

Types of Friction

Static Friction

When one body in contact with another is in equilibrium and is not on the point of sliding along the other, the force of friction is called static friction. The equilibrium in this case is non-limiting.

Limiting Friction

When one body is just at the point of sliding on the other; the friction in this case is maximum, known as the limiting friction. The equilibrium in the case is called the *limiting equilibrium.*

Dynamical Friction

When one body is sliding on the other, the friction in this case is slightly less than the limiting friction and is called dynamical friction, or kinetic friction.

Law of Friction

1. The direction of friction is opposite to the direction in which the body moves of tends to move.
2. The magnitude of friction is, up to certain extent, equal to the force tending to produce motion.
3. The magnitude of the limiting friction bears a constant ratio μ to the normal reaction between the surfaces. Here, μ is called the coefficient of friction.
4. The amount of friction is independent of the area and shape of the surface in contact, provided the normal pressure remains unaltered.
5. When the motion takes place, the friction still opposes the motion. It is independent of the velocity and is proportional to the normal reaction, however, is slightly less than the limiting friction.

Coefficient of Friction

The magnitude of the limiting friction F bears a constant ration μ to the normal reaction R between the surfaces. The constant μ is known as the coefficient of friction and

$$\frac{F}{R} = \mu.$$

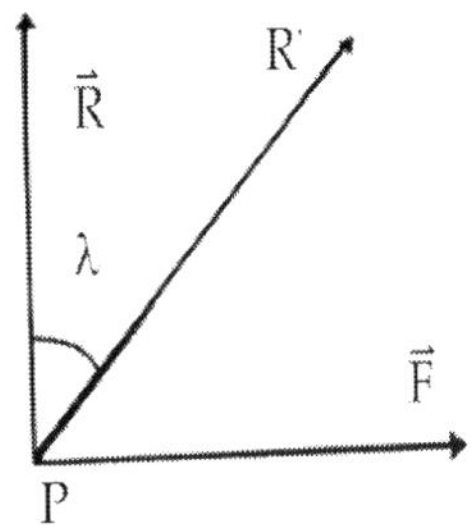

Angle of Friction

The angle λ, i.e. the direction of the resultant reaction R' when the body is just on the point of motion, is denoted as the angle of friction.

Now $R = R' \cos\lambda$ and $F = R' \sin\lambda$.

Since $\frac{F}{R} = \mu$, therefore,

$$\mu = \frac{R' \sin\lambda}{R' \cos\lambda} \quad \text{or} \quad \mu = \tan\lambda$$

Also $R^2 + F^2 = R'^2$

or $R' = \sqrt{R^2 + F^2} = \sqrt{R^2 + \mu^2 R^2}$

or $R' = R\sqrt{1 + \mu^2}$.

Cone of Friction

The cone of friction is a cone whose vertex is the point of contact P, with its axis along the common normal at P and whose semi – vertical angle is the angle of friction.

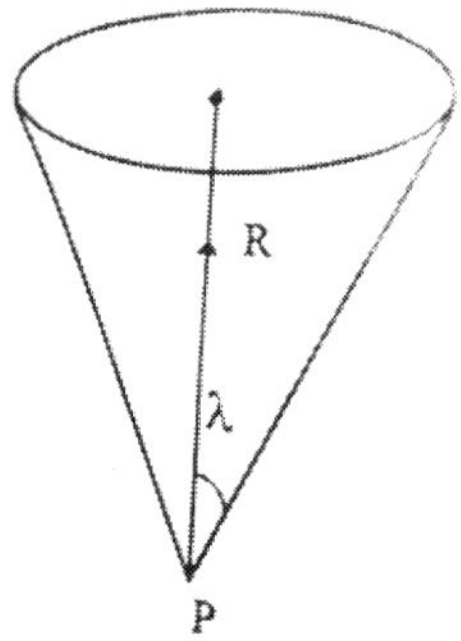

Equilibrium of a Particle on a Rough Horizontal Plane

We shall determine the least amount of force needed to send into motion a particle at rest on a rough horizontal plane. Let *P* be the required least amount of force that makes an angle with the horizontal plane. Let be the mass of the particle. For a limiting equilibrium

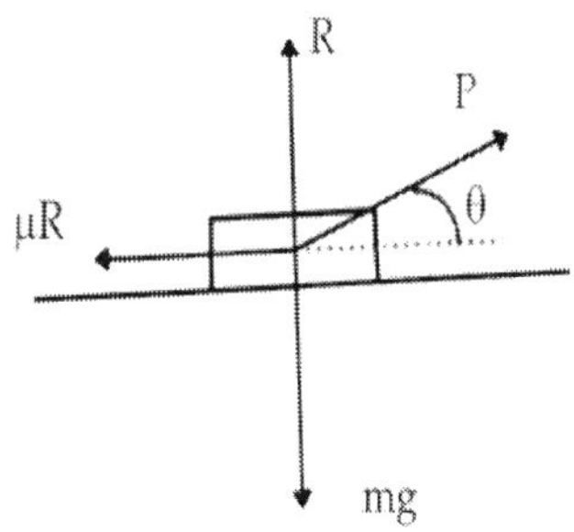

$$P\cos\theta = \mu R \qquad (1)$$

$$R + P\sin\theta = mg. \qquad (2)$$

From (2)

$$R = mg - p\sin\theta.$$

Putting this into eq. (1), we get

$$P(\cos\theta = \mu(mg - p\sin\theta)$$

$$P(\cos\theta + \mu\sin\theta) = \mu mg$$

$$P = \frac{\mu mg}{cos\theta + \mu\, sin\theta}.$$

Since $\mu = \tan\lambda$, therefore

$$P = \frac{mgtan\lambda}{cos\theta + tan\lambda\, sin\theta}$$

$$= \frac{mgsin\lambda}{cos\lambda + \tan\lambda\, sin\theta}$$

$$= \frac{mgsin\lambda}{\cos(\theta - \lambda)}.$$

P is at its minimum when $\cos(\theta - \lambda)$ is at its maximum i.e $\cos(\theta - \lambda) = 1$ or $\theta - \lambda = 0$ $or\ \theta = \lambda$.

Thus, $P_{least} = mg \sin\lambda$.

Equilibrium of a Particle on a Rough Horizontal Plane

Show that if a particle is in equilibrium on an inclined plane under its own weight, the inclination of the plane is equal to the angle of friction.

Proof.

Let be the inclination of the plane and *m* be the mass of the particle. For a limiting equilibrium, we have:

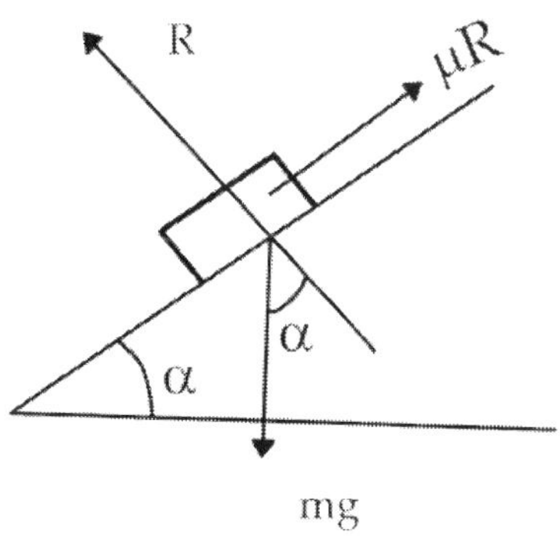

$R = mg\cos\alpha$ (1)

$\mu R = mg\sin\alpha$. (2)

From (1) and (2) we have

$$\mu = \frac{mg\sin\alpha}{mg\cos\alpha} = \tan\alpha$$

Since $\mu = \tan\lambda$ therefore

tan λ = tan α and

$$\alpha = \lambda,$$

where λ is the angle of friction.

Finding the least amount of force needed to send a particle into motion on a rough horizontal plane.

Let *P* be the required least amount of force which at angle θ with the inclined plane. Let α be the inclination of the plane and *m* be the mass of the particle. For a limiting equilibrium, we have,

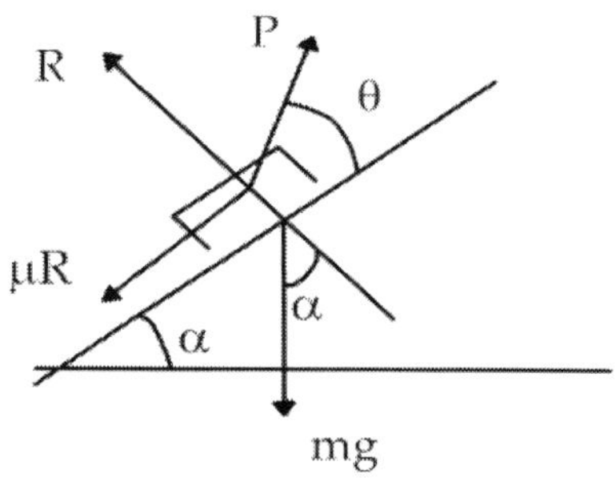

$P\cos\theta = \mu R + mg\sin\alpha$ (1)

$P\sin\theta + R = mg\cos\alpha.$ (2)

From (2)

$R = mg\cos\alpha - p\sin\theta$.

Then (1) becomes:

$P(\cos\theta) = \mu(mg\cos\alpha - P\sin\theta) + mg\sin\alpha$

$P(cos\theta+\mu sin\theta)=mg(cos\alpha+sin\alpha)$

$$P=\frac{mg\ (\mu\cos\alpha+sin\alpha)}{cos\theta+\mu sin\theta}.$$

Since $\mu=\tan\lambda$; with λ as the angle of friction, we have

$$P=\frac{mg\,(\sin\lambda\cos\alpha+\cos\lambda\sin\alpha)}{\cos\lambda\cos\theta+\sin\lambda\sin\theta}$$

$$=\frac{mg\ \sin(\alpha+\lambda)}{\cos(\theta-\lambda)}.$$

For minimum *P*, $cos(\theta-\lambda)$ is at its maximum, $i.e\ cos(\theta-\lambda)=1$.
or $\theta-\lambda=0$ or $\theta=\lambda$.

Thus,

$P_{least}=mg\sin(\alpha+\lambda)$.

Example

Consider the least amount of force *P* that will move a weight up an inclined plane. Show that this minimum of *P*, acting parallel to the plane, and moving it upwards, is $P\sqrt{1+\mu^2}$, where μ is the coefficient of friction.

Solution.

We know that the least amount of force required of *P* is

$P_{least}=mg\sin(\alpha+\lambda)$, (1)

where α is the inclination of the plane. Let P' be the magnitude of the least amount of force acting parallel to the plane. For a limiting equilibrium we have,

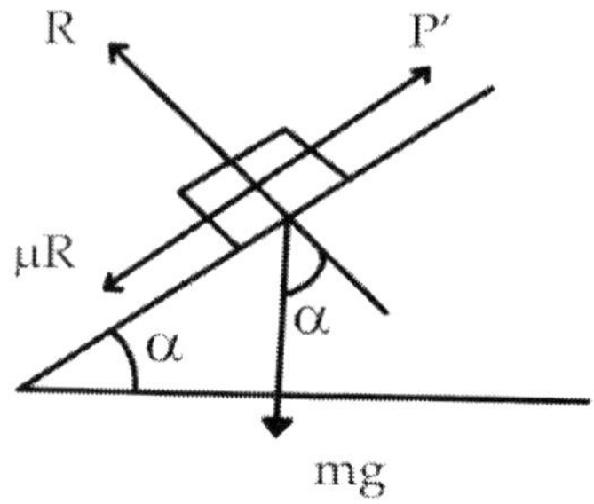

$P' = \mu R + mg \sin\alpha$ (2)

$R = mg cos\alpha$. (3)

From (2) and (3),

P'= $\mu mg\, cos\alpha + mg \sin\alpha$,

P' = mg ($\mu \cos\alpha + \sin\alpha$)

P' = mg ($tan\lambda\, cos\alpha + \sin\alpha$)

$= \left\{ mg\left(\frac{sin\lambda}{cos\lambda} cos\alpha + sin\alpha\right)\right\}$

$= mg\left(\frac{sin\lambda\, cos\alpha + cos\lambda\, sin\alpha}{cos\lambda}\right)$

$= \frac{mg\, sin(\alpha+\lambda)}{cos\lambda}$

$= \frac{P}{cos\lambda} = P\, sec\lambda = P\sqrt{1+\tan^2\lambda}$ = P $\sqrt{1+\mu^2}$.

To find the least amount of force needed to drag a particle on a rough inclined plane in a downward direction.

Let *P* be the force required for angle θ with the plane, and let α be the mass of the particle. For a limiting equilibrium, have:

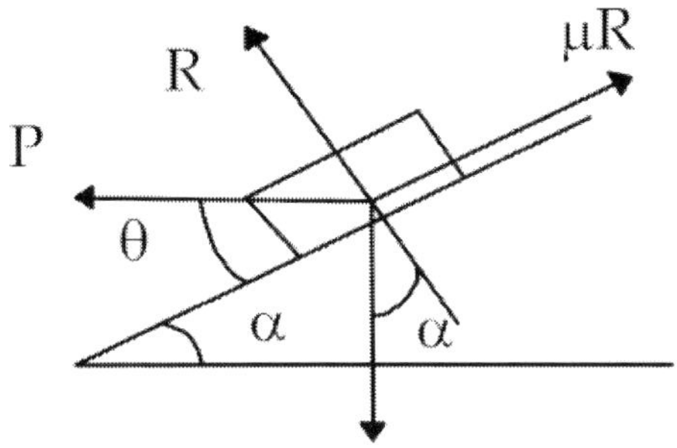

$$P\,cos\theta + mg\,sin\alpha = \mu R$$

$$P\,sin\theta + R = mg\,cos\alpha.$$

From eq. (2) we have

$$R = mg\,cos\alpha - p\,sin\theta.$$

Substituting this into eq. (1), we get

$$P\,cos\theta + mg\,\sin\alpha = \mu(mg\,cos\alpha - P\,sin\theta)$$

$$P(cos\theta + \mu\,sin\theta) = mg(\mu\cos\alpha - sin\alpha)$$

$$P = mg\frac{\mu\cos\alpha - sin\alpha}{cos\theta + \mu\,sin\theta}$$

$$P = mg\frac{tan\lambda\,cos\alpha - sin\alpha}{cos\theta + tan\lambda\ sin\theta}$$

$$P = mg\frac{sin\lambda\,cos\alpha - cos\lambda\,sin\alpha}{cos\lambda\,cos\theta + sin\lambda\,sin\theta}$$

$$P = mg\frac{\sin(\lambda - \alpha)}{\cos(\theta - \lambda)}$$

For minimal P, $\cos(\theta - \lambda)$ is maximum i.e.

$\cos(\theta - \lambda) = 1.$

Or $\theta - \lambda = 0$ or $\theta = \lambda$.

$P_{least} = mg\sin(\lambda - \alpha)$.

Find the force required to support a heavy body on an inclined plane with inclination $\acute{a};(a > \ddot{e})$.

Consider a particle at point A, the required force P acts along AB, making an angle θ with the plane. Let n be the mass of the particle. For a limiting equilibrium we have:

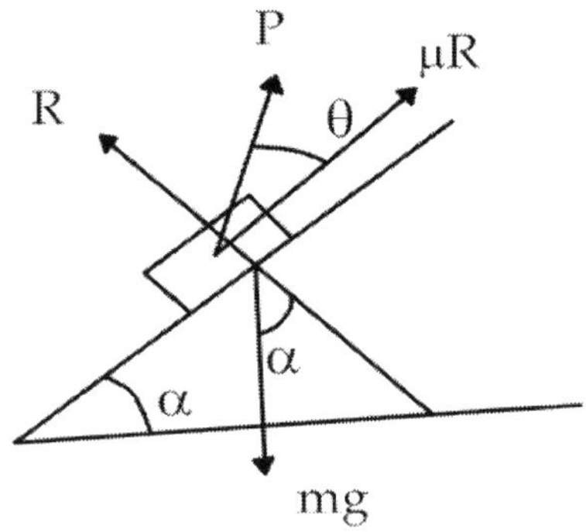

$$P\cos\theta + \mu R = mg\ \sin a$$

$$P\sin\theta + R = mg\ \cos a.$$

From eq. (2), we have

$$R = mg\cos a - P\sin\theta.$$

Substituting this into (1), we get:

$$P(\cos\theta + \mu\ (mg\cos a - p\sin\theta\ ) = mg\sin\alpha$$

$$P\left(\cos\theta - \mu\sin\theta\right) = mg(\sin\alpha - \mu\cos\alpha\ )$$

$$P = mg\frac{\sin\alpha\ - \mu\cos\alpha}{\cos\theta\ - \mu\ \sin\theta}$$

$$p = mg\frac{\sin\alpha - \tan\lambda\ \cos\alpha}{\cos\theta - \tan\lambda\sin\theta}$$

$$P = mg\frac{\cos\lambda\ \sin\alpha - \sin\lambda\ \cos\alpha}{\cos\lambda\cos\theta + \sin\lambda\ \ \sin\theta}$$

$$P = mg\frac{\sin\left(\alpha - \lambda\right)}{\cos\left(\theta + \lambda\right)}.$$

For minimal P, $\cos(\theta+\lambda)$ is at its maximum:

$$\cos(\theta+\lambda)=1$$

Or $\theta+\lambda=0$ or $\theta=-\lambda.$

Thus,

$$P_{least}=mg\,sin(\alpha-\lambda).$$

Example

A light ladder is supported on a rough floor and leans against an equally rough wall. Determine how far up the ladder a man can climb without slipping.

Solution

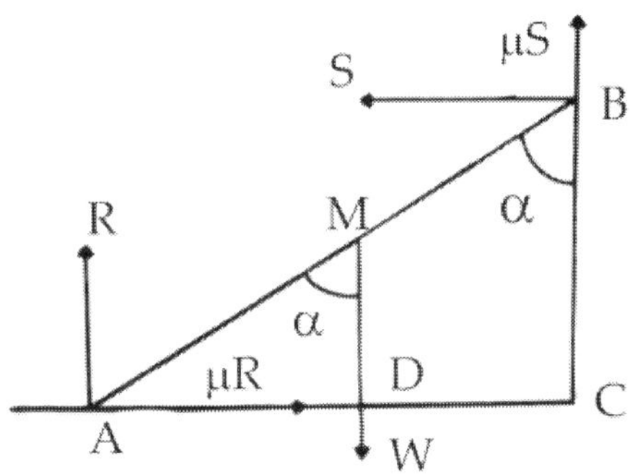

Let AB be the light ladder which makes an angle α with the wall. Let the man be at point M when the ladder is at a limiting equilibrium. Let W be the weight of the man. Suppose that $AM = x$, that R and S are the reaction of the ground and the wall, respectively, and $\mu R\, and\, \mu S$ be the force of friction. For a limiting equilibrium we have:

$$R+\ \mu S=W$$

$$S=\mu R.$$

From eq. (1), we have:

$$R=W-\mu S.$$

Substituting this into eq. (2) we get:

$S \; \mu(W - \mu S)$

$S + \mu^2 S = \mu W$

$$S = \frac{\mu W}{1+\mu^2}.$$

Taking the moment about A, have:

$$W(AD) = S(BC) + \mu S(AC)$$

$$W(AM\, sin\alpha) = S(AB\, cos\alpha) + \mu S(AB\, sin\alpha)$$

$$W(x\, sin\alpha) = S(l\, cos\alpha) + \mu S(l \sin\alpha)$$

$$W(x\, sin\alpha) = Sl(cot\alpha + \mu) \quad AB = l = \text{length of ladder}$$

$$Wx = \frac{\mu Wl}{1+\mu^2} = (cot\alpha + \mu)$$

$$x = \frac{\mu l}{1+\mu^2}(cot\alpha + \mu).$$

Example

A 4ft. long rod rests on a rough floor against the smooth edge of a table with a height of 3 ft. If the rod is on the point of slipping when inclined at an angle of to the horizontal, find the coefficient of friction.

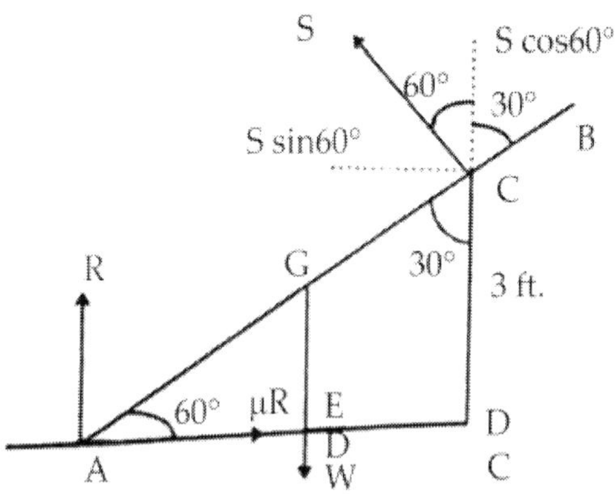

Solution.

Let AB be the rod of weight W acting at point G. Then we have

AB = 4ft, AG = 2ft, CD = 3ft.

Reaction R at A is perpendicular to the ground, and reaction S acts at C, perpendiculars to the rod. For a limiting equilibrium, we get:

$$R + S\cos 60^{\circ} = W$$

$$S\sin 60^{\circ} = \mu R.$$

From eq. (1), we have:

$$R = W - S\cos 60^{\circ}.$$

Substituting this into (2):

$$S\sin 60^{\circ}\,\mu\left(W - S\cos 60^{\circ\circ}\right)$$

$$S\,\frac{\sqrt{3}}{2} = \mu\left(W - \frac{S}{2}\right)$$

$$S = \frac{2\mu W}{\sqrt{3+\mu}}.$$

Let $AC = x$, then for ΔACD, we have:

$$\frac{DC}{AC} = \sin 60^{\circ}$$

$$\frac{3}{x} = \frac{\sqrt{3}}{2} \text{ or } x = 2\sqrt{3}. \qquad (4)$$

Taking the moment about A, we get:

$$W(AE) = S(AC)$$

$$W(AG\cos 60^{\circ}) = SX$$

Using (3) and (4), we have

$$W(2)\left(\frac{1}{2}\right) = \frac{2\mu W}{\sqrt{3}+\mu}(2\sqrt{3})$$

$$1=\frac{2\mu}{\sqrt{3}+\mu}(2\sqrt{3})$$

$$\sqrt{3}+\mu=4\sqrt{3}\,\mu$$

$$\mu=\frac{\sqrt{3}}{4\sqrt{3-1}}=0.292$$

Example

A uniform ladder rests in a limiting equilibrium with one end on a rough horizontal plane and the other against a smooth vertical wall. A man ascends the ladder. Show that he cannot go more than half way up.

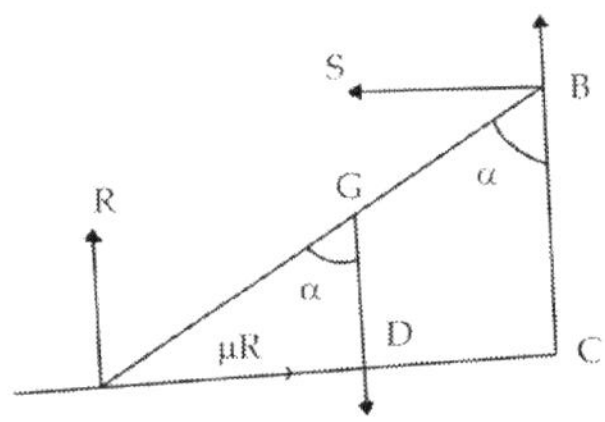

Solution.

Let AB be ladder of length I and weight W. If the ladder is in a limiting equilibrium, then we have:

$$R=W$$

$$S=\mu$$

From (1) and (2), we have:

$$S=\mu W$$

Taking the moment about A, we get:

$$W(AD)=S(BC)$$

$$W(AG\,sin\,\alpha)=\mu W(AB\cos\alpha),$$ using (3)

$$\frac{1}{2}=sin\alpha.\mu l\cos\alpha$$

$\tan\alpha = 2\mu$. (3a)

Let W_l be the weight of the man who ascend the ladder, and let the man be at point *M* when the ladder is in a limiting equilibrium. Let *AM* = *x* then we have:

(3a)

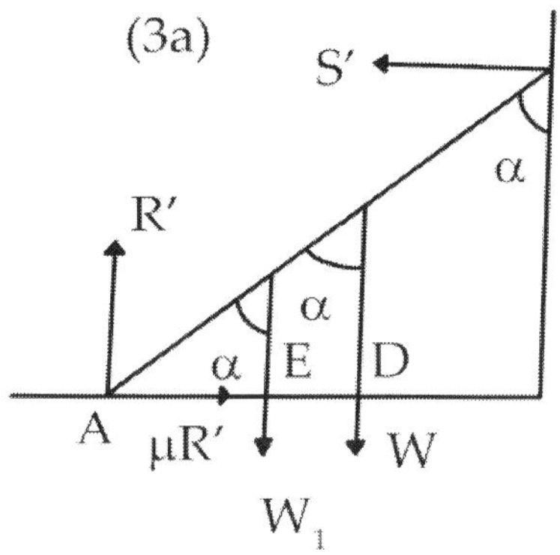

$R' = W + W_l$ (4)

$S' = \mu R'$. (5)

From (4) and (5)

$S' = \mu R'$. (6)

From (4) and (5)

$S' = \mu(W + W_l)$.

Taking the moment about A:

W(AD) + $W_l(AE) = S'(BC)$

W (AG sinα) + $W_l\ (AM\sin\alpha) = S'(AB\cos\alpha)$.

Using (6), we have:

$$W(\frac{1}{2}\sin\alpha) + W_l\ (AM\sin\alpha) = \mu(W + W_l)\ (l\cos\alpha)$$

$$W(\frac{l}{2}\tan\alpha) + W_l\ (x\tan\alpha) = \mu l(W + W_l).$$

Using (3a), we get:

$$W\left(\frac{l}{2}2\mu\right)+W_l\left(2\mu x\right)=\mu l W+\frac{\mu}{W_l}$$

$$W\left(l\mu\right)+2W_l\left(x\mu\right)=\mu l W+\mu l W_1$$

$$2W_l\left(x\mu\right)=\mu l w_1$$

$$2x=l \text{ or } x=\frac{1}{2}.$$

This shows that the man cannot go more than half way up the ladder.

Example

A uniform rod of weight W is placed with its lower end on a rough horizontal floor and its upper end against an equally rough vertical wall. The rod makes an angle α with the wall and is prevented from slipping down by a horizontal force P applied to its middle point. Prove that

$P=W,$

where λ is the angle of friction and $\lambda<\frac{1}{2}\alpha$.

Solution

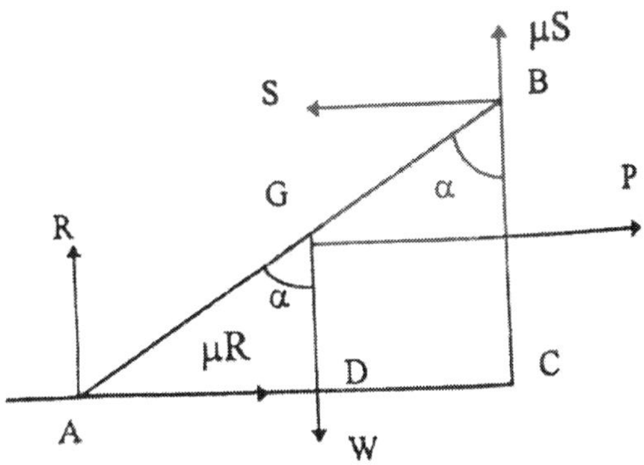

Let AB be rod of length l, with its weight W acting at its mid - point G. Then,

$AB=l$

$$AG = \frac{l}{2} = BG.$$

For a limiting equilibrium, we have:

$$P + \mu = S$$

$$R + \mu S = W.$$

From (1) and (2), we get:

$$\frac{P}{W} = \frac{S - \mu R}{R + \mu S}.$$

Taking the moment about *G*, we get:

$$\mu S(GE) + S(BE) + \mu R(GD) = R(DA)$$

$$\Rightarrow \mu S(BG \sin\alpha) + S\ (BG \cos\alpha) + \mu R(AG \cos\alpha) = R(AG \sin\alpha)$$

$$\Rightarrow \mu\left(\frac{1}{2}\right)\sin\alpha + S\left(\frac{1}{2}\right)\cos\alpha + \mu R\left(\frac{1}{2}\right)cos\alpha = R\left(\frac{1}{2}\right)sin\alpha$$

$$\Rightarrow \mu S \sin\alpha + S\cos\alpha + \mu R\, cos\alpha = R\sin\alpha$$

$$\Rightarrow S(\mu \sin\alpha + \cos\alpha) = R(\sin\alpha - \mu\, cos\alpha)$$

$$\Rightarrow S(\tan\lambda\, sin\alpha + cos\alpha) = R\ (sin\alpha - tan\lambda\, cos\alpha)$$

$$\Rightarrow S(\sin\lambda \sin\alpha + cos\lambda\, cos\alpha) = R(\cos\lambda\, sin\alpha - sin\lambda cos\alpha)$$

$$\Rightarrow S\cos(a - \lambda) = R\sin(\alpha - \lambda)$$

$$\Rightarrow S = R\,tan(\alpha - \lambda).$$

From (3) and (4), we get:

$$\frac{P}{W} = \frac{R\,tan(a - \lambda) - \mu R}{R + \mu \tan(\alpha - \lambda)}$$

$$\Rightarrow \frac{P}{W} = \frac{tan\ (a - \lambda) - \mu}{1 + \mu \tan(\alpha - \lambda)}$$

$$\Rightarrow \frac{P}{W} = \frac{\tan(a-\lambda) - \tan\mu}{1+\tan\lambda\,\tan(\alpha-\lambda)} = \tan(\alpha-\lambda-\lambda)$$

$$\Rightarrow P = \tan(\alpha - 2\lambda).$$

Since P is positive, $(\alpha - 2\lambda) > 0$, thus $\lambda < \frac{1}{2}\alpha.$

Example

A uniform semi-circular wire hangs on a rough peg, with the line joining its extremities makes an angle of 45° with the horizontal. If the wire is just on the point of slipping, find the coefficient of friction between the wire and the peg.

Solution

Let P be the peg and APB be the semi-circular wire with center at O and radius r, then

OR = r.

Let G be the center of gravity of wire, the:

$$OG = \frac{2r}{\pi}$$

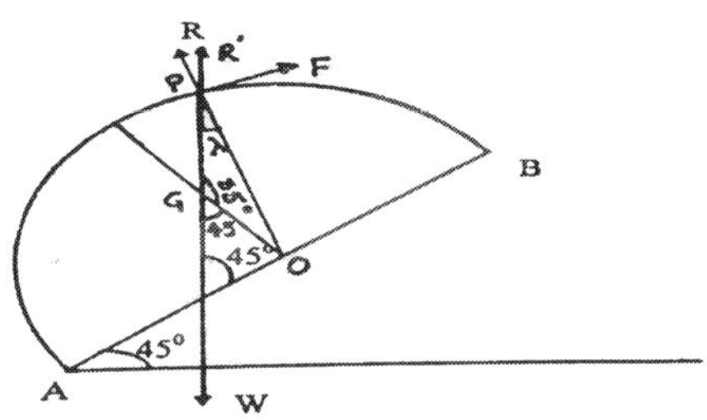

In equilibrium, the resultant reaction R' and W must be equal and opposite, and act along the same straight line. Therefore:

$< OPG = \lambda$ and $< OPG = 135°.$

For a limiting equilibrium, we have

$R = W sin\,\lambda$

and

Taking the moment about O, we get:

$OG\left(W\,sin45^0\right) = OP\left(\mu R\right)$

$$\frac{2r}{\pi} R \sec\lambda \left(\frac{1}{\sqrt{2}}\right) = r.\mu\ .R.$$

Using (1), we get:

$$\frac{sec\lambda}{\mu} = \frac{\pi}{\sqrt{2}}.$$

Squaring both sides, we get:

$$\frac{\sec^2\lambda}{\mu^2} = \frac{\pi^2}{2}$$

$$\frac{1+\mu^2}{\mu^2} = \frac{\pi^2}{2},$$

where $\sec^2 1+\mu^2 = 1+\tan^2 = 1+\mu^2$

$$\frac{1}{\mu^2} + 1 = \frac{\pi^2}{2}$$

$$\mu^2 = \frac{2}{\pi^2 - 2}$$

$$\mu = \sqrt{\frac{2}{\pi^2 - 2}} = 0.504.$$

Example

A hemispherical shell rests on a rough inclined plane with an angle of friction λ. Show that the inclination of the plane base to the horizontal cannot be greater than $\sin^{-1}\left(2sin\lambda\right)$.

Solution

Let G be the center of gravity of the hemisphere, with its weight W acting along GA. For a limiting equilibrium, the resultant of R and μR must balance the weight W, and so acts along AG. Therefore,

$< OAG = \lambda \; and < AGB = \theta$.

Since the angle between two lines is equal to the angle between their perpendiculars, we have:

$< OGA = 180^0 - \theta$.

Let r be the radius of the sphere, then

$OA = r \text{ and } OG = \frac{r}{2}$.

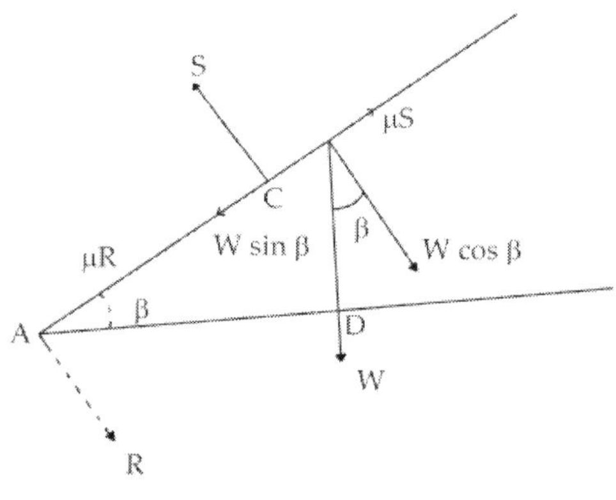

0For a limiting equilibrium:

$\mu r R = W \sin\lambda$

$R = W\cos\lambda$.

Taking the moment about 0:

$$(OG)W\sin\left(960^0 - \theta\right) + (OA)\mu R \sin\left(90^0\right) = 0$$

$$-\frac{r}{2}\frac{\mu R}{\sin\lambda}\sin\theta + r(\mu R) = 0$$

$$\sin\theta = 2\sin\lambda$$

$$\theta = \sin^{-1}(2\sin\lambda)$$

Example

A thin uniform rod passes over one peg and under another, with the coefficient of friction between each peg and the rod being μ. The distance between the pegs is a and the straight line joining them makes an angle β with the horizontal. Show that equilibrium is not possible unless the length of the rod is greater than $\frac{a}{\mu}(\mu + \tan\beta)$

Solution

Let *AB* be the rod of length x. Let the rod rest over the peg *C* and under the peg *A*. The rod will be of least length when its one is in contact with the lower peg. We know that *AC* = *a*. Denote the normal reaction at *A* and *C* is *R* and *S,* respectively. Resolving components along and perpendicular to the rod.

$$\mu .R + \mu.S = W \sin\beta \qquad (1)$$

$$S = R + W \cos\beta. \qquad (2)$$

Eliminating *R* from (1) and (2), We have:

$$\mu(S - W\cos\beta) + \mu S = W\sin\beta$$

$$2\mu S = W(\sin\beta + \mu\cos\beta)$$

$$S = \frac{W}{2\mu}(\sin\beta + \mu\cos\beta).$$

Taking the moment about *A*,

$$S(AC) = W(AD)$$

$$S(a) = W(AG\cos\beta)$$

$$\frac{aW}{2\mu}\sin\beta + \mu\cos\beta) = W\left(\frac{x}{2}\right)\cos\beta$$

$$x = \frac{a}{\mu}(\tan\beta + \mu)$$

.

This gives the least length of the rod. Under equilibrium, we have:

$$S\left|\overline{AC}\right| < W\left|\overline{AD}\right|$$

$$x > \frac{a}{\mu}\left(tan\beta + \mu\right)$$

Example

A uniform rod slides with its end on two equally rough, fixed rods, one being vertical and the other inclined at an angle α to the horizontal. Show that the angle θ to the horizontal of the movable rod, when it is on the point of sliding, is given by:

$$tan\theta = \frac{1-2\mu\ \tan\alpha - \mu^2}{2\ (\mu + \tan\alpha\ )}.$$

Solution

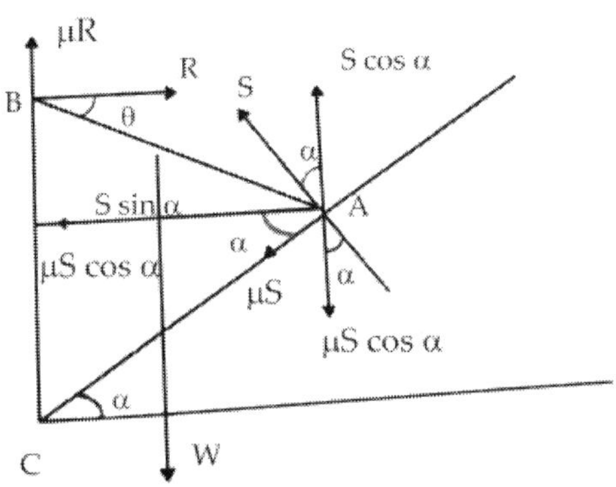

Let the movable rod be *AB*, under the action of three forces; the reactions of fixed rods and its weight.

Resolve the force horizontally and vertically, as shown in the figure above. Since the system is in equilibrium, we get:

$$R = \mu S\,cos\alpha + S\,sin\alpha \qquad (1)$$

$$\mu R + S\,cos\alpha = W + \mu\,sin\alpha. \qquad (2)$$

From (1) and (2), we have:

$$W = \frac{R\left(1+\mu^2\right)cos\alpha}{\mu\cos\alpha + sin\alpha}. \tag{3}$$

Taking the moment about A, we get:

$$\left(\frac{l}{2}\right)W\sin\left(90^0-\theta\right) = l\,R\sin\theta + l\,\mu\,R\,sin\left(90^0-\theta\right),$$

where l is the length of the movable rod AB,

$$\frac{Wcos\,\theta}{2} = R\,sin\,\theta + \mu\,R\cos\theta$$

$$R\,tan\theta = \frac{W}{2} - \mu R.$$

Using (3), we get:

$$R\tan\theta = \frac{R\left(1+\mu^2\right)cos\alpha}{2\,(\mu\,\cos\alpha + \sin\alpha\,)} = \mu R$$

$$\tan\theta = \frac{1-2\,\mu\,tan\alpha - \mu^2}{2\,(\mu + \tan\alpha\,)}.$$

Example

A weight W is laid upon a rough plane $\mu = 1\,/\sqrt{3}$, inclined at 45° to the horizon and is connected by string passing through a smooth ring A at the top of the plane with weight P hanging vertically. If W -= 3P, show that if θ is the greatest possible inclination of the string AW to the line of greatest slope on the plane, then:

$$\cos\theta = \frac{2\sqrt{2}}{3}.$$

Also find the direction in which W commences to move.

Solution

The line of greatest slope on a plane is the line perpendicular to the line if the intersection of the given plane with the horizontal plane. In the figure below, BC is the line of the greatest slope.

$$R = W\cos 45^\circ = \frac{W}{\sqrt{2}}$$

$$\mu R = \frac{1}{\sqrt{3}}\frac{W}{\sqrt{2}} = \frac{W}{6}.$$

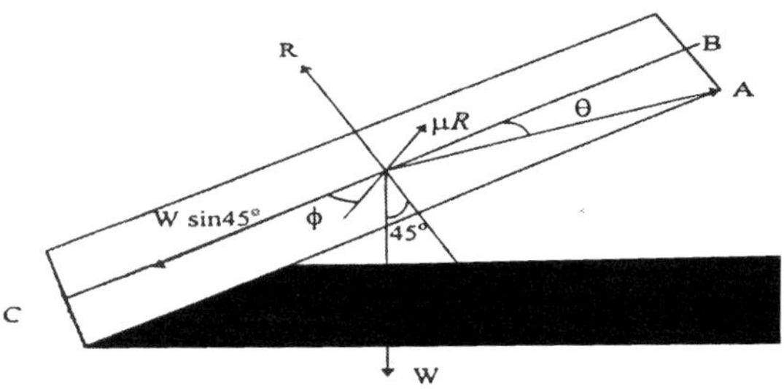

Suppose that the direction of motion of W makes an angle $\varnothing$ with the line of greatest slope, and hence, friction μR acts in the opposite direction. Now weight W is in equilibrium under the action of three forces: P, μR and W $\sin 45^\circ$. By Lami's theorem, we have:

$$\frac{P}{\sin\left(180^\circ - \varnothing\right)} = \frac{\mu R}{\sin\left(180^\circ - \theta\right)} = \frac{W\sin 45^\circ}{\sin(\theta + \varnothing)}$$

$$\frac{\frac{W}{3}}{\sin\varnothing} = \frac{\frac{W}{\sqrt{6}}}{\sin\theta} = \frac{\frac{W}{\sqrt{2}}}{\sin(\theta + \varnothing)}.$$

The relation $\frac{\frac{W}{3}}{\sin\varnothing} = \frac{\frac{W}{\sqrt{6}}}{\sin\theta}$ implies that

$$\sin\varnothing = \frac{\sqrt{6}}{3} = \sin\theta,$$

and the relation $\frac{\frac{W}{\sqrt{6}}}{\sin\theta} = \frac{\frac{W}{\sqrt{2}}}{\sin(\theta + \varnothing)}$ gives us

$$\sqrt{6}\, sin\theta = \sqrt{2}\ (\ \sin\theta \cos\varnothing + \cos\theta\, sin\varnothing).$$

Using (2), we get:

$$\sqrt{3}\, sin\theta = \left(sin\theta \cos\varnothing + \cos\theta \left(\frac{\sqrt{6}}{3} \right) \right)$$

$$cos\theta = \sqrt{3} - cos\,\theta \left(\frac{\sqrt{6}}{3} \right).$$

Squaring and adding (2) and (3), we get:

$$= \frac{2}{3}\sin^2\theta + 3 + \frac{2}{3}\cos^2\theta - 2\sqrt{2}\cos\theta$$

$$= \frac{2}{3} + 3 - 2\sqrt{2}\cos\theta$$

$$= 2\sqrt{2}\, cos\,\theta = \frac{4}{3} \quad \text{or} \quad cos\,\theta = \frac{2\sqrt{2}}{3}.$$

Putting this value into eq. (3), we get:

$$cos\varnothing = \sqrt{3} - \frac{2\sqrt{2}}{3}\frac{\sqrt{6}}{3} = \frac{5\sqrt{3}}{9}$$

$$\varnothing = \cos^{-1}\left(\frac{5\sqrt{3}}{9}\right).$$

This gives us the direction of *W*.

SUMMARY AND DISCUSSION

In this chapter, we discussed the important topics of friction, which play a very important role in mechanics, physics and engineering. Numerous real life problems are solved using the techniques used here.

Statistical Friction:

When one body in contact with another is in equilibrium and is not on the point of sliding along the other, the force of friction is called static friction. The equilibrium in this case is non-limiting.

Limiting Friction:

When one body is just at the point of sliding on the other; the friction in this case is maximum and is denoted as limiting friction, with a limiting equilibrium.

Dynamical Friction:

When one body is sliding on the other, the friction in this case is slightly less than the limiting friction and is denoted as the dynamical friction or kinetic friction.

Law of Friction

1. The direction of friction is opposite to the direction in which the body moves.
2. The magnitude of friction is, up to certain extent, equal to the force tending to produce motion.
3. The magnitude of the limiting friction bears a constant ratio μ to the normal reaction between the surfaces. The constant μ is called the coefficient of friction.
4. The amount of friction is independent of the area and shape of the surface in contact, provided the normal pressure remains unaltered.
5. When the motion takes place, the friction still opposes the motion, it is independent of the velocity and is proportional to the normal. It is also independent of the velocity and is proportional to the normal reaction, however, slightly less than the limiting friction.

Coefficient of Friction

The magnitude of the limiting friction F bears a constant ration μ to the normal reaction R between the surfaces. The constant μ is called the coefficient of friction and

$$\frac{F}{R} = \mu.$$

Angle of Friction

The angle λ of the direction of the resultant reaction R' when the body is just on the point of motion, is known as the angle of friction, and:

$R = R' \cos\lambda$ and $F - R' \sin\lambda$.

As $\frac{F}{R} = \mu$, we have

$\mu = \frac{R' \sin\lambda}{R' \cos\lambda}$ or $\mu = \tan\lambda$.

Also $R^2 + F^2 = R'^2$,

or $R' = \sqrt{R^2 + F^2} = \sqrt{R^2 + \mu^2 R^2}$

or $R' = R\sqrt{1 + \mu^2}$.

Cone of Friction

The cone of friction is a cone whose vertex is the point of contact *P*, with its axis along the common normal at *P* and whose semi – vertical angle is the angle of friction.

CHAPTER 4

VIRTUAL WORK

WORK DONE BY A FORCE

Consider a constant force F displacing a particle from point A to point B. Let the displacement $AB=S$, then work done W by the force F is defined as:

$W = F.S = FS\,cos\theta,$

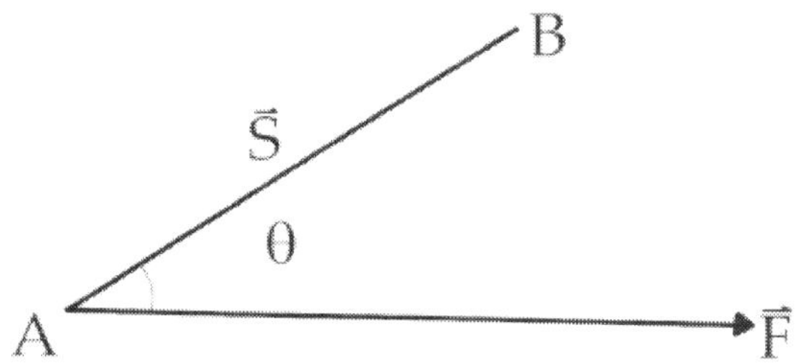

where θ is the angle between F and S.

Case (i) $\theta = 0$, then we have :

Work done $= W = FS\,cos\,\theta = FS.$

Case (ii) $\theta = \dfrac{\pi}{2}$, then we have :

Work done $= W = FS\cos\dfrac{\pi}{2} = FS(0) = 0.$

Therefore, when the applied force is perpendicular to the displacement the work done will be zero.

Case (iii) $\theta = \pi$, then we have:

Work done $= W = FS\cos\pi = FS(-1) = -FS.$

Case (iv) $0 \le \theta \le \frac{\pi}{2}$, then we have:

Work done $= W$ is positive, as cos $\theta \ge 0.$

Case (v) $\frac{\pi}{2} \le \theta \le \pi$, then we have:

Work done = W is negative, as $\cos\theta \le 0.$

Work done by the weight, *W*

If body of weight W falls down vertically through a distance h, the work done by gravity is positive and equal to Wh.

If the body rises vertically to height h. the work done is negative and is equal to -Wh. We say that the work is being done against gravity.

Work done by tension of an extensible light string

Let AB be a string with tension T in displaced position. Let A 'B' be another string and θ the angle which A 'B' makes with AB, as shown in the figure below:

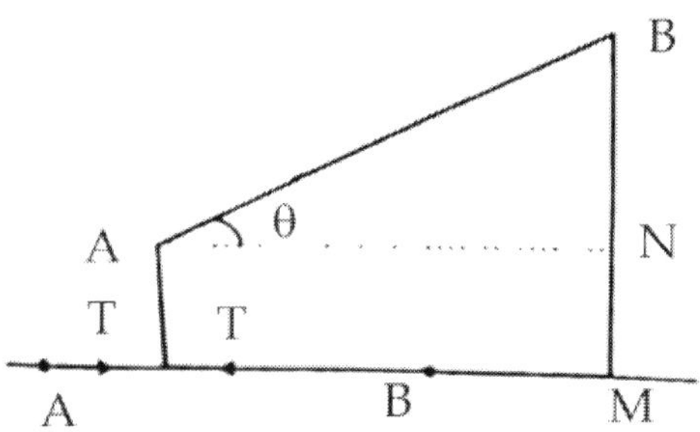

From the figure, we can see that:

Work done $= T(AL) - T(BM)$

$$=T(AB-LB)-T(LM-LB)$$

$$=T(AB-LB-LM+LB)$$

$$=T(AB-LM)$$

$$=T(AB-A'N)$$

$$=T(AB-A'B'cos\theta)$$

$$=T\left\{AB-A'B'\left(1=\frac{\theta^2}{2!}+\frac{\theta^4}{4!}-....\right)\right\}.$$

Neglecting θ^2 and higher powers of θ as being small, then:

Work done $=T(AB-A'B')$

$$=-T(A'B'-AB)$$

$$=-T\delta l.$$

Where δl is the extension in the length of the string. Similarly, the work done by thrust *T* of the light $T\delta l$.

Work Done by a force during a rotation

Suppose a force *F* acts on a body and rotates it about point *O* on the $xy-$ plane through an angle of θ.

Let a be the distance to force F from point O. Then we have:

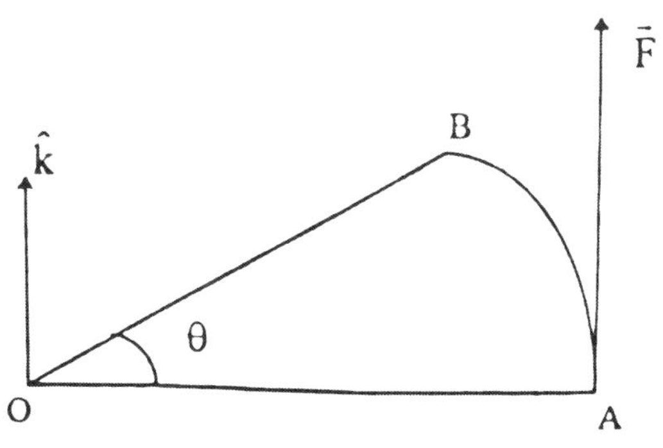

$G =$ Moment of F about $O = Fak,$

where k is a unit vector along the z-axis.

Displacement $= arc\ AB = \theta,\ (s = r\theta).$

Work done $= F\,a\theta = F\,a\theta k.k$

$= Fak\theta k = G\theta.$

Where $\theta = \theta K$ is the angular displacement?:

Work done by a couple

$Let\left(F, F\right)$ be a couple acting on a body on the xy plane whose forces F and $-F$ act at points A and B.

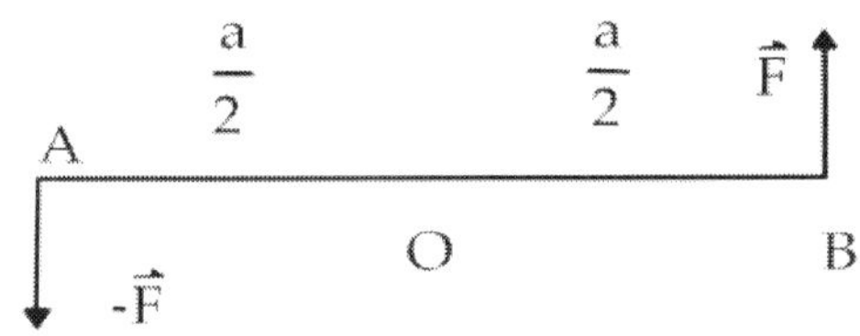

Let $AB = a$, with the axis of rotation passing through O, the midpoint of AB, then:

$G =$ moment of couple $= Fak.$

The work done for a small angular displacement will be

$$= F\left(\frac{a}{2}\right)\theta + F\left(\frac{a}{2}\right)\theta = Fa\theta = Fa\,\theta kk$$

$$= Fa\,k\theta K = G\theta.$$

The total work done by all the forces is equal to the work done by their resultant forces

Let R be the resultant of forces $F_1\ .F_2, F_3, \ldots..F_n$ acting on a particle then:

$R = F_1 + F_2 + F_3 + \ldots\ldots. + F_n.$

Let S be the displacement, then we have

$$RS = \left(F_1 + F_2 + F_3 + \ldots\ldots F_n\right)S$$

$$= F_1 S + F_2 S + F_3 S + \ldots\ldots\ldots + F_n S.$$

Virtual displacement and virtual work

If a set of particles or a body is in equilibrium under the action of forces then there is no motion and consequently there is no actual displacement.

Suppose that the set of particles or the body receives an imaginary displacement, the forces acting are regarded as constant during the displacement, which is known as the virtual displacement, and the work done by the forces during such a displacement is denoted as virtual work.

Constraint

A constraint is a restriction imposed on a mechanical system to get a desired position or motion. For the case of a particle moving down an inclined plane, the constraint on its motion is its contact with the plane. Any displacement along the plane is consistent with the constraint, and the displacement away from the plane is not consistent with the constraint. Due to the imposition of the constraint we get unknown forces acting on the system. These forced are called the reactions of the constraints.

Ideal constraint or workless constraints

If the forces of constraint do not virtual work, the constraints are said to be ideal or workless.

Example of workless constraints

(1) The reaction $\overline{R}$ *of* any smooth surface with which the body is in contact with is workless.

The surface is smooth and the reaction $\overline{R}$ is normal to the surface at the point of contact *p*. If p moves to a neighboring point p'/, the displacement $\overline{pp'}$ is at a right angle to $\overline{R}$. Hence, its virtual work is zero.

(2) The tension on an inextensible string is workless.

Let *AB* be a string whose tension is *T* and A 'B' be the displaced position of the string. The string is inextensible, therefore AB = A'B'/. Let θ be the angle which A'B' makes with AB, as shown in the figure below.

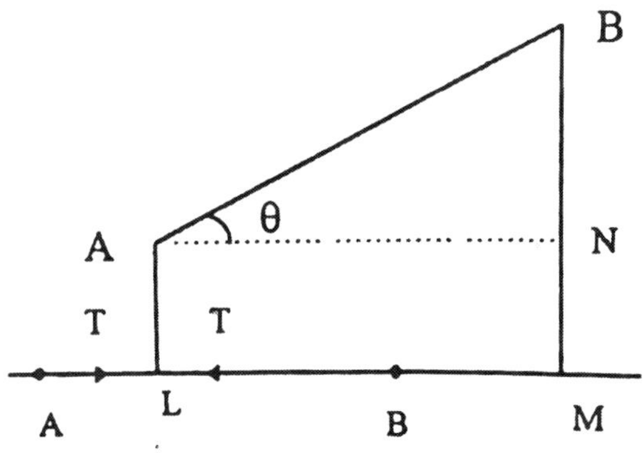

From the figure above, we have:

Work done = $T(AL)-T(BM)$

$=T(AB-LB)-T(LM-LB)$

$=A(AB-LB-LM+LB)$

$=T(AB-LM)$

$=T(AB-A'N)$

$=T(AB-A'B'cos\theta)$

$$=T\left\{AB-A'B'(1-\frac{\theta^2}{2!}+\frac{\theta^4}{4!}-\ldots.\right\}.$$

Neglecting θ^2 and higher powers of θ as being small, then we have:

Work done = $T\{AB-A'B'\}$

$=-T(A'B'-AB)=0; AB=A'B'.$

Similarly, the thrust of an inextensible rod is a workless constraint.

(3) The mutual reaction of two rigidly connected particles.

Let $\overline{r_1}$ and $\overline{r_2}$ be the position vectors of A and B respectively, as shown in the figure below.

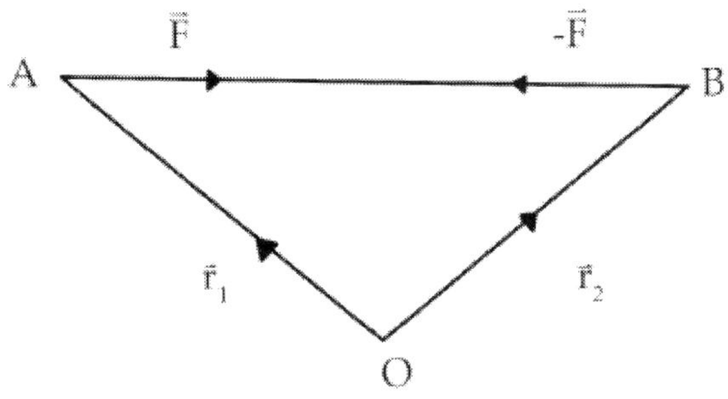

Let $\overline{F}$ and $-\overline{F}$ be forces of mutual reaction acting along AB and BA respectively, then $\overline{F}$ and AB are collinear . Therefore,

$$\overline{F} = k\overline{AB},$$

where k is constant.

$$\overline{F} = k\left(\overline{r_2} - \overline{r_1}\right).$$

Since the particle are rigidly connected, therefore:

$$\left|\overline{AB}\right| = \text{constant}$$

$$\left|\overline{r_2} - \overline{r_1}\right| = \text{constant}$$

$$\left|\overline{r_2} - \overline{r_1}\right|^2 = \text{constant}$$

$$\left(\overline{r_2} - \overline{r_1}\right)\left(\overline{r_2} - \overline{r_1}\right) = \text{constant}$$

$$\delta\left\{\left(\overline{r_2} - \overline{r_1}\right)\left(\overline{r_2} - \overline{r_1}\right)\right\} = 0$$

$$\left(\overline{r_2} - \overline{r_1}\right)\left(\delta\overline{r_2} - \delta\overline{r_1}\right) + \left(\delta\overline{r_2} - \delta\overline{r_1}\right) + \left(\overline{r_2} - \overline{r_1}\right) = 0$$

$$2\left(\overline{r_2} - \overline{r_1}\right)\left(\delta\overline{r_2} - \delta\overline{r_1}\right) = 0.$$

Multiplying by $\frac{k}{2}$:

$$K\left(\overline{r_2}-\overline{r_1}\right)\left(\delta\overline{r_2}-\delta\overline{r_1}\right)=0$$

$$\overline{F}\left(\delta\overline{r_2}-\delta\overline{r_1}\right)=0$$

$$\overline{F}(\delta\overline{r_2})-\overline{F}(\delta\overline{r_1})=0$$

$$\overline{F}(\delta\overline{r_2})+(-\overline{F})(\delta\overline{r_1})=0.$$

Therefore, the sum of the work done $\overline{F}$ and - $\overline{F}$ is zero. Hence, this constraint is workless.

(4) The reaction at a point or a fixed axis of rotation is workless as the displacement of the point of application of the force is zero.

Principal of virtual work for a single particle

A particle subject to workless constraints is in equilibrium if and only if zero virtual work is done by the applied forces in any arbitrary infinitesimal displacement consistent with the constraints.

Proof

Let the total force applied to the particle be $\overline{F}$ and the total forces applied by the constraint be $\overline{F}_c$. Let the particle be in equilibrium, then:

$$\overline{F}_a+\overline{F}_c=0.$$

Let $\delta\overline{r}$ be an infinitesimal displacement consistent with the constraints, then:

$$\left(\overline{F}_a+\overline{F}_c\right)\delta\overline{r}=0$$

$$\overline{F_a}\delta\overline{r}+\overline{F_c}\delta\overline{r}=0.$$

Since the constraints are workless, we get:

$$\overline{F_c}\delta\overline{r}=0.$$

Thus, $\overline{F_a \delta r} = 0,$ i.e. the virtual work done by the applied force is zero.

Conversely, let the virtual work done by the applied force be zero, then:

$$\overline{F_a \delta r} = 0.$$

Also, since the constraints are workless, we have:

$$\overline{F_c}\,\overline{\delta r} = 0:$$

Adding (1) and (2), we get:

$$\overline{F_a \delta r} + \overline{F_c}\,\overline{\delta r} = 0$$

$$(\overline{F_a} + \overline{F_c})\ \overline{\delta r} = 0.$$

Since $\delta\overline{r}$ is arbitrary, therefore $\delta\overline{r} \neq 0$ and hence

$$\overline{F_a} + \overline{F_c} = 0.$$

This shows that the particle is in equilibrium.

Principle of virtual work for a set of particles

A set of particles, subject to workless constraints, is in equilibrium if and only if zero virtual works is one by the applied forces in an arbitrary infinitesimal displacement consistent with constraints.

Let there be n particles in the set. Let $\overline{F_{ia}}$ be the total applied force and $\overline{F_{ic}}$ be the total force of constraints on the i^{tch} particle. The total force on the ith particle will be $\overline{F_{ia}} + \overline{F_{ic}}$. Thus:

Total force on all particles $= \sum_{i=1}^{n} \overline{F_{ia}} + \overline{F_c} = 0.$

Let the set of particles be in equilibrium, then:

$$\sum_{i=1}^{n} \overline{F_{ia}} + \overline{F_{ic}} = 0.$$

Let $\delta\overline{r}$ be the arbitrary infinitesimal displacement, then we get:

$$\sum_{i=1}^{n}(\overline{F_{ia}}+\overline{F_{ic}})\delta\overline{r}=0$$

$$\sum_{i=1}^{n}(\overline{F_{ia}}\delta\overline{r}+\overline{F_{ic}\delta\overline{r}})=0.$$

As the constraint is workless, thus:

$$\sum_{i=1}^{n}\overline{F_{ic}\delta\overline{r}}=0.$$

Hence,

$$\sum_{i=1}^{n}\overline{F_{ia}\delta\overline{r}}=0.$$

That is, zero virtual work is done by the applied forces. Conversely, let the virtual work done by the applied forces be zero, then:

$$\sum_{i=1}^{n}\overline{F_{ia}\delta\overline{r}}=0.$$

Also given that the constraints are workless:

$$\sum_{i=1}^{n}\overline{F_{ic}\delta\overline{r}}=0.$$

Adding (1) and (2), we get:

$$\sum_{i=1}^{n}\overline{F_{ia}\delta\overline{r}}+\sum_{i=1}^{n}\overline{F_{ic}\delta\overline{r}}=0$$

$$\sum_{i=1}^{n}(\overline{F_{ia}\delta\overline{r}}+\overline{F_{ic}\delta\overline{r}})=0$$

$$\sum_{i=1}^{n}(\overline{F_{ia}}+\overline{F_{ic}})\delta\overline{r}=0.$$

Since $\delta\overline{r}$ is arbitrary, therefore $\delta\overline{r} \neq 0$, hence,

$$\sum_{i=1}^{n}(\overline{F_{ia}} + \overline{F_{ic}}) = 0$$

.

Principal of virtual work of a single rigid body

A rigid body, subject to workless constraints, is in equilibrium if and only if zero virtual work is done by the applied forces and applied torques in any arbitrary infinitesimal displacement consistent with constraints.

Proof

A given system of forces acting on the rigid body can be reduced to a single $\overline{R}$ at any arbitrary point on the plane of forces, together with the couple $\overline{G}$. Let

$\overline{R} = \overline{R}_a + \overline{R}_c.$

Where $\overline{R}_a$ is the sum of the applied forces and $\overline{R_c}$ is the sum of the forces of the constraints. Let:

$$\overline{G} = \overline{G_a} + \overline{G_c},$$

where $\overline{G_a}$ is the sum of the applied torque and $\overline{G}_c$ is the sum of the torque of constraints. Let the body be in equilibrium, such that

$\overline{R} = 0$ and $\overline{G} = 0.$

Any arbitrary virtual displacement of the body will consist of a translation $\delta\overline{r}$ and rotation $\overline{\delta\theta}$. Thus,

$\overline{R}\delta\overline{r} = 0$ and $\overline{G}\delta\overline{\theta} = 0.$

Taking the sum, we have

$\overline{R}\delta\overline{r} + \overline{G}\delta\overline{\theta} = 0$.

Using (1) and (2), we get:

$\overline{R_a} + \overline{R_c})\delta\overline{r} + \left(\overline{G_a} + \overline{G_C}\right)\delta\theta = 0,$

$$\overline{R_a}\delta\overline{r}+\overline{R_c}\delta\overline{r}+\overline{G_a}\delta\theta+\overline{G_c}\delta\theta=0.$$

The constraints are workless, therefore,

$$\overline{R_c}.\delta\overline{r}+\overline{G_c}.\delta\theta=0.$$

And

$$\overline{R_a}\delta\overline{r}+\overline{G_a}\delta\theta=0.$$

Hence, the work done by the applied forces and applied torque is zero. Conversely, let the work done by the applied forces and applied torque be zero, then:

$$\overline{R_a}\,\delta\overline{r}+\overline{G_a}\delta\theta=0 \qquad (3)$$

The constraints are workless, therefore

$$\overline{R_c}\delta\overline{r}+\overline{G_c}\delta\theta=0 \qquad (4)$$

Adding (3) and (4), we get:

$$\overline{R_a}\,\delta\overline{r}+\overline{R_c}\delta\overline{r}+\overline{G_a}\delta\theta+\overline{G_c}\delta\theta=0$$

$$(\overline{R_a}+\overline{R_c})\delta\overline{r}+\left(\overline{G_a}+\overline{G_C}\right)\delta\theta=0$$

$$\overline{R_c}.\delta\overline{r}+\overline{G_c}.\delta\theta=0.$$

$\delta\overline{r}$ and $\delta\theta$ are arbitrary, therefore,

$\overline{R}=0$ and $\overline{G}=0$.

That is, the body is in equilibrium.

Principal of virtual work for a set of rigid bodies

A set of rigid bodies subject to workless constraints is in equilibrium if and only if zero virtual work is done by the applied forces and the applied torques in any arbitrary infinitesimal displacement consistent with the constraints.

Proof

Let there be n rigid bodies. The system of forces acting on the ith rigid body can be reduced to a single force $\overline{R}_i$ at an arbitrary point on the plane of forces together with a couple $\overline{G}_i$.

Let $\overline{R_i} = \overline{R}_{ia} + \overline{R}_{ic}$,

where $\overline{R}_{ia}$ is the sum of the applied forces and $\overline{R}ic$ is the sum of the forces of constraints on the ith rigid body. Let

$\overline{G_i} = \overline{G_{ia}} + \overline{G_{ic}}$,

where $\overline{G_{ia}}$ is the sum of is applied torques and $\overline{G_{ic}}$ is the sum of the torque of constraints on the ith rigid body. For an equilibrium, we have

$$\sum_{i=1}^{n} \overline{R}_i = 0 \quad \text{and} \quad \sum_{i=1}^{n} \overline{G}_i = 0 .$$

Any arbitrary virtual displacement of the body will consist of a translation $\delta\overline{r}$ *and* rotation $\delta\theta$. Thus

$$\sum_{i=1}^{n} \overline{R}_i \delta\overline{r} = 0 \quad \text{and} \quad \sum_{i=1}^{n} \overline{G}_i \delta\overline{\theta} = 0 .$$

Taking the sum:

$$\sum_{i=1}^{n} \overline{R}_i \delta\overline{r} = 0 + \sum_{i=1}^{n} \overline{G}_i \delta\overline{\theta} = 0 .$$

By using (1) and (2), we get:

$$\sum_{i=1}^{n} (\overline{R}_{ia} + \overline{R}_{ic}) \delta\overline{r} = 0 \sum_{i=1}^{n} (\overline{G}_{ia} + \overline{G}_{ic}) \delta\overline{\theta} = 0$$

$$\sum_{i=1}^{n} \overline{R}_{ia} \delta\overline{r} + \sum_{i=1}^{n} \overline{R}_{ic} \delta\overline{r} + \sum_{i=1}^{n} \overline{G}_{ia}\, \delta\overline{\theta} + \sum_{i=1}^{n} \overline{G}_{ic}\, \delta\overline{\theta} = 0.$$

Since the constraints are workless,

$$\sum_{i=1}^{n} \overline{R}_{ic} \delta \overline{r} + \sum_{i=1}^{n} \overline{G}_{ic} \delta \overline{\theta} = 0,$$

and therefore

$$\sum_{i=1}^{n} \overline{R}_{ia} \delta \overline{r} + \sum_{i=1}^{n} \overline{G}_{ia} \delta \overline{\theta}.$$

That is, the virtual work done by the applied forces and torque is zero. Conversely, let the virtual work done by the applied forces and applied torque be zero, then

$$\sum_{i=1}^{n} \overline{R}_{ia} \delta \overline{r} + \sum_{i=1}^{n} \overline{G}_{ia} \delta \overline{\theta}.$$

Also, given that the constraints are workless, we have:

$$\sum_{i=1}^{n} \overline{R}_{ic} \delta \overline{r} + \sum_{i=1}^{n} \overline{G}_{ic} \delta \overline{\theta} = 0.$$

Adding (3) and (4), we get:

$$\sum_{i=1}^{n} \overline{R}_{ia} \delta \overline{r} + \sum_{i=1}^{n} \overline{R}_{ic} \delta \overline{r} + \sum_{i=1}^{n} \overline{G}_{ia} \delta \overline{\theta} + \sum_{i=1}^{n} \overline{G}_{ic} \delta \overline{\theta} = 0,$$

$$\sum_{i=1}^{n} (\overline{R}_{ia} + \overline{R}_{ic}) \delta \overline{r} = 0 \sum_{i=1}^{n} (\overline{G}_{ia} + \overline{G}_{ic}) \delta \overline{\theta} = 0$$

$$\sum_{i=1}^{n} \overline{R}_{i} \delta \overline{r} + \sum_{i=1}^{n} \overline{G}_{i} \delta \overline{\theta} = 0.$$

Since $\delta \overline{r}$ and $\delta \overline{\theta}$ are arbitrary. Thus,

$$\sum_{i=1}^{n} \overline{R}_{i} \delta \overline{r} = 0 \text{ and} \sum_{i=1}^{n} \overline{G}_{i} \delta \overline{\theta} = 0.$$

i.e. the set of rigid bodies are in equilibrium.

Example

Four equally heavy uniform rods are freely joined to form a rhombus $ABCD$ that is freely suspended from A and kept in the shape of a square by an inextensible string connecting A and C. Show that the tension in the string is $2W$, where W is the weight of one rod.

Solution

Since A is a fixed point, we measure the distance from A. The weight of the four rods is $4W$, which act at the center of gravity G, where G is the point of intersection of the diagonals AC and BD.

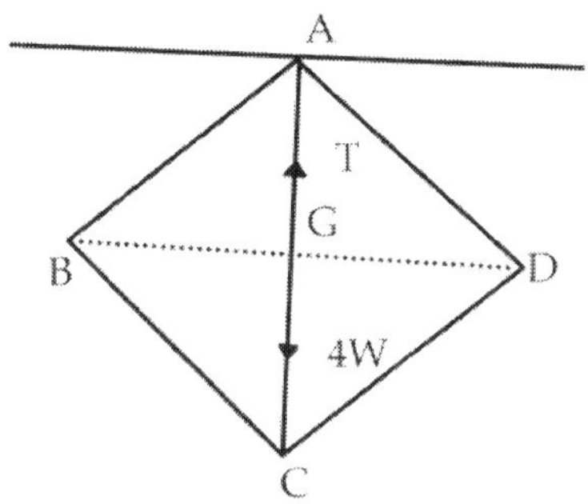

Suppose that the string is removed and a force equal to tension T is applied at C along AC. Let

$AG = y,\quad$ then $AC = 2y$.

The equation of virtual work:

$$4W\delta(AG) - T\delta(AC) = 2y$$

$$4W\delta(y) - T\delta(2y) = 0$$

$$4W - 2T = 0$$

$$T = 2W.$$

Example

A frame work $ABCD$ consists of four equal light rods smoothly joined together to form a square; it is suspended from a peg at A and a weight W is attached to C. The framework is kept in shape by a light rod connecting B and D. Determine the thrust of the rod.

Solution

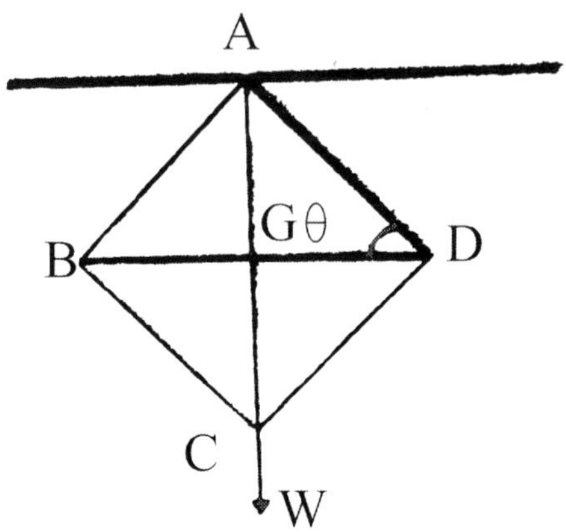

Let $\angle ADB = \theta$ and l be the length of each rod, then AD = l.

Thus $AG = AD\sin\theta = lsin\theta$

$$DG = ADcos\theta = lcos\theta$$

$$BD = 2(DG) = 2l\,cos\theta$$

$$AC = 2(AG) = 2l\sin\theta.$$

Let *T* be the thrust in rod *BD*. The equation of the virtual work is given by:

$$W\delta(AC) + T\delta(2lcos\theta) = 0$$

$$W\delta(2lsin\delta) + T\,\delta(2lcos\theta) = 0$$

$$W2lcos\theta\delta\theta - T\,2l\,sin\theta\delta\theta = 0$$

$$W\cos\theta - T\sin\theta = 0$$

$$T = Wcot\theta\,.$$

Put $\theta = 45°$, as $ABCD$ is a square, then we have:

$$T = W\,.$$

Note: The thrust in the rod T is positive when tension in the string is negative.

Example

Four uniform rods are freely joined at their extremities and from a parallelogram ABCS, which is suspended from the joint A and is kept in shape by an inextensible string AC. Prove that the tension of the string is equal to half of the whole weight.

Solution

Let W be the weight of the four rods. It acts at its center of gravity G, where G the point of intersection is of AC and BD. Suppose that the string is removed and a force equal to tension T is applied to C, along AC.

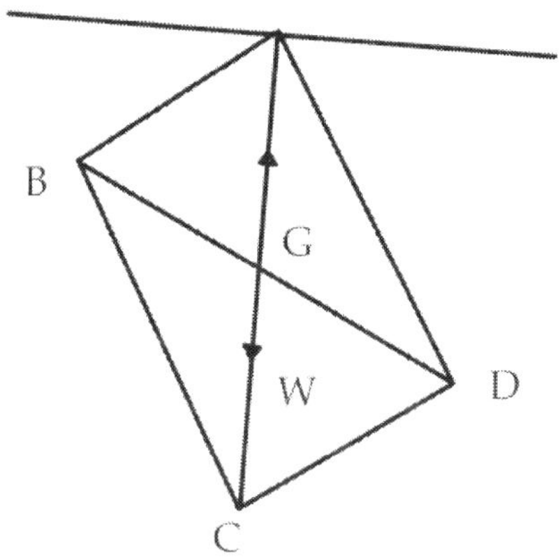

Let $AG=y$, then,

$AC=2y$

The equation of virtual work is given as:

$$W\delta(AG)-T\delta(AC)=0$$

$$W\delta(y)-T\delta(2y)=0$$

$$W\delta-2T\delta y=0$$

$$W-2T=0$$

$$T=\frac{W}{2}.$$

Example

Six equal bars are smoothly hinged at their extremities, forming a regular hexagon *ABCDEF,* kept in shape by vertical strings joining the middle points of BC, CD and AF, FE respectively. The side AB is held horizontally and at the upper most side. Prove that the tension of each string is three times the weight of the bar.

Solution

Let W be the weight of each bar, then the weight of six bars is $6W$ which acts at the center of gravity G. $Let\, P, Q, L, M$ and N be the mid points of the bars BC, CD, AB, AF and *EF* respectively. Let θ be the angle at which bar AF makes with the horizontal. Let l be the length of each bar. Let PQ and MN intersect CF at K and H, respectively.

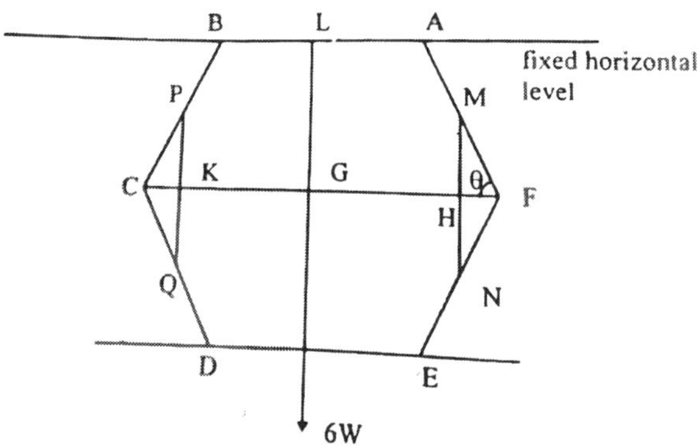

For the triangle MFH,

$$HF = MF\, cos\,\theta = \frac{1}{2}(AF)\theta = \frac{1}{2}l\cos\theta$$

$$MH = MF\sin\theta = \frac{1}{2}(AF)\sin\theta = \frac{1}{2}l\sin\theta.$$

Also,

$$MN = 2(MH) = lsin\theta$$

$$PQ = MN = l\sin\theta$$

$$LG = 2(MH) = l\,sin\theta$$

Let T be the tension in each string, then the equation of virtual work is given as:

$$6W\delta(LG)-T\delta(PQ)-(MN)=0$$

$$6W\delta(lsin\theta)-T\delta(lsin\theta)-T\delta(lsin\theta)=0$$

$$6W\,cos\theta-T\,cos\theta-T\,cos\theta=0$$

$$6W-2T=0,\ T=3W.$$

Example

Six equal uniform rods $AB, BC\,SD, SE, EF\ ,FA$, each of weight W, are freely jointed to form a regular hexagon. The rod AB is fixed in the horizontal position and the slope of the hexagon is maintained by a light rod joining C and F. Show that the thrust in the rod is $\sqrt{3}$ W.

Solution

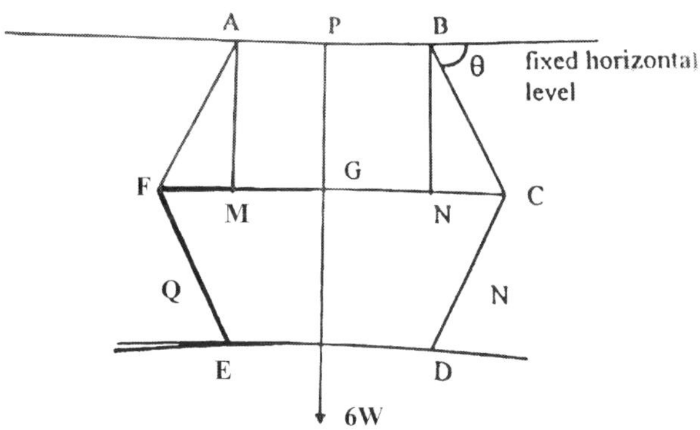

Let θ be the angle that rod BC makes with the horizontal. AM and BN are perpendicular to CF. Let P be the mid-point of AB, P be mid point of AB and l be the length of each rod. The weight of the six rods is $6W$, which acts at G, the center of gravity of six rods.

For triangle BCN

$$BN=BC\sin\theta=l\,sin\theta$$

$$CN=BC\cos\theta=l\,cos\theta.$$

Therefore,

$$PG = BN = l\,sin\theta$$

$$CF = CN + MN + MF$$

$$CF = l\,cos\theta + l + lcos\theta = 1 + 21\,cos\theta.$$

Let *T* be the thrust in rod CF, then the equation of virtual work is

$$6W\,\delta\left(PG\right) + T\delta\left(l + 2l\,cos\theta\right) = 0$$

$$6w\delta(l\sin\theta) + T\delta\left(l + 2lcos\theta\right) = 0$$

$$6wl\,cos\theta\delta\theta - T2l\,sin\theta\delta\theta = 0$$

$$6W\,cos\theta - 2T\,sin\theta = 0$$

$$T = 3W\,cot\theta.$$

For a regular hexagon, take $\theta = 60°$, then

$$T = \sqrt{3}\,W.$$

Example

A string of length *a* forms the shorter diagonal of a rhombus of four uniform rods hinged together, each of length *b* and weight *W*. If one of the rods be supported in a horizontal position, prove that the tension in the string is:

$$\frac{2W\left(2b^2 - a^2\right)}{b\sqrt{4b^2 - a^2}}.$$

Solution

The length of each rod equals *b*, so that

$$AB = BC = CD = DA = b.$$

Also, $AC = a$. Let $\angle ADC = 2\theta$, then $\angle ABD = \theta$ and $\angle BDC = \theta$. Since W is the weight of each rod, the weight of the four rods is 4W, which act at point *G*, the point of intersection of AC and *BD*.

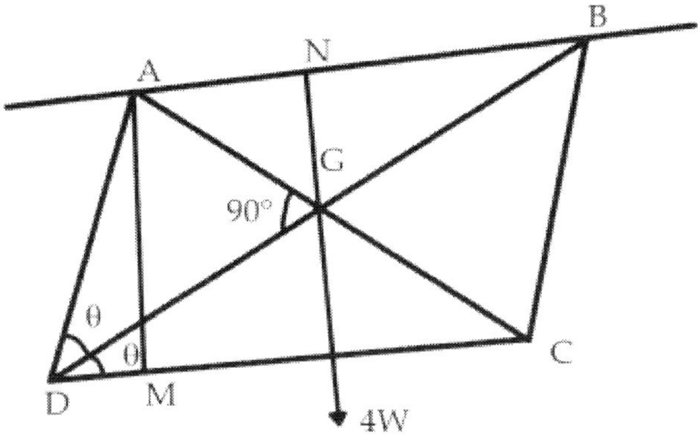

For the triangle ADM $AM = AD$ $\sin\theta = b\sin 2\theta.$

Also,

$NG =$ depth of G below $AB = \frac{1}{2}(AM) = \frac{1}{2}b\sin 2\theta$.

For the triangle $AGD,$

$AG = AD\, sin\theta = b\, sin\theta.$

Thus, $AC = 2(AG) = 2b\sin\theta.$

But $AC = a,$ therefore :

$$sin\theta = \frac{a}{2b}$$

Let T be the tension in the string, the equation of the virtual work is given as:

$$4w\delta(NG) - T\delta(AC) = 0$$

$$4W\delta\left(\frac{1}{2}b\sin 2\theta\right) - T\delta(2b sin\theta) = 0$$

$$4W\, cos2\theta - 2T\cos\theta = 0$$

$$T = \frac{2W\, cos2\theta}{cos\theta} = \frac{2W\,(1-\sin^2\theta)}{\sqrt{1-\sin^2\theta}}.$$

Using $\sin\theta = \frac{a}{2b}$, we have :

$$T = \frac{2\left(1-\frac{a^2}{2b^2}\right)}{\sqrt{1-\frac{a^2}{4b^2}}} = \frac{2W\left(2b^2-a^2\right)}{b\sqrt{4b^2-a^2}}.$$

Example

A uniform rod of length 2 α rests in equilibrium against a smooth vertical wall and upon a smooth peg at a distance b from the wall. Show that, in a position of equilibrium, the beam is inclined to the wall at an angle of $\sin^{-1}\left(\frac{b}{a}\right)^{\frac{1}{3}}$.

Solution

Let the rod be inclined at an angle θ to the wall. As $Ab=2a$, we have $,AG=a$ and:

$PN = b.$

The height of the center of gravity G above the peg is MN. Thus,

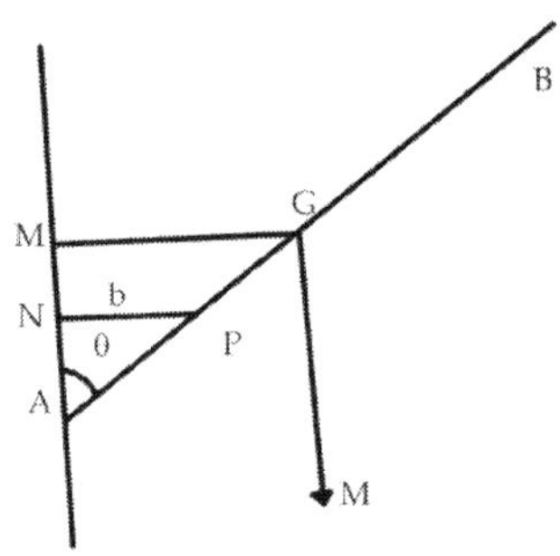

$$MN = AM - AN = AG\,cos\theta - NP\cot\theta$$

$$MN = a\,cos\theta - b\,cot\,\theta.$$

Let $z = acos\theta - b\,cot\theta$, Then the equation of virtual work is:

$$-W\,dz = 0$$

$$a\,sin\theta\,d\theta + b\,cosec^2\theta\,d\theta = 0$$

$$a\sin^3\theta = b \text{ or } \quad \theta = \sin^{-1}\left(\frac{b}{a}\right)^3.$$

Example

A heavy elastic string with a natural length of $2\pi a$ is placed around a smooth cone with a vertical axis and a semi vertical angle of α. If W is the weight and λ the modules of the string, prove that it will be in equilibrium in the form of a circle when radius is given by:

$$a\left(1+\frac{W}{2\pi a}\cot\alpha\right).$$

Solution

Let the string be in equilibrium at a depth y below the vertex of the cone, and the radius of circle as x. Then:

$$tan\alpha = \frac{x}{y}$$

$$x = y\tan\alpha.$$

The circumference of the circle is given by $2\pi x$.

By Hook's law the tension in the string is T.

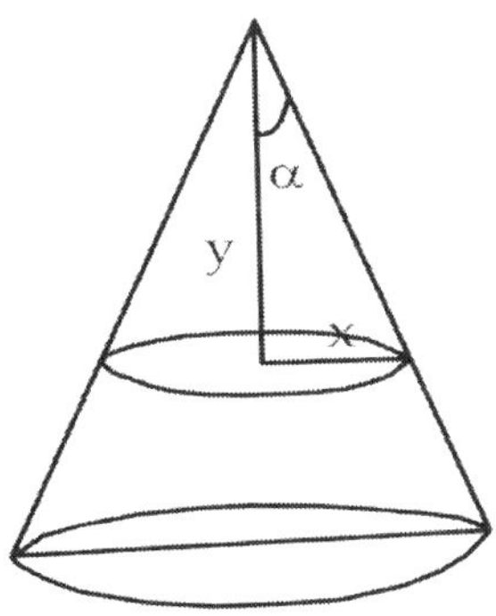

or $T = \frac{\lambda(\text{change in lenght})}{\text{orignal lenght}} = \frac{\lambda(2\pi x - 2\pi a)}{2\pi a} = \frac{\lambda(x-a)}{a}$.

Let the ring be displaced downward by displacement δy. The equation of virtual work is given by:

$$W\delta y - T\delta(2\pi x) = 0$$

$$W\delta - T\delta(2\pi \tan\alpha) = 0$$

$$W\delta y - 2\pi T \tan\alpha\, \delta y = 0$$

$$W = 2\pi T\alpha = 2\pi \frac{\lambda(x-a)}{a} tan\alpha$$

or $$x = a\left(1 + \frac{w}{2\pi a} cot\alpha\right).$$

Example

A rhombus is formed from of four uniforms rods joined together and suspended from point *A*. The rhombus is kept in position by a light rod joining the middle points of *BN* and *CD*; prove that if *T* is the thrust in the rod and *W* the weight of the rhombus, then:

$$T = Wtan\left(\frac{A}{2}\right).$$

Solution

The weight *W* of the rhombus acts at its center of gravity G. Let *P* and Q be midpoints of BC and CD respectively. Let l be the length of each rod and

$\angle CAD = \theta \, and \, \angle ACD = \theta.$

For the triangle $ADG,$ we have:

$AG = ADcos\theta = lcos\theta.$

Let PQ intersect AC at N, then for ΔCNQ, we have:

$QN = CQ sin\theta$

$PQ = 2QN = 2CQ\,sin\theta = CD\,sin\theta = l\,sin\theta$.

The equation of virtual work is given by:

$$W\delta(AG) + T\delta(PQ) = 0$$

$$W\delta(lcos\theta) + T\delta(l\,sin\theta) = 0$$

$$-W\,lsin\theta\delta\theta + Tl\,cos\theta\delta\theta = 0$$

$$Wsin\theta - Tcos\theta = 0$$

$$T = Wtan\theta = W\tan\left(\frac{A}{2}\right).$$

Four equally uniform rods, each of weight W, are connected at one end of each by means of a smooth joint and the other ends rest on a smooth table and are connected by equal strings. A weight W is suspended form the joint. Show that the tension at each of the strings is:

$$\frac{W+2w}{4}\frac{a}{\sqrt{4l^2-2a^2}},$$

where l is the length of each rod and a is length of each string.

Solution

Here $OA = OB = OC = OD = l$ and

$AB = BC = CD = AD = a$.

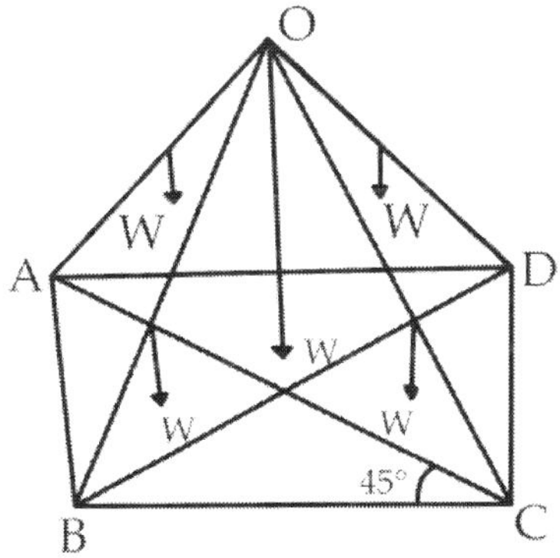

Let the suspended weight W make an angle with each rod. For the triangle OCG, we have:

$$CG = OC sin\theta = l sin\theta$$

$$OG = OC cos\theta = l\cos\theta$$

Also $AC = 2(CG) = 2l sin\theta.$

For the triangle ABC:

$$BC = AC cos45 = 2l sin\theta cos45 = \sqrt{2}\, l\, sin\theta.$$

Given that $BC = a,$ therefore,

$$sin\theta = \frac{a}{\sqrt{2l}}$$

$$tan\theta = \frac{a}{\sqrt{2l^2 - a^2}}$$

The height of the midpoint of each rod above the plane $ABCD$ is given as:

$$\frac{1}{2}(OG) = \frac{1}{2}(l cos\theta).$$

The equation of virtual work is:

$$-4w\delta\left(\frac{1}{2}cos\theta\right) - W\delta(l cos\theta) - 4T\delta\left(\sqrt{2}\, l\, sin\theta\right) = 0$$

$$2wl sin\theta\delta\theta + Wl\, sin\theta\delta\theta - 4\sqrt{2}\, l\, T\cos\theta\delta\theta = 0$$

$$2w + W - 4\sqrt{2}\, T\tan\theta = 0$$

$$T = \frac{W + 2w}{4\sqrt{2}} tan\theta = \frac{W + 2w}{4}\left(\frac{a}{\sqrt{4l^2 - 2a^2}}\right).$$

SUMMARY AND DISCUSSION

Work done by a force

Consider a constant force F that acts on a particle and causes a displacement from point A to point B. Let the displacement $AB = S$, then the work done, W, by force F, is defined as:

$W = F.S = FS\cos\theta,$

where θ is the angle between F and S.

If $\theta = 0$, then we have :

Work done $= W = FS\cos\theta = FS.$

If $\theta = \dfrac{\pi}{2}$, then

Work done $= W = FS\cos\dfrac{\pi}{2} = FS(0) = 0.$

Therefore, when the applied force is perpendicular to the displacement, the work done will be zero.

If $\theta = \pi$, then :

Work done $= W = FS\cos\pi = FS(-1) = -FS.$

If $0 \le \theta \le \dfrac{\pi}{2}$, then

Work done $= W$ is positive, as cos $\theta \ge 0.$

If $\dfrac{\pi}{2} \le \theta \le \pi$, then

Work done = W is negative, as $\cos\theta \le 0.$

Work done by the weight, W

If a body of weight W falls down vertically through a distance h, the work done by gravity is positive and equal to Wh .

If the body rises vertically at a height h, the work done is negative and is equal to - Wh. We say that work is being done against gravity.

The total work done by all the forces is equal to the work done by their resultants.

Virtual displacement and virtual work

If a set of particles or a body is in equilibrium under the action of some forces then there is no motion and consequently there is no actual displacement.

Suppose that the set of particles or body receives an imaginary displacement, the acting forces are regarded as constant during the displacement, denoted as the virtual displacement, and the work done by the forces during such a displacement is known as the virtual work.

Constant

A constraint is a restriction imposed on a mechanical system to get a desired position or motion. For the case of a particle moving down an inclined plane, the constraint on its motion is its contact with the plane. Any displacement along the plane is consistent with the constraint and any displacement away from the plane is not consistent with the constraint. Due to the imposition of constraint, we get unknown forces acting on the system. These forced are called reactions of the constraints.

Ideal constraint or workless constraints

If the forces of the constraint do not result in virtual work, the constraints are said to be ideal or workless.

The tension on an inextensible string is workless.

Principal of virtual work for a single particle

A particle subject to workless constraints is in equilibrium if and only if zero virtual work is done by the applied forces in any arbitrary infinitesimal displacement consistent with the constraints.

Principle of virtual work for set particles

A set of particles, subject to workless constraints, is in equilibrium if and only if zero virtual work is one by the applied forces in an arbitrary infinitesimal displacement consistent with the constraints.

Principal of virtual work of a single rigid body

A rigid body, subject to workless constraints, is equilibrium if and only if zero virtual work is done by the applied forces and applied torques in any arbitrary infinitesimal displacement consistent with the constraints.

Principal of virtual work for a set of rigid bodies

A set of rigid bodies, subject to workless constraints, is in equilibrium if and only if zero virtual work is done by the applied forces and applied torques in any arbitrary infinitesimal displacement consistent with the constraints.

CHAPTER 5

KINEMATICS

INTRODUCTION

In this chapter, we will discuss the important topic of mechanics.

Dynamics

The branch of mechanics that deals with the motion of objects. It is divided into two branches, namely kinematics and kinetics.

Kinematics

Kinematics deals with the geometry of motion, without any reference to the forces acting on the body.

Kinetics

Kinetics relate the motion of the body with the forces causing the motion.

CARTESIAN COMPONENT OF VELOCITY AND ACCELERATION

Consider the motion of the particle in the -plane, as shown in the figure below.

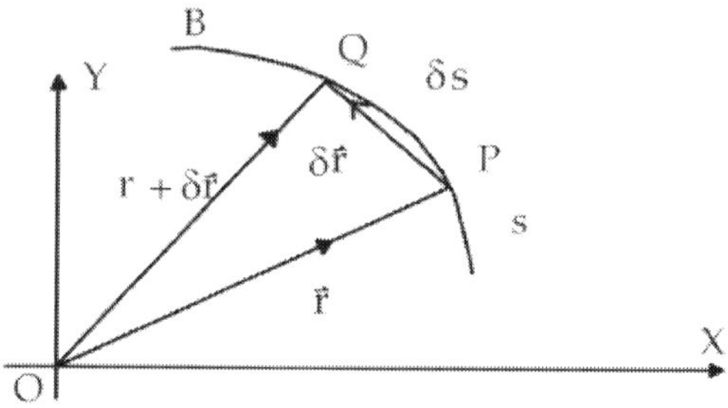

- Let $P(\bar{r})$ and $Q(\bar{r}+S\bar{r})$ be the position of the particle at time t and $t+\delta t$ respectively, then we have: $\overline{PQ} = \delta\bar{r}$.

Let S be the length of arc AB measured from some fixed point on the curve. Let PQ = δs, then $\frac{\delta\bar{r}}{\delta t} Lim \frac{\delta\bar{r}}{\delta t}$ is the instantaneous velocity of the particle at P and denoted by $\bar{a}$.Let $P(x, y)$ be a point and $\bar{r} = xi + yj$, then

$$\bar{v} = \frac{d\bar{r}}{dt} = \frac{dx}{dt}i + \frac{dy}{dt}j$$

and $$\bar{a} = \frac{d^2\bar{r}}{dt^2} = \frac{d^2x}{dt^2}i + \frac{d^2y}{dt^2}j$$, where

$v_x = x -$ Component of velocity $= \frac{dx}{dt}$

$v_y = y -$ Component of velocity $= \frac{dy}{dt}$

$a_x = x -$ Component of acceleration $= \frac{d^{2x}}{dt^2}$

$a_y = x -$ Component of acceleration $= \frac{d^{2y}}{dt^2}$

$$v = |\bar{v}| = \sqrt{\left(\frac{dx}{dt}\right)^2 + \left(\frac{dy}{dt}\right)^2} = \frac{ds}{dt}.$$

Example

Describe the path of a particle whose position vector is $\overline{r} = a\,cosh\omega t\overline{i} + b\,sinh\omega t\overline{j}\ (0 \le t < \infty)$, where a, b and ω are constants. Find its velocity and acceleration.

Solution

As $\overline{r} = a\,cosh\omega\overline{i} + bsinh\omega\,\overline{j}$, we have

$x = a\cosh\omega t, \quad y = b\,sinh\omega t$

$\frac{x}{a} = cosh\omega t \qquad \frac{y}{b} = sinh\omega t.$

Thus

$\frac{x^2}{a^2} - \frac{y^2}{b^2} = \cosh^2(\omega t) - \sinh^2(\omega t) = 1$ is the required path carried out by the particle.

Now $\overline{v} = \frac{d\overline{r}}{dt} = a\omega sinh\omega\overline{i} + b\omega\cosh\omega t\overline{j}$

$\overline{a} = \frac{d^2\overline{r}}{dt^2} = a\omega^2 cosh\overline{i} + b\omega^2 sinh\omega t\overline{j}$

$= \omega^2\left(a\,cosh\omega\overline{i} + bsinh\omega\overline{j}\right) = \omega^2\overline{r}$.

Thus, $\overline{a} = \omega^2\overline{r}$.

Example

The position of the particle moving along an ellipse is given by $\overline{r} = a\cos\overline{i} + b\sin t\,\overline{j}$. If $a > b$, find the position of the particle when its velocity has a maximum or minimum magnitude.

Solution

Given $\bar{r} = a\cos t\,\bar{i} + b\sin t\,\bar{j}$, we have:

$$\bar{v} = \frac{d\bar{r}}{dt} = -a\,sint\,\bar{i} + b\cos t\,\bar{j}$$

$$v = |\bar{v}| = \sqrt{a^2\sin^2 t + b^2\cos^2 t}$$

$$= \sqrt{a^2\sin^2 t + b^2(1 - \sin^2 t)}$$

$$= \sqrt{a^2\sin^2 t + b^2 - b^2\sin^2 t)}$$

or $v = \sqrt{b^2 + \left(a^2 - b^2\right)\sin^2 t}$,

where v is maximum when $\sin^2 t = 1$. Therefore,

$sint = \pm 1$.

When $\sin = 1, \quad t = \pi/2$.

When $\sin = -1 \quad t = 3(\pi/2)$.

Thus, for $t = \frac{\pi}{2}, 3\left(\frac{\pi}{2}\right)$, $\bar{r} = a\cos\left(\frac{\pi}{2}\right)\bar{i} + b\sin\left(\frac{\pi}{2}\right)\bar{j} = b\bar{j}.$

For $t = \frac{3\pi}{2}$, $r = a\cos\left(\frac{3\pi}{2}\right)\bar{i} + b\sin\left(\frac{3\pi}{2}\right)\bar{j} = -b\bar{j}$.

Therefore,

$\bar{r} = \pm b\bar{j}.$

v is minimum when $\sin^2 t = 0$, hence $t = 0, \pi$.

For $t = 0$ $\bar{r} = a\cos(0)\bar{i} + b\sin(0)\bar{j} = a\bar{i}$.

For $t = \pi$, $\bar{r} = a\cos(\pi)\bar{i} + b\sin(\pi)\bar{j} = a\bar{i}$.

Hence, the velocity has a minimum magnitude when $\bar{r} = \pm a\bar{i}$.

Radial and transverse components of velocity and acceleration

Let *P* be the position of the particle at time t and let $P(r, o)$ be its polar coordinates. The direction following an increasing θ and perpendicular to the direction of the particle is denoted as the transverse direction.

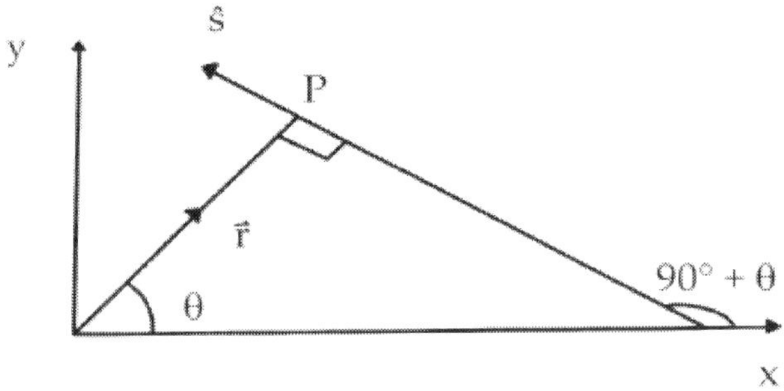

$\overline{r} = cos\theta\overline{i} + sin\theta\overline{j}$

$\overline{s} = \cos(90+\theta)\overline{i} + \sin(90+\theta)\overline{j}$

$= -\sin\theta + \cos\theta\overline{j}$.

Now $\frac{d\overline{r}}{dt} = \frac{d\overline{r}}{d\theta}$

$= (-\sin\theta\overline{i} + cos\theta\overline{j})\frac{d\theta}{dt} = \overline{s}\theta = \overline{s}\theta; \frac{d\theta}{dt} = \theta$ and

$\frac{d\overline{s}}{dt} = \frac{d\overline{s}}{d\theta}\frac{d\theta}{dt}$

$= -(\cos\theta\overline{i} + sin\theta\overline{i})\frac{d\theta}{dt} = \overline{r}\theta = -\theta\overline{r}$.

Now $\overline{v} = \frac{d\overline{r}}{dt} = \frac{d}{dt}(r.\overline{r}) = \frac{dr}{dt}\overline{r}$ + r $\frac{d\overline{r}}{dt} + r\frac{d\overline{r}}{dt} = r\overline{r} + r\theta\overline{s}$.

Thus, we have:

The radial component of the velocity = $v_r = r$

The transverse component of velocity = $v_\theta = r\theta.\overline{s}$.

Also, $\overline{a} = \frac{d\overline{v}}{dt} = \frac{d}{dt}\left(s\overline{r} + r\theta\overline{s}\right)$

$\ddot{r}\hat{r} + \dot{r}\dot{\theta}\hat{s} + \dot{r}\dot{\theta}\hat{s} + \overline{r}\ddot{\theta}\overline{s} + r\dot{\theta}(\dot{\theta}\widehat{r})$

$= \left(\ddot{r} + r\theta^2\right)\dot{\hat{r}} + \left(2\dot{r}\dot{\theta} + r\ddot{\theta}\right)\hat{s}.$

Thus, we have:

The radial component of the acceleration = $a_r = r + \ddot{r}\theta^2$

The transverse component of acceleration = $a_0 = 2\dot{r}\dot{\theta} + r\ddot{\theta}.$

Note that $a_0 = \frac{1}{r}\frac{d}{dt}\left(r^2\dot{\theta}\right)$.

Example

Find the radial and transverse components of the velocity of a particle moving along the curve $ax^2 + by^2 = 1$ at any time t with a polar angle $\theta = ct^2$.

Solution

Given that $\theta = ct^2$, we have $\theta = 2ct$. Changing the equation of the given curve into polar coordinates using $x = r\cos\theta$ and $y = r\sin\theta$, we get:

$ar^2 cos^2\theta + br^2\sin^2\theta = 1$,

or $r = \left(a cos^2\theta + b\sin^2\theta\right)^{\frac{1}{2}}$.

Thus, $r = \frac{dt}{dt} = \frac{dr}{d\theta}\frac{d\theta}{dt}$

$= -\frac{1}{2}\left(a\cos^2\theta + b\sin\theta\right)^{-\frac{3}{2}}(2)(-a+b)\sin\theta cps\theta\theta$

$$=-\left(a\cos^2\theta+b\sin\theta\right)^{\frac{3}{2}}\left(-a+b\right)\left(2ct\right)\sin\theta\cos\theta,$$

$$v_r=r=\frac{ct\left(a-b\right)sin2\theta}{\left(a\cos^2\theta+b\,sin^2\theta\right)^{\frac{3}{2}}},$$

$$v_\theta=r\theta=\frac{2ct}{\left(a\cos^2\theta+b\,sin^2\theta\right)^{\frac{1}{2}}}.$$

Example

The velocities of a particle along the perpendicular to the radius vector are λr^2 and $\mu\theta^2$, where λ and μ are constants. Find its radial and transverse components of acceleration in terms of *r* and θ.

Solution

Here $v_r=r\lambda r^2$ and $v_\theta=r\theta=\mu\theta^2$,

or $r=\lambda r^2$ and $\theta=\dfrac{\mu\theta^2}{r}$.

Therefore, by differentiating, we get:

$$r=2\lambda rr=2\lambda r\left(\lambda^2 r\right)=2\lambda^2 r^3.$$

Thus, $a_r=r-r\theta^2=2\lambda^2 r^3-r\left(\dfrac{\mu\theta^2}{r}\right)^2$

$$=2\lambda^2 r^3-\frac{\mu^2\theta^4}{r},$$

$$a_\theta=\frac{1}{r}\frac{d}{dt}\left(r^2\theta\right)=\frac{1}{r}\frac{d}{dt}\left(r^2\theta\right)=\frac{1}{r}\frac{d}{dt}\left\{r^2\left(\frac{\mu\theta}{r}\right)\right\}$$

$$=\frac{1}{r}\frac{d}{dt}\left\{r\left(\mu\theta^2\right)\right\}$$

$$=\frac{\mu}{r}\frac{d}{dt}\left(r\mu^2\right)$$

$$=\frac{\mu}{r}\left(r\theta^2+2r\theta\right)$$

$$=\frac{\mu}{r}\left(\lambda r^2\right)\theta^2+2\frac{\mu}{r}r\theta\left(\frac{\mu\theta}{r}\right)$$

$$=\mu\lambda r\theta+2\frac{\lambda^1\theta^3}{r}.$$

Tangential and normal components of velocity and acceleration

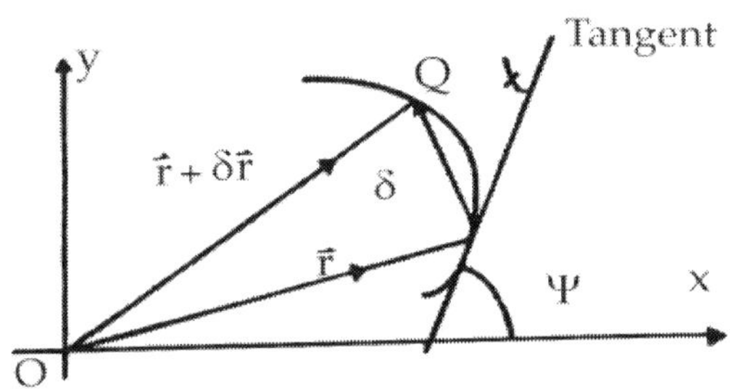

Let $P(\bar{r})$ and Q $(\bar{r}+\delta\bar{r})$ be the positions of the particles at time t and $t+\delta t$ respectively. Then $\overline{PQ}=\delta\bar{r}$...... Let arc $PQ=\delta s,$then :

$$\bar{v}=\frac{d\bar{r}}{dt}=\frac{d\bar{r}}{ds}\frac{ds}{dt}=\frac{d\bar{r}}{dt}=\frac{d\bar{r}}{ds}v=v\frac{d\bar{r}}{ds}.$$

Now $\left|\frac{d\bar{r}}{dt}\right|=Lim_{\delta\to 0}\frac{\left|\overline{\delta r}\right|}{\delta s}=Lim_{\delta\to 0}\frac{\left|\overline{PQ}\right|}{arc\,PQ}=1.$

Also $\frac{d\bar{r}}{dt}=Lim_{\delta\to 0}\frac{\delta\bar{r}}{\delta s}=Lim_{\delta\to 0}\frac{\delta\bar{r}}{\delta s}=Lim_{\delta\to 0}\frac{1}{\delta s}(\delta\bar{r}),$ which is a vector along the tangent at P Therefore, $\frac{d\bar{r}}{ds}$ is a unit tangent at P. Let $\bar{t}$ be a unit vector along the tangent at P . Therefore,

$$\frac{d\overline{r}}{ds} = \hat{t}$$

Substituting this into eq. (1):

$$\overline{v} = v\hat{t} = v\hat{t} + 0\hat{n}.$$

Where $\overline{n}$ is a unit vector along normal at *P*. Thus,

$v_t =$ Tangential component of the velocity = v, and

$v_n =$ Normal component of velocity = 0.

Hence, the velocity is determined along the tangent to the path.

Let $\overline{a}$ be the acceleration, then:

$$\overline{a} = \frac{d\overline{v}}{dt} = \frac{d}{dt}\left(v\hat{t}\right) = \frac{dv}{dt}\hat{t} + v\frac{d\hat{t}}{dt}$$

$$= \frac{dv}{dt}\hat{t} + v\frac{d\hat{t}}{d\psi}\frac{d\psi}{ds}\frac{ds}{dt}$$

$$= \frac{dv}{dt}\hat{t} + v\frac{d\hat{t}}{d\psi}Kv$$,

where K = $\frac{d\psi}{ds}$ is a curvature, thus:

$$\overline{a} = \frac{dv}{dt}\hat{t} + \frac{v^2}{p}\frac{d\hat{t}}{d\psi}.$$

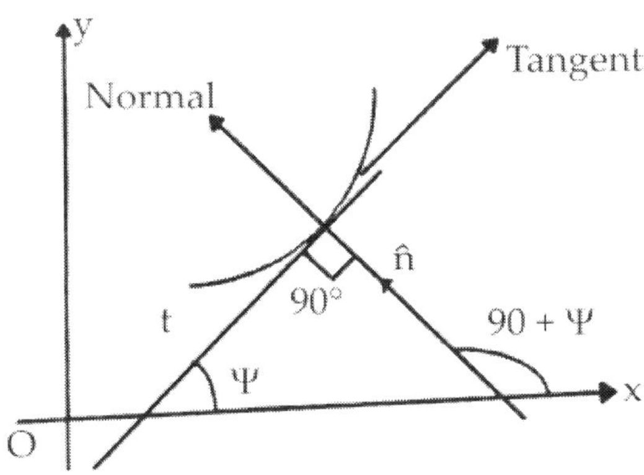

$$K = \frac{1}{\rho}$$

Where ρ is the radius of the curvature? Let $\hat{t}$ and $\hat{n}$ be the unit vectors along the tangent and normal at *P*, then:

$$\hat{t} = cos\psi\hat{i} + sin\psi\hat{j}$$

$$\hat{n} = \cos(\psi + 90)\hat{i} + \sin(\psi + 90)\hat{j},$$

or

$$\hat{t} = cos\psi\hat{i} + sin\psi\hat{j}$$

$$\hat{n} = sin\psi\hat{i} + cos\psi\hat{j} = \hat{n}.$$

Therefore,

$$\frac{d\hat{t}}{d\psi} = -sin\psi\hat{i} + cos\psi\hat{j} = \hat{n}.$$

Substituting this into eq. (2), we get:

$$\bar{a} = \frac{dv}{dt}\hat{t} + \frac{v^2}{p}\hat{n}.$$

Hence,

$a_t =$ tangential component of acceleration $= \frac{dv}{dt}$

$a_n =$ Normal component of acceleration $= \frac{v^2}{p}$.

Example

Find the tangential and normal components of the acceleration of a particle describing the ellipse $\frac{x^2}{a^2} + \frac{y^2}{b^2} = 1,$ with uniform speed *v*, when the particle is at (0, b).

Solution

Given that $\frac{x^2}{a^2}+\frac{y^2}{b^2}=1,$ differentiating $w.r.t.$ 'xx', we get:

$$\frac{2x}{a^2}+\frac{2y}{b^2}y'=0 \qquad \text{or} \quad y'=-\frac{b^2x}{a^2y}.$$

Differentiating again,

$$y''=\frac{b^2}{a^2}\left(\frac{y-xy'}{y^2}\right).$$

Thus at $(0,b)$:

$$y''=\frac{b}{a^2}$$

$$y'=0.$$

Now, from differential calculus, we know that:

$$p=\frac{\left[1+\left(\frac{dy}{dx}\right)^2\right]^{\frac{3}{2}}}{\frac{d^2y}{dx^2}},$$

so that

$$p=\frac{\left[1+(0)^2\right]^{\frac{3}{2}}}{\frac{b}{a^2}}=\frac{a^2}{b}.$$

p is always positive, and velocity is uniform, therefore,

$\frac{dv}{dt}=0$, and

$$a_t=\frac{dv}{dt}=0,\ a_n=\frac{by^2}{a^2}.$$

Example

A particle is moving with uniform speed along the curve $x^2 y = a\left(x^2 + \frac{a^2}{\sqrt{5}}\right)$, show that its acceleration has a maximum value of $\frac{10v^2}{9a}$.

Solution

Given that $x^2 y = a\left(x^2 + \frac{a^2}{\sqrt{5}}\right)$, or $y = a + \frac{a^3 x^{-2}}{\sqrt{5}}$, and differentiating $w.r.t.'x'$:

$$y' = \frac{2a^3 x^{-3}}{\sqrt{5}}$$

$$y'' = \frac{6a^3 x^{-4}}{\sqrt{5}}.$$

From the differential calculus, we know that:

$$\frac{\left[1+\left(\frac{dy}{dx}\right)^2\right]^{\frac{3}{2}}}{\frac{d^2y}{dx^2}},$$

so that $p = \frac{\left[1+\left(\frac{4a^6}{5x^6}\right)^2\right]^{\frac{3}{2}}}{\frac{6a^3}{\sqrt{5x^4}}}$.

Since speed is uniform, we have $\frac{dv}{dt} = 0$:

$$\bar{a} = \frac{dv}{dt}\hat{t} + \frac{v^2}{p}\hat{n} = \frac{v^2}{p}\hat{n},$$

$$|\bar{a}| = \frac{v^2}{p}\hat{n}$$

$|\bar{a}|$ is maximum when p is minimum, $i.e.\ \frac{dp}{dx} = 0$.

Differentiating (1), then:

$$\frac{dp}{dx} = \frac{30a^3x^3\left(\frac{3}{2}\right)\left(5x^6+4a^6\right)^{\frac{1}{2}}\left(30x^5\right) - \left(5x^6+4a^6\right)^{\frac{3}{2}}\left(150a^3x^4\right)}{\left(30a^3x^5\right)^2}$$

$$= \frac{30a^3x^4\left(5x^6+4a^6\right)^{\frac{1}{2}}\left(45x^6-25x^6-20a^6\right)}{\left(30a^3x^5\right)^2}$$

$$= \frac{20\left(5x^6+4a^6\right)^{\frac{1}{2}}\left(x^6-a^6\right)}{30a^3x^6}$$

$$= \frac{20\left(5x^6+4a^6\right)^{\frac{1}{2}}}{30a^3x^6}(x+a)(x-a)\left(x^4+a^2x^2+a^4\right).$$

Thus $\frac{dp}{dx} = 0$ implies that $x = a, x = -a$. Hence,

$\frac{dp}{dx} < 0$ before $x = a$.

$\frac{dp}{dx} > 0$ after $x = a$.

Therefore, p is minimum when $x = a$. Thus for $x = a$, eq. (1) gives us

$$p = \frac{\left(5a^6+4a^6\right)^{\frac{3}{2}}}{30a^3a^5} = \frac{9a}{10}.$$

e maximum value of magnitude of the acceleration is $\frac{v^2}{p_{min}}$,

or $a = \frac{10v^2}{9a}$.

Example

A point describes a cycloid $s = 4a\ sin\psi$ with uniform speed v. Find the acceleration at any point.

Solution

Given that $s = 4a\,sin\psi$,

$$\frac{ds}{d\psi} = 4a\,cos\psi = p,$$

$$\bar{a} = \frac{dv}{dt}\hat{t} + \frac{v^2}{p} = \hat{n} = \frac{v^2}{p} = \hat{n}.$$

We know that v is uniform, $i.e\frac{dv}{dt} = 0.$

Using eq. (1) we have:

$$\bar{a} = \frac{v^2}{4acos\psi}\hat{n} = \frac{v^2}{4a\sqrt{1-\sin^2\psi}}\hat{n},$$

or

$$\bar{a} = \frac{v^2}{4a\sqrt{1-\left(\frac{s^2}{16a^2}\right)}}\hat{n} = \frac{v^2}{4a\sqrt{16a^2-s^2}}\hat{n}.$$

Relative velocity

Let $+\left(\overline{r_p}\right) and\, Q\left(\overline{r_Q}\right)$ be the positions of two particles moving on the $xy-$ plane.

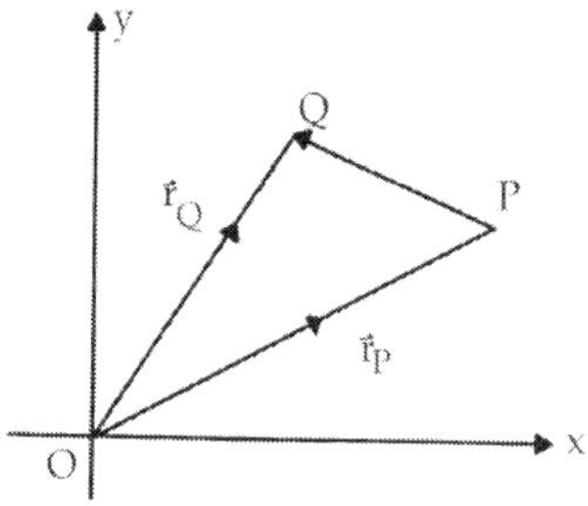

Then $\left(\overline{PQ}\right) = \left(\overline{r}_Q - \overline{r}_p\right)$.

The rate of change of the displacement vector $\overline{PQ}$ is called the relative velocity of Q $w.r.t\ p$, and is given as:

$$\frac{d}{dt}\left(\overline{PQ}\right),$$

or $$v_{rel} = \frac{d}{dt}\left(\overline{r}_Q - \overline{r}_p\right) = \overline{r}_Q - \overline{r}_p$$

= (velocity of Q) – (velocity of P).

If the velocity of two particles is equal them relatives' velocity is zero.

Angular Velocity

The angular velocity ω is the rate of change of the angular displacement θ *w.r.t.* time t. Let p be the position of the particles moving about O. Let

$$\angle XOP = \theta \text{ then } \omega = \frac{d\theta}{dt}$$

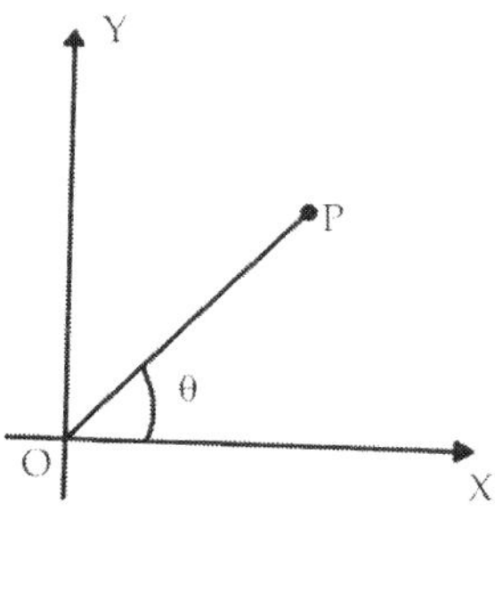

gives us the magnitude of angular velocity. The angular velocity is a vector quantity and its direction is perpendicular to the plane rotation of particles. Thus,

$$\bar{\omega} = \frac{d\theta}{dt}\bar{k}.$$

If the rotation is clockwise then the angular velocity is taken as positive, otherwise it is negative.

Consider a particle moving along a circle of radius r and center O. The linear velocity v at p is taken along the tangent at p. Let $\overline{OP} = \bar{r}, \angle POQ = \theta, arc\ PQ = s,$ then :

$$s = r\theta, \text{or} \frac{ds}{dt} = r\frac{d\theta}{dt},$$

as r is constant. Thus,

$$v = r\omega.$$

In vector form, this relation is written as:

$$v = rx\omega.$$

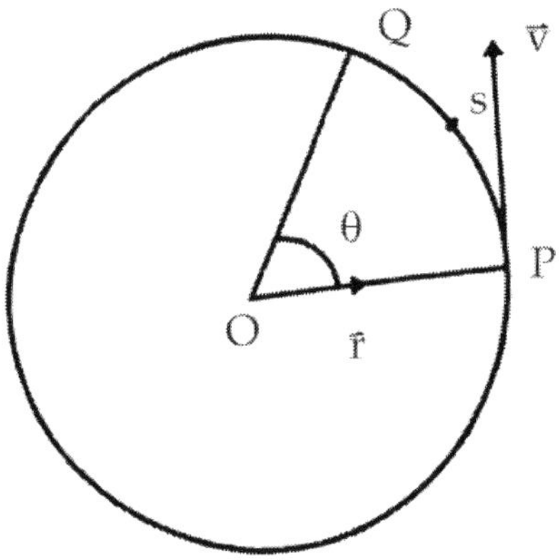

SUMMARY AND DISCUSSION

Dynamics

The branch of mechanics that deals with the motion of objects. It is divided into two branches, namely kinematics and kinetics.

Kinematics

Kinematics deals with the geometry of motion, without any reference to the forces acting on the body.

Kinetics

Kinetics relate the motion of the body with the forces causing the motion.

Cartesian components of velocity and acceleration

If $\bar{r} = xi + yj$ then

$$\bar{v} = \frac{d\bar{r}}{dt} = \frac{dx}{dt}i + \frac{dy}{dt}j,$$

and $$\bar{a} = \frac{d^2\bar{r}}{dt^2} = \frac{d^2x}{dt^2}i + \frac{d^2y}{dt^2}j$$

$v_x = x -$ Component of velocity $= \frac{dx}{dt}$

$v_y = y -$ Component of velocity $= \frac{dy}{dt}$

$a_x = x -$ Component of acceleration $= \frac{d^{2x}}{dt^2}$

$a_y = x -$ Component of acceleration $= \frac{d^{2y}}{dt^2}$

$$v = |\bar{v}| = \sqrt{\left(\frac{dx}{dt}\right)^2 + \left(\frac{dy}{dt}\right)^2} = \frac{ds}{dt}.$$

Radial and transverse components of velocity and acceleration

Radial component of the velocity = $v_r = r$.

Transverse component of velocity = $v_\theta = r\theta\overline{s}$.

Radial component of the acceleration = $a_r = \ddot{r} + r\dot{\theta}^2$.

Transverse component of acceleration = $a_0 = 2\dot{r}\dot{\theta} + r\ddot{\theta}$.

$$a_0 = \frac{1}{r}\frac{d}{dt}(r^2\dot{\theta}).$$

Tangential and normal components of velocity and acceleration

$v_t =$ Tangential component of the velocity = v.

$v_n =$ Normal component of velocity = 0.

Hence velocity is along the tangent to the path.

Let $\bar{a}$ be the acceleration, then

$$\bar{a} = \frac{d\bar{v}}{dt} = \frac{d}{dt}\left(v\hat{t}\right) = \frac{dv}{dt}\hat{t} + v\frac{d\hat{t}}{dt}$$

$$= \frac{dv}{dt}\hat{t} + v\frac{d\hat{t}}{d\psi}\frac{d\psi}{ds}\frac{ds}{dt}$$

$$= \frac{dv}{dt}\hat{t} + v\frac{d\hat{t}}{d\psi}Kv$$

Where K = $\frac{d\psi}{ds}$ *is curvature.Thus*

$$\bar{a} = \frac{dv}{dt}\hat{t} + \frac{v^2}{p}\frac{d\hat{t}}{d\psi}$$

Hence, $a_t =$ tangential component of acceleration $= \frac{dv}{dt}$,

$a_n =$ Normal component of acceleration $= \frac{v^2}{p}$.

Angular velocity

The angular velocity ω is the rate of change of the angular displacement θ $w.r.t.$ time t.

$$\omega = \frac{d\theta}{dt},$$

$$\bar{\omega} = \frac{d\theta}{dt}\bar{k}.$$

If the rotation is clockwise then the angular velocity is taken as positive, otherwise it is negative.

$v = r\omega.$

In vector form this relation is written as:

$v = rx\omega.$

CHAPTER 6

RECTILINEAR MOTION

INTRODUCTION

The motion of a particle along a straight line is denoted as rectilinear motion. Let the particle start from point O along the $x-axis$. After time t, the particle is at point P, a distance x from O. Let $\overline{r}$ be the position vector of point $P\ w.r.t.$ the origin O, then we have:

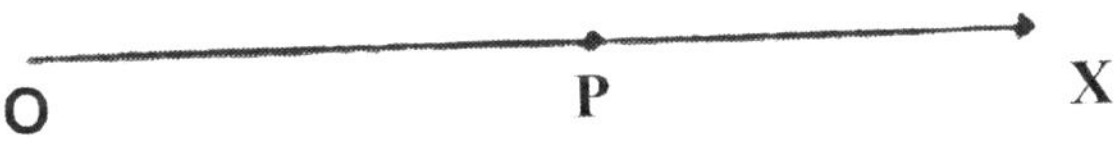

$$\overline{r} = \overline{OP} = x\hat{i}$$

$$\overline{v} = \frac{d\overline{r}}{dt} = \frac{dx}{dt}\hat{i}.$$

Let $v = |\overline{v}|$ and $a = |\overline{a}|$, then

$$v = \frac{dx}{dt},$$

$$a = \frac{d^2x}{dt^2}.$$

Also, $a = \frac{dv}{dt} = \frac{dv}{dx}\,\frac{dx}{dt} = \frac{dv}{dx}v = v\frac{dv}{dx}$

Motion with constant acceleration

Suppose a particle starts from O with velocity u at time $t = 0$, and constant acceleration a. Let the particle be at a distance x from O after time t. Then we have:

$a = \frac{dv}{dt}$, where a is constant.

Integration yields:

$v = at + A$.

Since velocity $v = u$ at $t = 0$, we have, $A = u$, and thus,

$v = at + u = u + at$.

Now, since $v = \frac{dx}{dt}$, therefore eq. (1) gives us:

$$\frac{dx}{dt} = u + at.$$

By integrating:

$$x = ut + a\left(\frac{t^2}{2}\right) + B.$$

Using $x = 0,\ t = 0,$ we get, $B = 0,$ hence:

$$x = ut + \frac{at^2}{2}.$$

Also, $a = v\frac{dv}{dx}$, where a is constant, then $ax = vdv$.

Integrating

$$\frac{v^2}{2} = ac + C.$$

Since $v = u$, at $x = 0$, therefore, $C = \frac{u^2}{2}$, and hence

$$v^2 - u^2 = 2ax.$$

The equations (1),(2) and (3) are called equations of motion. If the particle moves with constant retardation, the acceleration is taken as negative.

Vertical motion under gravity

For a falling body, the acceleration is constant. It is called acceleration due to gravity, and is denoted by g. In an FPS system, $g = 32\,\text{ft/sec}^2$. In a CGS system $g=981\,\text{cm/sec}^2$ and in MKS system $g = 9.81\,m/sec^2$. If the body is projected vertically upward, then the acceleration g is negative. For falling bodies, the equation of motion is given as:

$$v = u + gt$$

$$x = ut + \frac{1}{2}gt^2$$

$$v^2 - u^2 = 2gx.$$

Example

A particle is projected vertically upward with velocity $\sqrt{2gh}$ and another falls from a height h at the same time. Find the height of the point where they meet each other.

Solution

Let both particles meet at point P at height x. Then for the first particle:

$$x = ut - \frac{1}{2}gt^2.$$

Letting $u = \sqrt{2gh}$, we have:

$$x = \sqrt{2gh}\,t - \frac{1}{2}gt^2 \quad (1)$$

For the second particle,

$$x = ut + \frac{1}{2}gt^2$$

$$x = ut + \frac{1}{2}gt^2 .$$

Letting $u = 0$,$x = h - x$, we have:

$$h - x = 0 + \frac{1}{2}gt^2 . \qquad (2)$$

Using eqs. (1) and (2), we get:

$$u = \sqrt{2gh}t$$

or $t = \sqrt{2gh}t.$

Thus equation (1) gives us:

$$x = \sqrt{2gh}t\left(\frac{h}{\sqrt{2gh}}\right) - \frac{1}{2}g\left(\frac{h^2}{2gh}\right) = h - \frac{h}{4} = \frac{3h}{4} .$$

Example

A particle is projected vertically upward. After time *t,* another particle is sent up from the same point with the same velocity and meets the first at height *h* during the downward flight of the first. Find the velocity of the projection.

Solution

Let *u* be the velocity of the projection and v the velocity at height *h*, then we have:

$$v^2 - u^2 = -2gh$$

or $v = \sqrt{u^2 - 2gh}.$

Now, the time taken by the first particle from height *h* to the highest point and back to the height *h* is *t*. Therefore, the time taken from the highest point is *t/2*.

Velocity at highest point $= 0.$

Velocity at height $h = v$.

Here, we use the formula:

$$v = u - gt$$

Letting $v = 0, u = v, t =$ t/2, we get:

$$0 = v - \frac{gt}{2}.$$

From (1) and (2),

$$\sqrt{u^2 - 2gh} = \frac{gt}{2}$$

$$u^2 - 2gh = \frac{g^2t^2}{4}$$

$$u^2 = 2gh + \frac{g^2t^2}{4}$$

$$u = \sqrt{\frac{8gh + g^2t^2}{4}}.$$

Distance traveled in the nth second

Let x_1 and x_2 be the distance travelled in the first *n* and *n–1* seconds respectively, then from the equation of motion:

$$x = ut + \frac{1}{2}at^2.$$

Then $x_1 = un + \frac{1}{2}an^2,$

and $x_2 = u(n-1) + \frac{1}{2}a(n-a)^2$

The distance travelled in the nth second = $x_1 - x_2$,

$$x_1 - x_2 = un + \frac{a}{2}n^2 - \left\{u(n-a) + \frac{a}{2}(n-1)^2\right\}$$

$$= un + \frac{a}{2}n^2 - \left\{ un - u + \frac{a}{2}\left(n^2 - 2n + 1\right) \right\}$$

$$= u + \frac{1}{2}a\left(2n - 1\right).$$

Motion with variable acceleration

(i) Time dependent acceleration

As the acceleration is time dependent, it is function of time. Therefore, we can write

$$a = f\left(t\right)$$

or $$\frac{dv}{dt} = f\left(t\right)$$

or $$dv = f\left(t\right)dt.$$

Integrating

$$\int dv = \int f\left(t\right)dt$$

or $$v = g\left(t\right) + A,$$

where g(t) = $\int f\left(t\right)dt,$

and A is a constant of integration. This gives us the velocity. Also since $v = \frac{dx}{dt}$, we have:

$$\frac{dx}{dt} = g\left(t\right) + A,$$

or $$dx = g\left(t\right)dt + A\,dt.$$

By integrating, we get

$$x = \int g\left(t\right)dt + At + B.$$

where B is another constant that denotes the distance travelled by the particle. By using the initial conditions, we can determine the constants A and B.

Example 3

Discuss the motion of the particle moving in a straight line if it starts from rest a $t = 0$ and its acceleration is equal to $ac\cos t + b\sin t$.

Solution

Given that acceleration is $a = ac\cos t + b\sin t$, and by integrating, we have:

$v = a\sin t - b\cos t + A.$

As $v = 0$ at $t = 0$, we have $A = b$, thus

$$C = a\sin t - b\cos t + b.$$

Also, $v = \dfrac{dx}{dt} = a\sin t - b\cos t + b$.

Integrating again,

$x = -a\cos t - b\sin t + bt + B,$

where B is a constant of integration. Since $x = 0$, at $t = 0$, therefore $B = a$. Thus:

$x = -a\cos t \ \ \text{-}b\sin t + bt + a$

$= a(1 - \cos t) + b(t - \sin t).$

(ii) Velocity dependent acceleration

Let the acceleration a be function of velocity v, then we have

$a = f(v).$

Since $a = v\dfrac{dv}{dx}$, we have:

$$v\frac{dv}{f(v)} = dx.$$

Integrating:

$$\int v\frac{dv}{f(v)} = \int dx,$$

or $\quad x = g(v) + A,$

where $g(v) = \int v\frac{dv}{f(v)}$ and A is a constant of integration, giving the distance travelled by the particles.

Again consider $a = f(v)$. Since

$$a = \frac{dv}{dt}, \text{therefore:}$$

$$\frac{dv}{f(v)} = dt \quad \text{or} \quad \int\frac{dv}{f(v)} = \int dt,$$

or $\quad t = h(v) + B,$

where $h(v) = \int\frac{dv}{f(v)}$ and B is a constant of integration, giving the time travelled by the particles.

Example

A particle moves in a straight line with an acceleration kv^{3}. If its initial velocity is u, find the velocity and the time spent when the particle has travelled a distance of x.

Solution

Given that $a = kv^3$ and $a = v\frac{dv}{dx}$, we have:

$$v\frac{dv}{dx} = kv^3$$

or $v^{-2}\,dv = k\,dx$

or $\int v^{-2}\,dv = kdx$

or $v^{-1} = kx + A.$

Given that $v = u, x = 0 \text{ at } t = 0, \text{therefore } A = -\frac{1}{u}$. Thus

$\frac{1}{v} = \frac{1}{u} - kx \quad or \quad c = \frac{u}{1-kux}$.

Again since $v = \frac{dx}{dt}, \text{then}$

$\frac{dx}{dt} = \frac{u}{1-kux}$,

or $(1-kux)\,dx = u\,dt, \text{thus}:$

$\int(1-kux)\,dx = u\int dt$

$$x - ku\left(\frac{x^2}{2}\right) = ut + B.$$

Using $x = 0, \text{at } t = 0, \text{we get } B = 0.$ Hence,

$$x - ku\left(\frac{x^2}{2}\right) = ut \text{ or } t = \frac{x}{2u}(2-kux).$$

Example

A particle starts with a velocity u and move in a straight line. If the particle suffers a retardation equal to the square of the velocity, find the distance travelled by the particle in time t.

Solution. The retardation is given as $v^2, \text{therefore}:$

$a = -vt^2$.

We know that:

$$v\frac{dv}{dx} = a \text{ ,therefore:}$$

$$\frac{dv}{dx} = -v$$
,

or $$\frac{dv}{v} = -dx$$

or $$\int\frac{dv}{v} = -\int dx$$

or $$In\,v = -x + A,$$

where A is the constant of integration. Using $v = u, x = 0 \text{ at } t = 0,$ we have: $A = In\,u,$ thus,

$$\ln\left(\frac{u}{v}\right) = x \quad or \ \frac{u}{v} = e^{x} \quad or \ v = \frac{u}{e^{x}}.$$

Since $v = \dfrac{dx}{dt},$ we have $\dfrac{dx}{dt} = \dfrac{u}{e^{x}}$

or $e^{x} = u\,dt$, and hence:

$$e^{x} = ut + B,$$

where B is a constant of integration.

Using $x=0 \text{ at } t=0,$ we get $B=1$. Hence

$$e^{x} = ut + 1 \text{ or } x = in(ut + 1).$$

(iii) Distance dependent acceleration

Let the acceleration be a function of distance x, then we have $a = f(x)$.

Since $v\dfrac{dv}{dx} = f(x)dx,$

or $\int v dv = \int f(x)dx,$

$$\frac{v^2}{2} = g(x) + A,$$

where g(x) $\int f(x)\,dx$ and A is the constant of integration,

or $v = \pm\sqrt{2g(x)+2A}.$

We know that $\frac{dx}{dt} = v$, therefore

$$\frac{dx}{dt} = \pm\sqrt{2g(x)+2A},$$

$$dt = \pm\frac{dx}{\sqrt{2g(x)+2A}}.$$

Integrating, we get:

$$t = \pm\int\frac{dx}{\sqrt{2g(x)+2A}} + B,$$

where B is the constant of integration.

Example

Discuss the motion of the particle moving in a straight line if it starts at a distance a from a point O and moves with an acceleration equal to times its distance from O.

Solution

Let x be the distance of the particle from O, then:

Acceleration $= a = \mu x.$

Since $v\frac{dv}{dx} = a$ or $v\frac{dv}{dx} = \mu x$ or $v dv = \mu x dx$

or $\int v dv = \mu \int x dx$

$$\frac{v^2}{2} = \mu\frac{x^2}{2} + A.$$

Given $v = 0, x = a \text{ at } t = 0,$ such that

$A = -\mu \frac{a^2}{2}$, thus,

$$\frac{v^2}{2} = \mu\left(\frac{x^2}{2} - \frac{a^2}{2}\right),$$

or $v = \pm\sqrt{\mu}\sqrt{x^2 - a^2}$.

Take $v = \sqrt{\mu}\sqrt{x^2 - a^2}$, since $v = \frac{dx}{dt}$, then

$$\int \frac{dx}{\sqrt{x^2 - a^2}} = \sqrt{\mu} \int dt$$

or $\quad \cosh^{-1}\left(\frac{x}{a}\right) = \sqrt{\mu} \int dt.$

Using $x = a \text{ at } t = 0, B = 0,$ thus

$$\cosh^{-1}\left(\frac{x}{a}\right) = \sqrt{\mu}\, t$$

or $\quad x = a\cosh(\sqrt{\mu t}).$

If we take $v = -\sqrt{\mu}\sqrt{x^2 - a^2}$, as $v = \frac{dx}{dt}$, then

$$\int \frac{dx}{\sqrt{x^2 - a^2}} = -\sqrt{\mu} \int dt + C$$

$$\cosh^{-1}\left(\frac{x}{a}\right) = -\sqrt{\mu}\, t + C$$

Using $x = a, t = 0, C = 0,$ thus

$$\cosh^{-1}\left(\frac{x}{a}\right) = -\sqrt{\mu}\, t$$

or $= a \cosh\left(-\sqrt{\mu}\, t\right) = a\cosh\left(\sqrt{\mu}\, t\right).$

Graphical solutions of rectilinear problems

Consider the graph of the curve $y = f(x)$ in the $xy-$ plane, then:

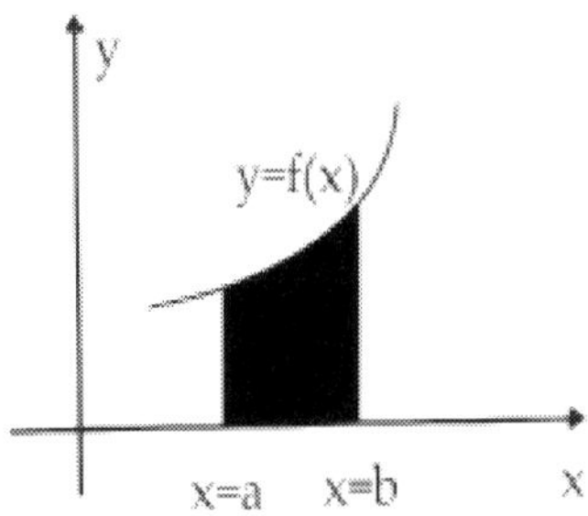

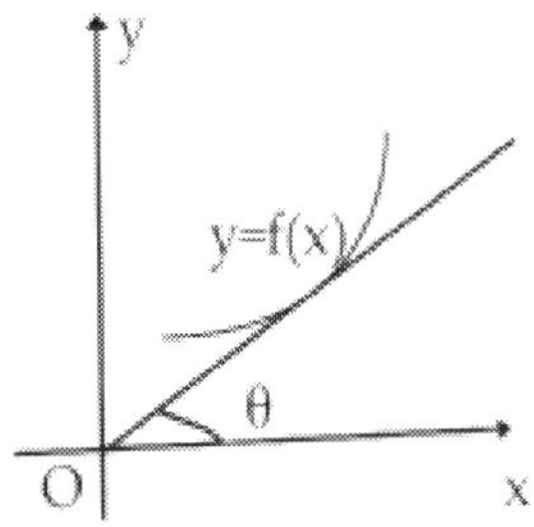

(a) $\frac{dy}{dx}$ is slope of the tangent at any point.

(b) The area bounded by the curve $y = f(x)$, from $x = a$ to $x = b$ is $= \int_a^b y\,dx$.

(i) Velocity-time curve

Consider the velocity time graph *v = v(t)*, then acceleration = a = $\frac{dv}{dt}$

or $a =$ slope of velocity – time curve = $\tan\theta$.

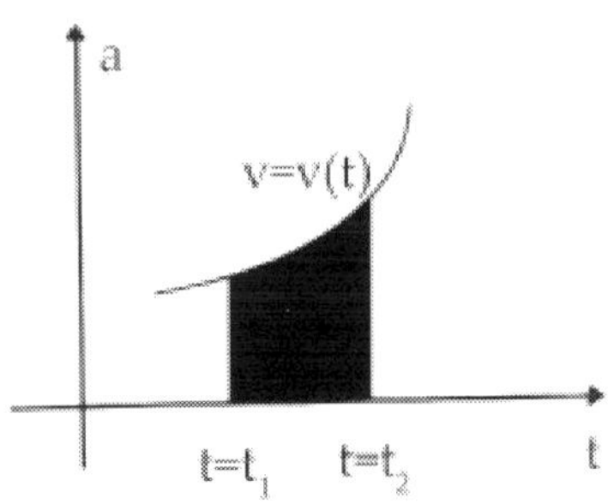

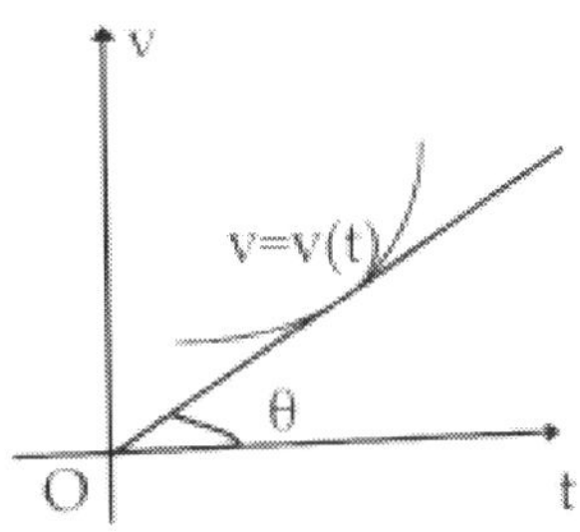

We know that $v = \frac{dx}{dt}$, thus,

$$dx = vdt$$

or $\int_{t=t_1}^{t=t_2} dx = \int_{t=t_1}^{t=t_2} v dt$

or $|x|_{t=t_1}^{t=t_2} = \int_{t=t_1}^{t=t_2} v dt$

or $x(t_1) - x(t_2) = \int_{t=t_1}^{t=t_2} v dt.$

Change in distance $=$area bounded by velocity – time curve, on the t-axis at $t = t_1$ and $t = t_2$.

(ii) Space – time curve

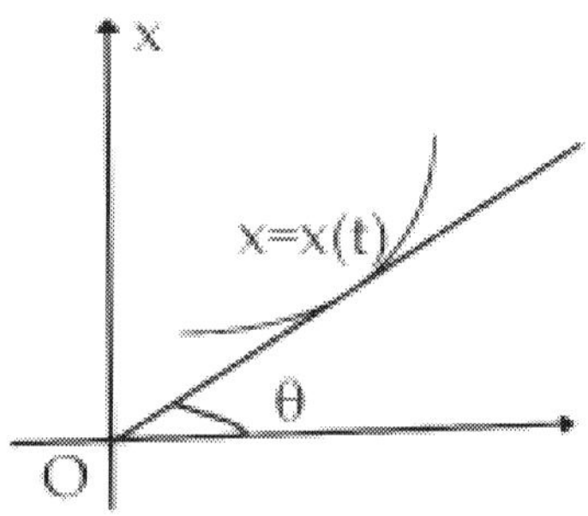

Consider the space- time curve $x = x(t)$, we have

velocity $= v = \frac{dx}{dt}$, or

velocity$=$ slope of the space-time curve

$= tan\theta$.

(ii) Acceleration-time graph

Consider the acceleration – time graph $a = a(t)$, then, we have:

$a = \frac{dv}{dt}$ or $dv = adt$

or $\int_{t=t_1}^{t=t_2} dv = \int_{t=t_1}^{t=t_2} adt$

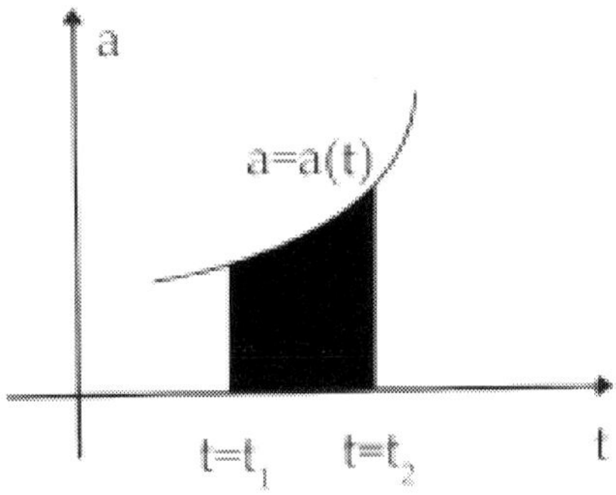

or $|v|_{t=t_1}^{t=t_2} = \int_{t=t_1}^{t=t_2} adt$

or $v(t_1)-v(t_2)=\int_{t=t_1}^{t=t_2} adt.$

Change in velocity = area bounded be the acceleration- time curve, on the t-axis at $t = t_1$ and $t = t_2$.

(iv) Velocity-space curve

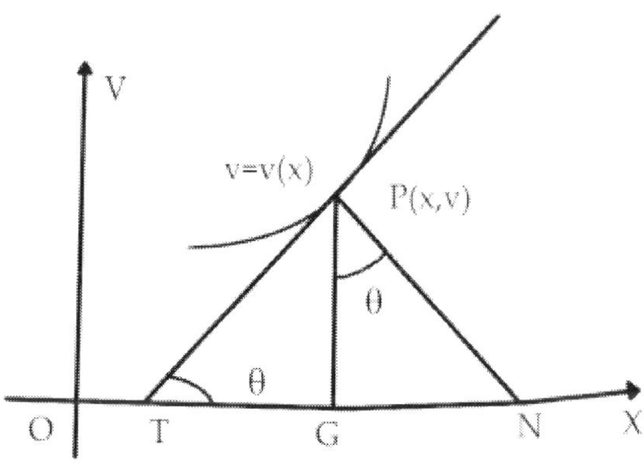

Consider the velocity- space curve $v = v(x)$. We have:

$a = v\frac{dv}{dx}$, where,

Slope of tangent = $\frac{dv}{dx}$.

Therefore, $tan\theta = \frac{dv}{dx}$

for triangle PNG

$$\tan\theta = \frac{GN}{GP} = \frac{GN}{v}.$$

From (1) and (2), we have:

$\frac{dv}{dx} = \frac{GN}{v}$ or $v\frac{dv}{dx} = GN.$

Therefore,

Acceleration $= a =$ Length of subnormal at P.

Example

A particle starts from rest with a constant acceleration a. When its velocity acquires a certain value v, it moves uniformly and then its velocity starts decreasing with a constant retardation of $2a$, until it comes to rest. Find the distance traveled by the particle if the time taken from rest to rest is t.

Solution

Let t_1, t_2 and t_3 be the times for the accelerated, uniform and decelerated motion, respectively, then

$$t = t_1 + t_2 + t_3.$$

Acceleration $= a =$ slope of OA $= \frac{v}{t_1}$

$$\text{or } t_1 = \frac{v}{a}.$$

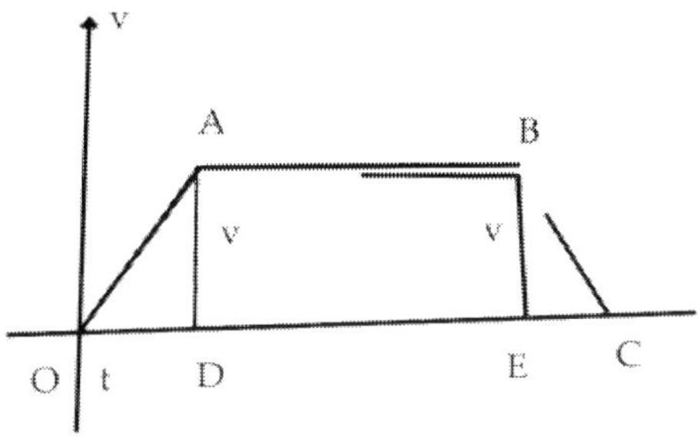

Similarly, retardation = $2a = \frac{v}{t^3}$,

or $t_3 = \frac{v}{2a}$.

Thus, $t = t_1 + t_2 + t_3$

$= \frac{v}{a} + t_2 + \frac{v}{2a}$.

This implies that

$t_2 = t - \frac{3v}{2a}$.

Distance = area under the velocity- time curve

= area of trapezium OABC

$= \frac{1}{2}(OC + AB)(AD)$

$= \frac{1}{2}(t_1 + t_2 + t_3 + t_2)v$

$= \frac{1}{2}(t + t_2)v$

$= \frac{1}{2}\left(t + t - \frac{3v}{2a}\right)v = vt - \frac{3v^2}{4a}$.

Example

A particle moving along a straight line starts from rest and accelerates uniformly until it attains the velocity *v*. The motion slows down and the particle come to rest after traversing a total distance *x*. If the acceleration is *f*, find the retardation and the total time taken for the particle to move from rest to rest.

Solution

Let t_1 and t_2 be the time for the acceleration and deceleration motion respectively, then

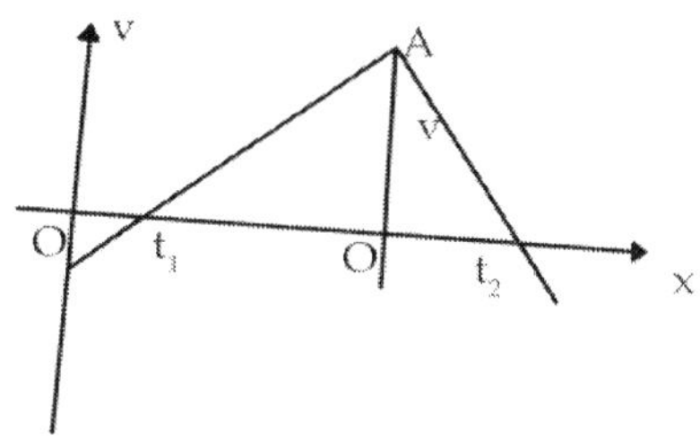

acceleration = f

$= slope\, of\ OA$

or $f = \frac{v}{t_1}$ or $t_1 = \frac{v}{f}$.

Let *g* be the retardation, then

g = slope of *AB*

$= \frac{v}{t^2}$

or $t^2 = \frac{v}{g}$.

Distance = x = area of triangle OAB $= \frac{1}{2}(t_1 + t_2)v$

or $t_1 + t_2 = \frac{2x}{v}$.

Thus total time = $\frac{2x}{v}$.

Letting $t_1 = \frac{v}{f}$ and $t_2 = \frac{v}{g}$ in (1), then

$$x = \frac{1}{2}\left(\frac{v}{f}+\frac{v}{g}\right)v = \frac{1}{2}\left(\frac{1}{f}+\frac{1}{g}\right)v^2$$

or $$\frac{2x}{v^2} = \frac{1}{f}+\frac{1}{g}$$

or $$\frac{1}{g} = \frac{2x}{v^2} - \frac{1}{f} = \frac{2xf - V^2}{fv^2}$$

or $$g = \frac{fv^2}{2xf - v^2}$$

SUMMARY AND DISCUSSION

Rectilinear motion

The motion of a particle along a straight line is called the rectilinear motion.

Vertical motion under gravity

For a falling body, the acceleration is constant. It is called acceleration due to gravity, and is denoted by *g*. In an FPS system, $g = 32\,\text{ft}/\text{sec}^2$. In a CGS system $g=981\,\text{cm}/\text{sec}^2$ and in MKS system $g = 9.81m/sec^2$. If the body is projected vertically upward, then the acceleration *g* is negative. For falling bodies, the equation of motion is given as:

$$v = u + gt$$

$$x = ut + \frac{1}{2}gt^2$$

$$v^2 - u^2 = 2gx.$$

Distance traveled in the nth second

Let x_1 and x_2 be the distance travelled in the first n and $n - 1$ seconds, respectively. From the equation of motion, we have:

$$x = ut + \frac{1}{2}at^2.$$

Then, $x_1 = un + \frac{1}{2}an^2,$

and $x_2 = u(n-1) + \frac{1}{2}a(n-a)^2.$

Distance travelled in the nth second = $x_1 - x_2$:

$$x_1 - x_2 = un + \frac{a}{2}n^2 - \left\{u(n-a) + \frac{a}{2}(n-1)^2\right\}$$

$$= un + \frac{a}{2}n^2 - \left\{un - u + \frac{a}{2}(n^2 - 2n + 1)\right\}$$

$$= u + \frac{1}{2}a(2n-1).$$

Motion with variable acceleration

(i) Time dependent acceleration

(ii) Velocity dependent acceleration

(iii) Distance dependent acceleration.

CHAPTER 7

PROJECTILES

PROJECTILES

An object thrown into the air with a certain velocity from a gun or dropped from a moving plane under the action of gravity is called a projectile. Thus, a projectile moves with a constant horizontal velocity and falls freely under the action of gravity. The path of a projectile is called the trajectory.

Equation of the trajectory of a projectile

Let the particle *m* be projected from point *O* with an initial svelocity making an angle with the horizontal. Take *O* as the origin and the horizontal and vertical lines through *O* as and respectively.

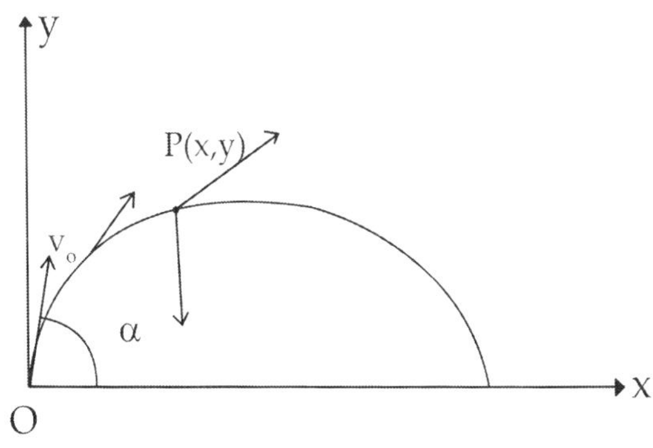

Suppose that after time t, the particle is at point $P(x, y)$ whose position vector is $\overline{r}$. *i.e*

$$\overline{r} = x\hat{i} + y\hat{j}$$

$$\dot{\overline{r}} = \dot{x}\hat{i} + \dot{y}\hat{j}$$

$$\ddot{\overline{r}} = \ddot{x}\hat{i} + \ddot{y}\hat{j}.$$

The gravitational force $\overline{F}$ acting on the particle at P is:

$$\overline{F} = mg\hat{j}\ \mathrm{S}$$

$$m\ddot{\overline{r}} = -mg\hat{j}$$

$$or\ \ddot{\overline{r}} = g\hat{j}$$

$$\ddot{x}\hat{i} + \ddot{y}\hat{j} = -g\hat{j}, \text{then}$$

$$\ddot{x} = 0 \quad \text{and} \quad \ddot{y} = -g.$$

Integrating, we get

$$\dot{x} = A \quad \text{and}\ \dot{y} = -gt + B.$$

At $t = 0,$ the particle is at O, therefore

at $t = 0,\ \dot{x} = v_0\ cos\alpha, \dot{y} = v_o\ sin\alpha$.

Hence,

$A = v_o\ cos\alpha, B = v_o\ sin\alpha$ and

$$\dot{x} = v_0 \cos\alpha,\ \dot{y} = -gt + v_o\ sin\alpha.$$

This gives us the horizontal and vertical component of velocity at any time t. Integrating again,

$$x = (v_0 \cos\alpha)t, \text{ and } y = -g\frac{t^2}{2} + t + D$$

at $t = 0, x = 0, y = 0,$ therfore $C = 0, D = 0.$ Hence

$$x = (v_0 \cos\alpha)t$$

$$y = v_0\, sin\alpha - \frac{1}{2}gt^2$$

The eqs (1) and (2) are parametric equation of the trajectory. To find the Cartesian equation, eliminate t from the eqs. (1) and (2), to get

This is the Cartesian equation of trajectory · from (3), we have

$$\frac{gx^2}{2v_o^2\cos^2\alpha} - x\frac{sin\alpha}{cos\alpha} = -y$$

$$\text{or} \quad x^2 - x\frac{sin\alpha}{cos\alpha}\left(\frac{2v_o^2\cos^2\alpha}{g}\right) = -y\frac{2v_o^2\cos^2\alpha}{g}$$

$$\text{or} \quad x^2 - \frac{2xv_0^2\, sin\alpha\cos\alpha}{g} = -2y\frac{v_o^2\cos^2\alpha}{g}$$

$$\text{or} \quad x^2 - 2x\frac{v_0^2 sin\alpha\cos\alpha}{g} + \left(\frac{v_0^2 sin\alpha\cos\alpha}{g}\right)^2$$

$$= -2y\frac{v_o^2\cos^2\alpha}{g} + \left(\frac{v_0^2 sin\alpha\cos\alpha}{g}\right)^2$$

$$\text{or} \left(x - \frac{v_0^2 sin\alpha\cos\alpha}{g}\right)^2 = \frac{2v_o^2\cos^2\alpha}{g}\left(y - \frac{v_o^2\sin^2\alpha}{2g}\right).$$

(1). Vertex

$\left(\frac{v_0^2 sin\alpha\cos\alpha}{g}, \frac{v_o^2\sin^2\alpha}{2g}\right)$ is the vertex of the parabola.

(ii) Latus rectum

Latus Rectum = $|4a| = \frac{2v_o^2\cos^2\alpha}{g}$.

(iii) Maximum height

Maximum height attained above x- axis

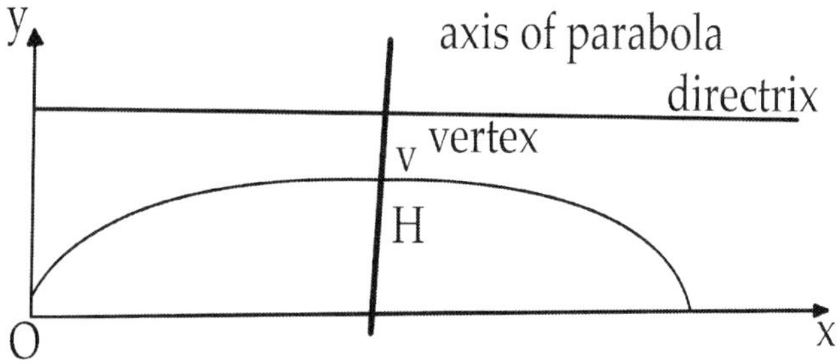

H = Maximum height attained above x – axis

$= y -$ Coordinate of vertex

$$= \frac{v_o^2 \sin^2 \alpha}{2g}.$$

This equation can also be obtained by putting $\frac{dy}{dx} = 0$ into the equation of the trajectory.

(iv) Equation of directrix

Height of Directrix above x –axis

$$= H + \frac{1}{4}(\,latus\,rectum)$$

$$= \frac{v_o^2 \sin^2 \alpha}{2g} + \frac{v_o^2 \cos^2 \alpha}{2g}$$

$$= \frac{v_o^2}{2g}\left(\sin^2 \alpha + \cos^2 \alpha\right) = \frac{v_o^2}{2g}.$$

The equation of directrix is

$$y = \frac{v_o^2}{2g}.$$

(v) Focus

x – Coordinate of focus = x – coordinate of vertex

$$=\frac{v_0^2 sin\alpha \cos\alpha}{g}$$

y – Coordinate of focus = H - $\frac{1}{4}(latus\, rectum)$

$$=\frac{v_o^2 \sin^2\alpha}{2g}+\frac{v_o^2 \cos^2\alpha}{2g}$$

$$=\frac{v_o^2}{2g}\left(\sin^2\alpha+\cos^2\alpha\right)$$

$$=-\frac{v_o^2}{2g}cos2\alpha$$

Thus, the focus is

$$=\left(\frac{v_0^2 sin\alpha \cos\alpha}{g},-\frac{v_o^2}{2g}cos2\alpha\right)=\left(\frac{v_o^2}{2g}sin2\alpha-\frac{v_o^2}{2g}cos2\alpha\right).$$

(vi) Time of flight

The time taken by the particle to reach the horizontal plane through the point of projection is called its time of flight. In this case, let y = 0, in the equation

$$y=g\frac{t^2}{2}+(v_o \sin\alpha)t,$$

then $0=-g\frac{t^2}{2}+(v_o \sin\alpha)t$.

As $t\neq 0$, therefore

$$v_o \sin\alpha-\frac{1}{2}gt=0,$$

or $t = \frac{2v_o\, sin\alpha}{g}$ is the time of flight.

(vii) Range

(a) Horizontal range

The horizontal range of the projectile is the horizontal distance described by the particle during the time of flight. Therefore,

Range R = (horizontal velocity) × (time of flight),

$$v_o \cos\alpha \left(\frac{2v_o\, sin\alpha}{g} \right),$$

$$= \frac{v_o^2\, 2(\sin\alpha\, cos\alpha)}{g} = \frac{2v_o^2 sin2\alpha}{g}.$$

This range can also be obtained by inserting $y = 0$ into the cartesian equation of the trajectory:

$$y = xtan\alpha - \frac{gx^2}{2v_o^2}\sec^2\alpha.$$

Also, it is notable that this range will remain the same if we replace α by $\frac{\pi}{2} - \alpha$.

(b) Maximum range

Range $R = \frac{v_o^2 sin2\alpha}{g}$

For maximum range $\frac{dr}{d\alpha} = 0,$

$$\Rightarrow \frac{2v_o^2}{g}\cos 2\alpha = 0$$

$$\Rightarrow cos2\alpha = 0$$

$$\Rightarrow 2\alpha = \frac{\pi}{2}$$

$$\Rightarrow \alpha = \frac{\pi}{4}.$$

Therefore $R_{max} = \frac{v_o^2}{g}$.

Range in inclined plane

Let β be the inclination of the inclined plane. A particle is projected from point O with velocity v_o by making an angle α to the horizontal, where $\beta < \alpha$. Let the projectile meet the inclined plane at $P(x, y)$, then OP = r is called the range of the inclined plane.

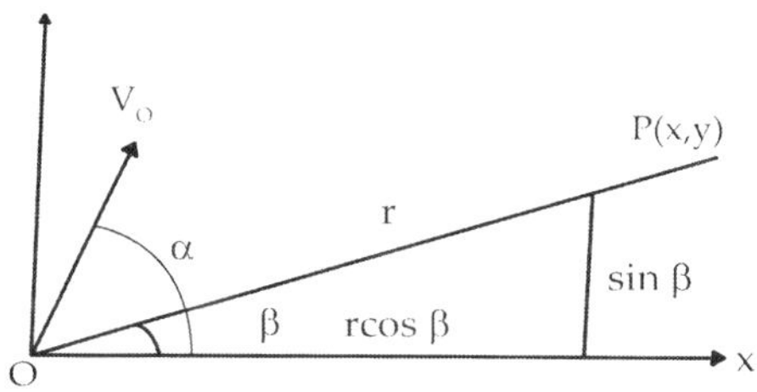

$$x = rcos\,\beta,\ y = rsin\beta$$

$$P(x, y) = (rcos\beta, rsin\beta).$$

The equation the of trajectory is:

$$y = \tan\alpha - \frac{1}{2}g\frac{x^2}{v_o^2}\sec^2\alpha.$$

The point lies on the trajectory. Thus,

we have: $r\,sin\beta = r\beta tan\alpha - \frac{1}{2}g\frac{r\cos^2\beta}{v_o^2\cos^2\alpha}$

or $\frac{grcos^2\beta}{2v_o^2\cos^2\alpha} = \cos\beta\frac{sin\alpha}{cos\alpha} - sin\beta$

$$= \frac{sin\alpha \, cos\beta - cos\alpha \, sin\beta}{cos\alpha}$$

$$= \frac{\sin(\acute{a} - \hat{a})}{\acute{a}os}$$

Thus, we have:

$$r = \frac{2v_o^2 \cos^2 \alpha}{g \cos^2 \beta} \times \frac{\sin(\alpha - \beta)}{cos\alpha}$$

$$= \frac{2v_o^2 \cos^2 \alpha \sin(\alpha - \beta)}{g \cos^2 \beta}.$$

For the range down the inclined plane, change $\beta \, to - \beta.$

Maximum Range

Equation (1) written, by using

$$\sin(\alpha + \beta) - \sin(\alpha - \beta) = 2 \cos \alpha sin\beta$$

$$r = \frac{v_o^2}{gcos^2 \beta} (\sin(2\alpha - \beta) = 1, \text{such that}:$$

$$2\alpha - \beta = \frac{\pi}{2} \; or \; \alpha = \frac{\beta}{2} + \frac{\pi}{4}.$$

Thus $r_{max} = \frac{v_o^2}{gcos^2 \beta}(1 - \sin \beta)$

$$\frac{v_o^2(1 - sin\beta)}{g(1 - \sin^2 \beta)} = \frac{v_o^2}{g(1 - \sin \beta)},$$

or, for the maximum range, use $\frac{dr}{dx} = 0$

or $\cos(2\alpha - \beta) = 0$

or $2\alpha - \beta = \frac{\pi}{2}.$

Time of flight

$$\text{Time of flight} = \frac{Horizontol\ distance}{Horizontol\ velocity}$$

$$= \frac{x}{v_o \cos\alpha} = \frac{r\cos\beta}{v_o \cos\alpha}.$$

Using (1), we get:

$$\text{Time of flight} = \frac{2v_o^2 \cos^2\alpha \sin(\alpha-\beta)}{g\cos^2\beta}\frac{\cos\beta}{v_o \cos\alpha}$$

$$= \frac{2v_o^2 \sin(\alpha-\beta)}{g\cos\beta}.$$

For the time of flight down the plane, change β to $-\beta$.

Show that the speed at a point *P* of the projectile is the same as that attained by a particle of the same mass at *P* falling freely from the point vertically above *P* on the Directrix of the trajectory.

Let v_o be the velocity of the projection and v be the velocity at point $P(x, y)$ of trajectory, then

$$\bar{v} = \dot{x}\hat{i} + \dot{y}\hat{j}$$

$$\bar{v} = (v_0 \cos\alpha)\hat{i} + (v_o \sin\alpha - gt)\hat{j}$$

$$v = |\bar{v}| = \sqrt{(v_0 \cos^2\alpha) + (v_o \sin\alpha - gt)^2}$$

$$v = \sqrt{v_o^2 \cos^2\alpha + v_o^2 \sin^2\alpha - 2gtv_o \sin\alpha + g^2t^2}$$

$$= \sqrt{v^2 - 2g(v_o t \sin\alpha - \frac{1}{2}gt^2}.$$

Since

$$y = v_o t \sin\alpha - \frac{1}{2} g t^2$$

Therefore,

$$v = \sqrt{v_0^2 - 2gy}$$

Let N be the point on the Directrix vertically above $P(x, y)$. From the figure:

$$MN = MP + PN$$

$$\frac{V_0^2}{2g} = y + PN$$

Let v' be the velocity of the particle failing freely at a distance NP. Using the equation of motion, we have:

$v^2 - U^2 = 2gx$ where u is initial velocity and v is final velocity.

Here $u = 0, x = PN, v = v'$, then:

$$v'^2 - 0 = 2g(PN)$$

$$v'^2 = 2g\left(\frac{v_o^2}{2g} - y\right)$$

$$v'^2 = v_o^2 - 2gy$$

$$v' = \sqrt{v_o^2 - 2gy}$$

From (1) and (2), we have:

$$v' = v$$

Parabola of safety

The equation of the projectile is:

$$y = x\,tan\alpha - \frac{1}{2} g \frac{x^2}{v^2} \sec^2\alpha,$$

which shows that for different values of α we have different trajectories. Thus, we define the locus of all points on different trajectories lying at a maximum distance from the point of projection O as the parabola of safety. All the trajectories from O must touch the parabola of safety at some point. Thus, it is envelope of the family of all trajectories with the same velocity of protestation $v_o, \alpha.$ The equation of the trajectory is:

$$y = x\tan\alpha - \frac{1}{2}g\frac{x^2}{v_o^2}\sec^2\alpha$$

$$y = x\tan\alpha - \frac{1}{2}g\frac{x^2}{v^2}$$

$$\frac{gx^2}{2v_o^2}\tan^2\alpha - x\tan\alpha + \left(\alpha + \frac{gx^2}{2v_o^2}\right) = 0$$

,

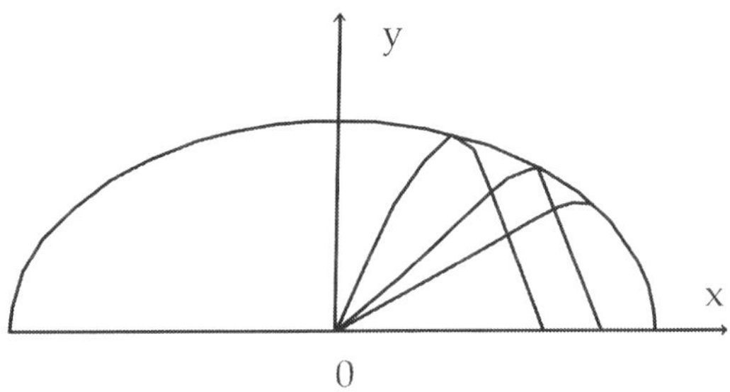

which is quadratic in $tan\alpha.$ For the equation of the envelope, let the discriminate equal 0:

$$x^2 - 4\left(\frac{gx^2}{v_o^2}\right)\left(y + \frac{gx^2}{v_o^2}\right) = 0.$$

Since $x \neq 0,$ divided by x^2, to get:

$$-2\left(\frac{g}{v_o^2}\right)\left(y + \frac{gx^2}{v_o^2}\right) = 0$$

$$-\frac{2gy}{v_o^2} - \frac{g^2x^2}{v_o^2} = 0$$

$$\frac{g^2x^2}{v_o^2} = \frac{2gy}{v_o^2} + 1$$

$$x^2 = -2\frac{v_o^2}{g}\left(y - \frac{v_o^2}{2g}\right).$$

This is the equation of parabola of safety with vertex $\left(0, \frac{v_o^2}{2g}\right)$ and latus rectum = $\frac{2v_o^2}{g}$.

The equation of the Directrix is given as:

$$y = \frac{v_o^2}{2g} + \frac{1}{4}(latus\, rectum)$$

$$y = \frac{v_o^2}{2g} + \frac{1}{4}\left(\frac{2v_o^2}{g}\right)$$

$$y = \frac{v_o^2}{2g} + \frac{v_o^2}{2g} = \frac{v_o^2}{g}$$

$$x - coordinate\, of\;\; focus = x - coordinate\, of\;\; vertex = 0$$

$$y - coordinate\;\; of\;\; focus\; = y - \;\; coordinate\;\; of\;\; vertex\; = 0$$

$$-\frac{1}{4}(latus\, rectum)$$

$$= \frac{v_o^2}{2g} - \frac{1}{4}\frac{2v_o^2}{g} = 0$$

Therefore, the focus is $(0,0)$.

Projection with air resistance

Let a particle of mass m be projected with an initial velocity v_o, making an angle α with the horizontal, as shown in the figure.

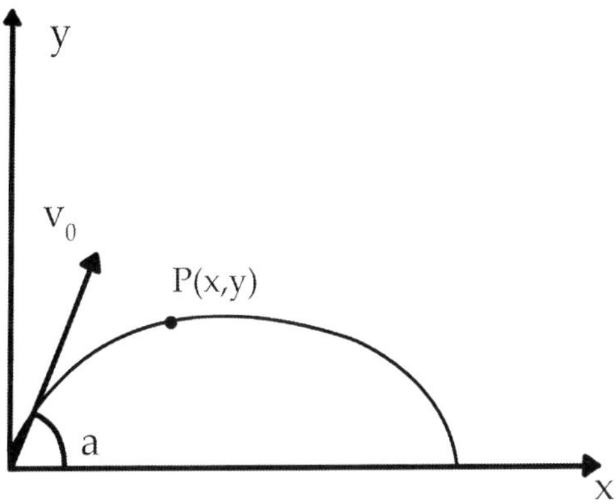

Here, we restrict ourselves to the simple case of air resistance proportional to the first power of velocity. We suppose that after some time t the particle is at P(x, y). The equations of motion, with mass canceled, are:

$$\ddot{x} = -k\dot{x}$$

$$\ddot{y} = -g - k\dot{y}.$$

From(1),

$$\frac{dx}{dt} = -kx$$

$$\frac{dx}{x} = -kdt$$

.

Integrate:

$$In(x) = -kt + c_1$$

At $t = 0\, x = v_o \cos\alpha$, therefore $c_1 = in(v_o cos\alpha)$

$\text{In(x)} = -kt + in(v_o cos\alpha)$

$$x = v_o \cos\alpha\, e^{-kt} \qquad (3)$$

$$x = -\frac{v_o}{k} cos\alpha\, e^{-kt} + c_2$$

at $t=0, x=0,$ therefore

$$c_2 = \frac{v_o}{k} cos\alpha, \text{thus}$$

$$x = \frac{v_o}{k} \cos\alpha\, e^{-kt} + \frac{v_o}{k} cos\alpha$$

Or $$x = -\frac{v_o}{k}\left(1 - e^{-kt}\right) cos\alpha \qquad (4)$$

Now, consider equation (2):

$$\frac{dy}{dt} - -g - ky$$

$$or \frac{dy}{g + ky} = -dt$$

Integrate

$$in\left(g + ky\right) = -kt + c_3$$

$$at\, t = 0, y = v_o \sin\alpha, \text{this implies that}$$

$$c_3 = in\ (g + kv_o \sin\alpha)$$

$$In\ \left(g + ky\right) = -kt + In\left(g + kv_o sin\alpha\right)$$

$$In \frac{g + ky}{g + kv_o \sin\alpha} = -kt$$

$$g + ky = \left(g + kv_o \sin\alpha\right) e^{-kt}$$

$$y = -\frac{g}{k} + \left(\frac{g}{k} + kv_o\ sin\alpha\right) e^{-kt}. \qquad (5)$$

Again integrating,

$$y = -\frac{g}{k} - \frac{1}{k}\left(\frac{g}{k} + v_o sin\alpha\right) e^{-ky} + c_4$$

at $t = 0, y = 0.$ This implies that:

$$c_4 = \frac{1}{k}\left(\frac{g}{k} + v_o\, sin\alpha\right).$$

Thus:

$$y = -\frac{g}{k} - \frac{1}{k}\left(\frac{g}{k} + v_o\, sin\alpha\right)e^{-kt} + \frac{1}{k}\left(\frac{g}{k} + v_o\, sin\alpha\right)$$

$$y = -\frac{g}{k}t - \frac{1}{k}\left(\frac{g}{k} + v_o\, sin\alpha\right)\left(1 - e^{-kt}\right) \qquad (6)$$

The equation of the path is obtained by eliminating t from equation $(4)\text{and}(6)$ so that we have:

$$y = -\frac{xcos\alpha}{v_o}\left(\frac{g}{k} + v_o\, sin\alpha\right) + \frac{g}{K^2} In\left(1 - \frac{xcos\alpha}{v_o}\right).$$

Example

A shell burst on contact with the ground, with its pieces flying in all directions at a speed of up to 80 ft./sec. Prove that a man at 100 ft. from where the shell burst is in danger for $\frac{5}{\sqrt{2}}$ seconds.

Solution

We know that the range of the particle is

$$R = \frac{v_o^2 sin2\alpha}{g}.$$

Here $R = 100\, ft., v_o = 80\, ft\, /\, sec., g = 32\, ft\, /\, sec^2$, therefore:

$$100 = \frac{6400 sin2\alpha}{32}$$

$$\text{or } \sin 2\alpha = \frac{1}{2}, \qquad \text{or } \alpha = 15^\circ, 75^\circ$$

For the range of $100\, ft,$ there are two directions of projections. Let

t_1 and t_2 be the times of flight respectively, then we have:

$$t_1 = \frac{2v_o sin15^o}{g}$$

and $t_2 = \dfrac{2v_o sin75^o}{g}$.

The man is in danger for time = $t_2 - t_1$

$$= \frac{2v_o sin15^o}{g} - \frac{2v_o sin75^o}{g}$$

$$= \frac{2v_o}{g}\left(sin15^o - sin75^o\right)$$

$$= \frac{160}{32}\left(2\cos\left(\frac{75+15}{2}\right)\sin\left(\frac{75-15}{2}\right)\right)$$

$$=5\left(2cos45^o\, sin30^o\right) = \frac{5}{\sqrt{2}} sec.$$

Example

Find the range of a rifle bullet when α is the elevation of the projection and v_o is the speed. Show that, if the rifle is filled with the same elevation and speed from a car travelling at v towards the target, the range will increase by $\dfrac{2vv_o sin\alpha}{g}$.

Solution

We know that the range is given by:

$$R = (horizontal\, velocity)(time\, of\ flight),$$

$$R = v_o cos\alpha\left(\frac{2v_o sin\alpha}{g}\right) = \frac{2v_0^2 \sin\alpha cos\alpha}{g}.$$

When the shell is fired from a car moving with speed v towards the target, the horizontal velocity is increased by v.

Horizontal velocity =

New $range + R'$

$= (new\, horizontal\, time\, of\ flight)$

$$= (v_o cos\alpha + v)\left(\frac{2v_o sin\alpha}{g}\right)$$

$$R' = \frac{2v_0^2 \sin\alpha cos\alpha}{g} + \left(\frac{2v_o v\, sin\alpha}{g}\right).$$

Increase in range = $r' - R = \left(\frac{2v_o v\, sin\alpha}{g}\right)$

Example

A battleship is moving with speed v, mounted with a gun pointing straight behind and with an angle of elevation α. If v_o is the speed of projection (relative to the gun), show that the range is:

$$\left(\frac{2v_o sin\alpha}{g}\right)(v_o\, cos\alpha - v).$$

Prove also that the angle of elevation for the maximum range is

$$\cos^{-1}\left(\frac{v + \sqrt{v^2 + 8v_o^2}}{4v_o}\right).$$

Solution

The ship is moving with speed v away from the target, so the horizontal velocity of the projectile will decrease by v.

Actual horizontal velocity = $v_o \cos\alpha - v$

Actual range = R = (horizontal velocity) (time of flight)

$$\left(v_o cos\alpha - v\right)\left(\frac{2v_o sin\alpha}{g}\right)$$

$$\left(\frac{2v_o sin\alpha}{g}\right)\left(v_o \, cos\alpha - v\right)$$

Since we have

$$R = \left(\frac{2v_o sin\alpha}{g}\right)\left(v_o cos\alpha - v\right)$$

$$\frac{v_o}{g}\,(v_o \sin 2\alpha - 2v \, sin\alpha),$$

for maximum range, we get:

$$\frac{dR}{d\alpha} = \frac{2v_o}{g}\left(v_o \, cos 2\alpha - v \, cos\alpha\right) = 0$$

$$v_o \cos 2\alpha - v cos\alpha = 0$$

$$v_o \; (2\cos^2 \alpha - 1) - v cos\alpha = 0$$

$$v_o (2\cos^2 \alpha - 1) - v \, cos\alpha = 0$$

$$2v_0 \cos^2 \alpha - v cos\alpha - v_0 = 0$$

$$cos\alpha = \frac{v \pm \sqrt{v^2 + 8v_o^2}}{4V_o}$$

$$\frac{d^2R}{d\alpha^2} = \frac{2v_o}{g}\left(-2v_o \, sin 2\alpha + v sin\alpha\right)$$

$$= -\left(\frac{2v_o sin\alpha}{g}\right) 4v_o \, cos\alpha -$$

when $\cos\alpha = \frac{v + \sqrt{v^2 + 8v_o^2}}{4V_o}$

$$\frac{dR}{d\alpha} = \frac{2v_o sin\alpha}{g}(\sqrt{v^2 + 8v_0^2}) = -ve$$

For 0 <

Thus, R is maximum when

$$cos\alpha = \frac{v + \sqrt{v^2 + 8v_o^2}}{4V_o}$$

$$\text{or } \alpha = \cos^{-1}\left(\frac{v + \sqrt{v^2 + 8v_o^2}}{4V_o}\right)$$

$$\text{or } \alpha = \cos^{-1}\left(\frac{v + \sqrt{v^2 + 8v_o^2}}{4V_o}\right).$$

Example

Prove that the speed required to project a particle from a height h to fall a horizontal distance *a* from the point of projection is at least

$$\sqrt{g(\sqrt{a^2 + h^2} - h)}$$

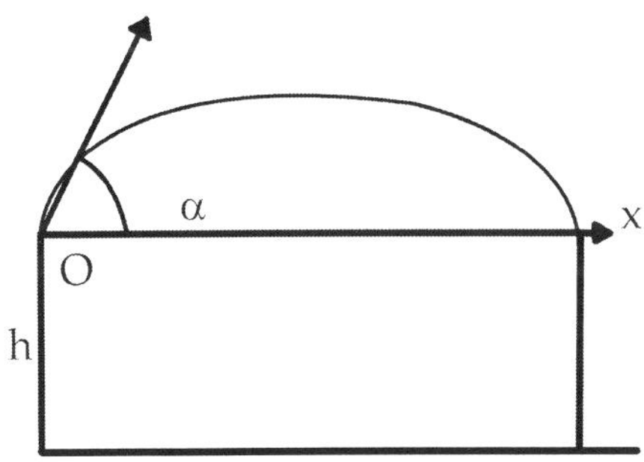

Solution

The equation of the trajectory is

$$y = xtan\alpha - \frac{gx^2}{2v_o^2}\sec^2\alpha$$

.

Point $(a,-h)$ lies on the trajectory, such that

$$-h = a\tan\alpha - \frac{ga^2}{2v_0^2}\sec^2\alpha$$

$$-h = a\tan\alpha - \frac{ga^2}{2v_0^2}(1+\tan^2\alpha)$$

$$ga^2\tan^2\alpha - 2av_o tan\alpha + ga^2 - 2hv_o^2) \geq 0.$$

Since $tan\alpha$ is real, the discriminant $\geq 0, i.e.$

$$4a^2v_o^4 - 4ga^2\left(ga^2 - 2hv_o^2\right) \geq 0$$

$$\left(v_0^2 + gh\right)^2 \geq g^2\left(a^2 + h^2\right)$$

$$v_0^2 + gh \geq g\sqrt{a^2 + h^2}$$

$$v_0^2 \geq g\left(\sqrt{a^2 + h^2} - h\right)$$

$$v_o \geq \sqrt{g\left(\sqrt{a^2 + h^2} - h\right)}.$$

Therefore, the least value of v_o is $\sqrt{g\left(\sqrt{a^2 + h^2} - h\right)}$.

Example

A particle is launched at an angle *a* from a cliff of height H above sea level. If it falls into the sea at a distance D from the base of the cliff, prove that the maximum height above the sea level is

$$H + \frac{D^2\tan^2\alpha}{4(H + D\tan\alpha)}.$$

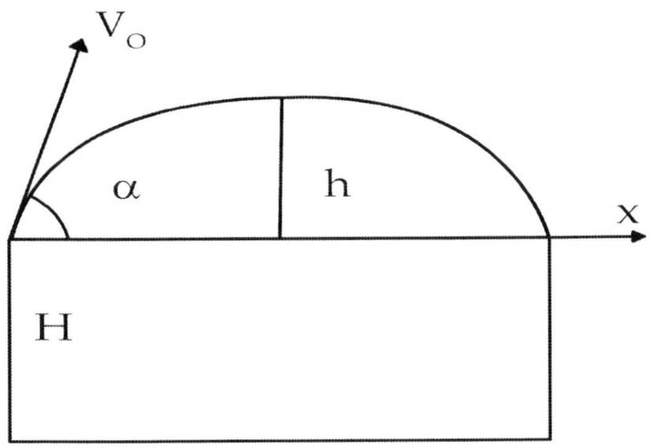

Solution

The equation of the trajectory is

$$y = xtan - \frac{gx^2}{2v_0^2}\sec^2\alpha$$

Point (D, -H) lies on the trajectory, therefore

$$-H = D\tan\alpha - \frac{gD^2}{2v_o^2}\sec^2\alpha$$

$$H + D\tan\alpha - \frac{gD^2}{2v_o^2\cos^2\alpha}$$

$$v_o^2 = \frac{gD^2}{2\cos^2\alpha(H + D\tan\alpha)}$$

Now, $\text{h} = \text{height above x} - \text{axis, such that}$

$$h = \frac{v_o^2\sin^2\alpha}{2g}.$$

Substituting the value of v_o^2, then

$$h = \frac{gD^2}{2\cos^2\alpha\left(H + Dtan\alpha\right)}\frac{\sin^2\alpha}{2g}$$

$$=\frac{D^2\tan^2\alpha}{4(H+D\tan\alpha)}.$$

Height above sea level $=H+h$

$$=H+\frac{D^2\tan^2\alpha}{4(H+D\tan\alpha)}.$$

Example

A fort and a ship are both armed with guns which give their projectile a muzzle velocity $\sqrt{2gk}$. The guns in the fort are at height h above the guns in the ship. If d_1 and d_2 are the greatest horizontal ranges at which the fort and ship respectively can engage, prove that:

$$\frac{d_1}{d_2}=\sqrt{\frac{k+h}{k-h}}$$

Solution

Let S be the ship and F be the fort.

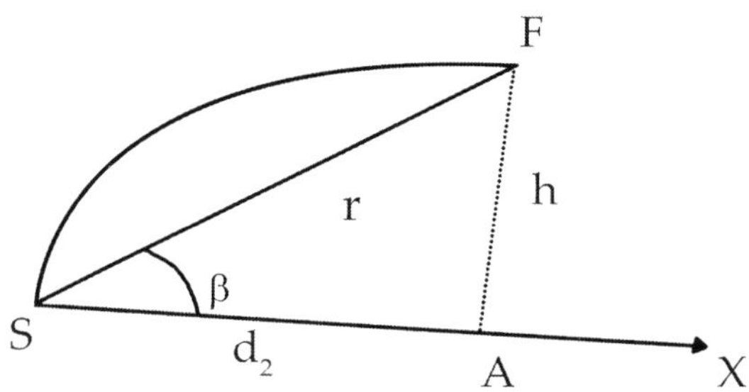

Let $\angle XSF=\beta$ and $SF=r$

d_2 is the greatest horizontal range for the gun on the ship, hence r is the maximum range on an inclined plane with inclination β. Thus,

$$r=\frac{v_o^2}{g(1+\sin\beta)}.$$

Insert $v_o=\sqrt{2gk}$ into the above equation:

$$r = \frac{2gk}{g(1+sin\beta)} = \frac{2k}{1+\sin\beta}$$

$$r + r\sin\beta = 2k,$$

From figure $h = r \ \ sin\beta$, therfore

$$r + h = 2k.$$

Also, from right angle triangle SAF, we have

$$r^2 = h^2 + d_2^2$$

Such that:

$$h^2 + d_2^2 = (2k-h)^2 = 4k^2 - 4kh + h^2$$

$$d_2^2 = 4k^2 - 4kh = 4k(k-h) \qquad (1)$$

For the gun in the fort, change h to – h:

$$d_1^2 = 4k(K+h) \qquad (2)$$

Dividing (2) by (1):

$$\frac{d_1}{d_2} = \sqrt{\frac{k+h}{k-h}}$$

Example

The horizontal range of a ball for two different projections is the same and equal to R. If h_1 *and* h_2 are the greatest heights in the two projections, prove that:

$$R^2 = 16h_1h_2.$$

Solution

Let R be the range, α the angle of the projection and v_o the velocity of the projection, then

$$R = \frac{2v_o^2 \, sin\alpha cos\alpha}{g}.$$

Let R' be the range when angle of the projectile is $\frac{\pi}{2}-\alpha$, then

$$R' \frac{2v_o^2 \sin\left(\frac{\pi}{2}-\alpha\right) cos\left(\frac{\pi}{2}-\alpha\right)}{g}$$

$$\text{or } R' = \frac{2v_o^2\, sin\alpha cos\alpha}{g}.$$

This shows that R is the same when the angle of projection is α and $\frac{\pi}{2}-\alpha.$

Let $h_1 and\, h_2$ be the maximum height in these cases respectively, therefore

$$h_1 = \frac{v_o^2 \sin^2 \alpha}{2g}$$

$$h_2 = \frac{v_o^2 \sin^2\left(\frac{\pi}{2}-\alpha\right)}{2g} = \frac{v_o^2 \cos^2 \alpha}{2g}$$

Hence,

$$16h_1h_2 = 16\frac{v_o^2 \sin^2 \alpha}{2g}\left(\frac{v_o^2 \cos^2 \alpha}{2g}\right)$$

$$= 4\frac{v_o^4 \sin^2 \alpha \cos^2 \alpha}{g} = R^2$$

$$i.e\;\; R^2 = 16h_1h_2.$$

Example

Prove that if the time of flight of a bullet over a horizontal range R is t seconds, then the inclination of the direction to the horizontal is:

$$tan^{-1}\left(\frac{gt^2}{2R}\right)$$

Solution

We know that

$$T = \frac{2_o \sin\alpha}{g}$$

$$R = \frac{2v_o^2 \, sin\alpha cos\alpha}{g}$$

$$\frac{gT^2}{2R} = \frac{4v_o^2 \sin^2\alpha}{g} \frac{g}{4v_o^2 \, sin\alpha cos\alpha} = tan\alpha.$$

Hence

$$\alpha = \tan^{-1}\left(\frac{gT^2}{2R}\right).$$

Example

A cricket ball thrown from a height of a 6 ft. at an angle of with the horizontal speed/ caught by another player at a height 2 ft from the ground. How far apart were the two men?

Solution

Let the two men b x feet apart. As the ball is caught at a height of 2 ft. from the ground, the distance fallen vertically is 6-2 = 4ft. The point $(x,-4)$ lies on the trajectory. Now

$$y = (v_o \, sin\alpha)t - \frac{1}{2}gt^2.$$

Here $y = -4$, $v_0 = 6 - ft\ /sec.,\ \alpha = 3 =^0, t = ?$

$$-4 == (60 sin 30)t - \frac{1}{2}(32)t^2$$

$$8t^2 - 15t - 2 = 0$$

Or $($t-2) (8t+1) = 0

$$t = 2 \ or \quad t = -\frac{1}{8}.$$

Since $t \neq -\frac{1}{8}$, therefore $t = 2$.

$x = Horzontal\ distance$

$= (Horizontal\ velocity)(time)$

$= (v_o cos\alpha)t = (60 cos 30^0)(2)$

or $x = 60\sqrt{3}\ ft.$

SUMMARY AND DISCUSSION

Projectile

An object thrown into the air with a certain velocity from a gun or dropped from a moving plane under the action of gravity is called a projectile. Thus, a projectile moves with a constant horizontal velocity and at the same falls freely under the action of gravity. The path of the projectile is called the trajectory.

The horizontal and vertical components of the velocity at any time t are $\dot{x} = v_0 \cos\alpha,\ \dot{y} = -gt + v_o\ sin\alpha$, and

$x = (v_0 \cos\alpha)t$, $y = v_0\ sin\alpha - \frac{1}{2}gt^2$ are the parametric equations of the trajectory.

$$\left(x - \frac{v_0^2 sin\alpha \cos\alpha}{g}\right)^2 = \frac{2v_o^2 \cos^2\alpha}{g}\left(y - \frac{v_o^2 \sin^2\alpha}{2g}\right).$$

(i) *Vertex*

$$\left(\frac{v_0^2 sin\alpha \cos\alpha}{g}, \frac{v_o^2 \sin^2\alpha}{2g}\right).$$

(ii) Latus Rectum

Latus Rectum = $|4a| = \dfrac{2v_o^2 \cos^2 \alpha}{g}$.

(iii) Maximum Height

$$= \frac{v_o^2 \sin^2 \alpha}{2g}.$$

(iv) Equation of Directrix

$$y = \frac{v_o^2}{2g}.$$

(v) Focus

$$= \left(\frac{v_0^2 sin\alpha \cos\alpha}{g}, -\frac{v_o^2}{2g} cos2\alpha \right) = \left(\frac{v_o^2}{2g} sin2\alpha - \frac{v_o^2}{2g} cos2\alpha \right).$$

(vi) Time of flight

The time taken by the particle to reach the horizontal plane through the point of projection is called its time of flight, *t*:

$$t = \frac{2v_o \, sin\alpha}{g}.$$

(vii) Range

(a) Horizontal Range

The horizontal range of the projectile is the horizontal distance described by the particle during the time of flight.

Range R = (horizontal velocity) (time of flight)

$$v_o \cos\alpha \left(\frac{2v_o \, sin\alpha}{g} \right)$$

$$= \frac{v_o^2 2(\sin \alpha \, cos\alpha)}{g} = \frac{2v_o^2 sin2\alpha}{g}.$$

The range can also be obtained by putting y = 0 into the Cartesian equation of the trajectory:

$$y = xtan\alpha - \frac{gx^2}{2v_o^2}\sec^2 \alpha.$$

Also, it is notable that this range will remain the same if we replace α with $\frac{\pi}{2} - \alpha$.

(b) Maximum Range

Range R = $\frac{v_o^2 sin2\alpha}{g}$

For the maximum range, $\frac{dr}{d\alpha} = 0$

$$\Rightarrow \frac{2v_o^2}{g}\cos 2\alpha = 0$$

$$\Rightarrow cos2\alpha = 0$$

$$\Rightarrow 2\alpha = \frac{\pi}{2}$$

$$\Rightarrow \alpha = \frac{\pi}{4}$$

Therefore, $R_{max} = \frac{v_o^2}{g}$.

Time of Flight

Time of flight = $\frac{Horizontol\,distance}{Horizontol\,velocity}$

$$= \frac{x}{v_o \cos\alpha} = \frac{r cos\beta}{v_o \cos\alpha}.$$

Using (1), we get

Time of flight = $\frac{2v_o^2 \cos^2\alpha \sin(\alpha-\beta)}{g\cos^2\beta} \frac{cos\beta}{v_o cos\alpha}$

$$= \frac{2v_o^2 \sin(\alpha-\beta)}{g\cos\beta}.$$

For the time of flight down the plane, change $\beta\, to - \beta.$

Parabola of Safety

The equation of the projectile is

$$y = x\, tan\alpha - \frac{1}{2} g \frac{x^2}{v^2} \sec^2\alpha.$$

This shows that for different values of α we have different trajectories. Thus, we define the locus of all points on different trajectories that lie at a maximum distance from the point of projection O as the parabola of safety.

CHAPTER 8

DAMPED HARMONIC OSCILLATOR

SIMPLE HARMONIC MOTION

Simple harmonic motion is defined as the motion of a particle moving in a straight line with an acceleration that is always directed towards a fixed point in the line and proportional to the distance of the particle from that point.

A	o	p	B
$x=-a$	$x=0$	$x=x$	$x=a$

Let O be the fixed point on the line along the moving particle. Let the particle be at point P. Let OP = x. Then equation of motion is:

accleration $=-\lambda x$,

where λ is the constant of proportional. The negative sign shows that acceleration is directed in the opposite direction in which x increases.

Since acceleration $=a=v\frac{dv}{dx}$, therefore

$$v\frac{dv}{dx}=-\lambda x$$

$$or\ \ vdv=-\lambda xdx$$

$$or\ \int vdv=-\lambda\int xdx$$

$$or \frac{v^2}{2} = -\lambda\left(\frac{x^2}{2}\right) + B,$$

where B is the constant of integration.

Let the velocity of a particle be zero at point A, whose distance from O is a. Thus, when $x = a$, we get $v = 0$ so that

$$0 = -\lambda \frac{a^2}{2} + B$$

$$\text{or } B = \lambda \frac{a^2}{2}.$$

Thus, $\frac{v^2}{2} = -\lambda\left(\frac{x^2}{2} - \frac{a^2}{2}\right)$,

$$\text{or } v^2 = \lambda\left(a^2 - x^2\right)$$

$$\text{or } v = \pm\sqrt{\lambda}\sqrt{a^2 - x^2}.$$

This gives us the velocity for a given displacement. If the particle is moving to the right, then as t increase, x also increase and so the velocity is positive,

$$v = \sqrt{\lambda}\sqrt{a^2 - x^2}.$$

Since $v = \frac{dx}{dt}$, we get

$$\frac{dx}{dt} = \sqrt{\lambda}\sqrt{a^2 - x^2}$$

$$\text{or } \frac{dx}{\sqrt{a^2 - x^2}} = \sqrt{\lambda}\, dt$$

$$\text{or } \int \frac{dx}{\sqrt{a^2 - x^2}} = \sqrt{\lambda} \int dt$$

$$\text{or } \sin^{-1}\left(\frac{x}{a}\right) = \sqrt{\lambda}\, t + C. \qquad (1)$$

Where C is the constant of integration.

Case(i)

Let the particle be at point A at t = 0, then x = a at t = a, so that (1) gives us

$$\sin^{-1}\left(\frac{a}{a}\right) = \sqrt{\lambda}\,(0) + C$$

or $C = \frac{\pi}{2}$.

Then $\frac{x}{a} = \sin\left(\sqrt{\lambda}\,t + \frac{\pi}{2}\right)$

or $x = a\sin\left(\sqrt{\lambda}t + \frac{\eth}{2}\right)$

or $x = a\cos(\sqrt{\lambda}t)$.

Which gives us the displacement at time t.

Case (ii)

Let the particle be at O at t = 0, then x = 0 at t = 0, so that from (1)

$$\sin^{-1}\left(\frac{O}{a}\right) = \sqrt{\lambda}\,(0) + C, \text{or } C = 0,$$

then $\sin^{-1}\left(\frac{x}{a}\right) = \sqrt{\lambda}t$

or $x = a\sin\left(\sqrt{\lambda}t\right)$.

General solution

The differential equation of simple harmonic motion is given as:

$a = -\sqrt{\lambda}\,x$.

Since $a = \frac{d^2x^2}{dt}$

or $\frac{d^2x}{dt^2} = -\sqrt{\lambda}x$,

the characteristics equation is $D^2 + \lambda = 0$, with roots:

$D = \pm i\sqrt{\lambda}$

Thus, the general solution is

$x = c_1 \cos\sqrt{\lambda}\,t + c_2 \sin\sqrt{\lambda}\,t$.

Nature of simple harmonic motion

Consider the equation:

$x = a\cos\sqrt{\lambda}\,t$.

Since -1 $\leq \cos\left(\sqrt{\lambda}t\right) \leq 1$, we have $-a \leq x \leq a$, and thus the maximum displacement from the fixed-point O is called the center of the motion. The maximum displacement from the center is called the amplitude of the motion.

A	o	p	B
$xs = -a$	$x = 0$	$x = x$	$x = a$

Let $x = a\cos\sqrt{\lambda}t$. (1)

Then $x = a\cos\left(\sqrt{\lambda}\,t + 2\pi\right) = a\cos\sqrt{\lambda}\ (t + \frac{2\pi}{\sqrt{\lambda}})$. (2)

From (1) and (2), the distance at time t and $t + \frac{2\pi}{\sqrt{\lambda}}$ is same.

Differentiating (1), we get:

$\frac{dx}{dt} = -a\sqrt{\lambda}\sin\left(\sqrt{\lambda}t\right)$ (3)

or $\frac{dx}{dt} = -a\sqrt{\lambda}\sin\left(\sqrt{\lambda}t + 2\pi\right)$

or $\frac{dx}{dt} = -a\sqrt{\lambda}\sin\sqrt{\lambda}\left(t+\frac{2\pi}{\sqrt{\lambda}}\right).$

(4)

From (3) and (4), the velocity at time $t+\frac{2\pi}{\sqrt{\lambda}}$ is also the same.

Therefore, the motion is repeated after time $t=\frac{2\pi}{\sqrt{\lambda}}$ and the particle oscillates between x = -a and

x = a.

velocity $= v = \sqrt{\lambda}\sqrt{a^2-x^2} = \sqrt{\lambda}\sqrt{a^2-x^2} = 0$,

and acceleration = a = - $\sqrt{\lambda}x = -\sqrt{\lambda}\,a$, which has maximum magnitude. Thus, at point B, x = a, its velocity is zero but its acceleration is maximum and directed towards O. Due to the acceleration, the particle moves towards O. At point O, x = 0, therefore its acceleration is zero but its velocity is maximum and due to the velocity, the particle moves towards point A.

At point A, x = -a, velocity is zero and acceleration is maximum due to acceleration. The particle moves towards O and finally comes to rest at A. Thus, the particle completes an oscillation or vibration. The motion is then repeated after time $2\pi/\lambda$, and the time taken to complete one oscillation is known as the time period and denoted by T, therefore,

$$T=\frac{2\pi}{\sqrt{\lambda}}$$

The number of oscillations completed in a unit of time is called the frequency and is denoted by υ, thus:

$$\upsilon=\frac{1}{T}=\frac{\sqrt{\lambda}}{2\pi}.$$

If a particle describes a cycle with a constant angular velocity, prove that the foot of the perpendicular from the particle on the diameter moves with a simple harmonic motion.

Proof

Let the particle move along a circle with center O and radius a and ω be the angular velocity. Let the center O be the origin and diameter AA' on the x-axis. Let $Q(x, y)$ be the position of the particle at time t and $p(x,0)$ be the foot of the perpendicular from Q to diameter AA', with $\angle AOQ = \theta$, then :

$$\frac{d\theta}{dt} = \omega, \qquad x = a\,cos\theta.$$

Differentiate $w.r.t.t,$ then

$$x = -a\sin\frac{d\theta}{dt} = -\omega sin\theta.$$

Differentiating again $w.r.t$.'t', we get:

$$x = -a\omega\cos\frac{d\theta}{dt} = -a\omega^2\cos\theta = -a\theta^2 x$$

The acceleration is proportional to the distance. Hence the foot of the perpendicular P describes a simple harmonic motion,

$\lambda = \omega^2$ ù, $or\sqrt{\lambda} =$

for time period = T = $\frac{2\pi}{\sqrt{\lambda}} = \frac{2\pi}{\omega}$

Example

If a P moves with a velocity v given by $v^2 = -n^2\left(ax^2 + 2bx + c\right),$ show that P describes a simple harmonic motion. Find the centre, the amplitude and time period of motion.

Solution

We have $v^2 = -n^2\left(ax^2 + 2bx + c\right).$

Differentiate $w.r.t\,x$ to get

$$2v\frac{dv}{dx}=n^2\left(2ax+2b\right)$$

$$\text{or } v\frac{dv}{dx}=-n^2\left(2ax+2b\right)$$

$$\text{or } v\frac{dv}{dx}=-n^2\left(ax+b\right)$$

$$\text{or } v\frac{dv}{dx}=-an^2\left(x+\frac{b}{a}\right).$$

Let X = x + $\frac{b}{a}$, then the above equation gives us

$$v\frac{dv}{dx}=-an^2X$$

$$i.e\ v\frac{dv}{dx}\propto X$$

Therefore, P describes a simple harmonic oscillation.

To find the center let X = 0, then $x+\frac{b}{a}=0$, so that

$x=-\frac{b}{a}$ is centre of motion.

Also, here $\lambda = an^2$ $.so\sqrt{\lambda}=\sqrt{a}\,n$ and

the time period = $\frac{2\pi}{\sqrt{\lambda}}=\frac{2\pi}{\sqrt{a}n}$.

To find the amplitude put v = 0, then

$$0=-n^2\left(ax^2+2bx+c\right)$$

$$or\ ax^2+2bx+c=0$$

$$or\ \ x=\frac{-b\pm\sqrt{b^2-ac}}{a}.$$

Let O be the origin, then

$$OA = \frac{-b+\sqrt{b^2-ac}}{a}$$

and $OA' = \dfrac{-b-\sqrt{b^2-ac}}{a}$.

Let C be the center, then

OC = $-\dfrac{b}{a}$ and therefore, the amplitude equals CA. Thus

$$Amplitude = CA == \frac{-b+\sqrt{b^2-ac}}{a} - \left(\frac{b}{a}\right) = \frac{\sqrt{b^2-ac}}{a}$$

$$\underline{A' \qquad\qquad C \qquad\qquad O \qquad\qquad A}$$

$$x = \frac{-b-\sqrt{b^2-ac}}{r} \qquad x = -\frac{b}{a} \qquad x = \frac{-b+\sqrt{b^2-ac}}{a}$$

Example

A particle describing a simple harmonic motion has velocities 5 ft./sec and 4 ft./sec when its distance from the center is 12m and 13m, respectively. Find the time period of motion.

Solution

We know that

$$v^2 = \lambda\left(a^2 - x^2\right),$$

therefore, when $x = 12\,\text{ft}, v = 5\,\text{ft}/\text{sec},$ we have $25 = \lambda\left(a^2 - 144\right)$.

When $x = 13\,\text{ft}, v = 4\,\text{ft}/\text{sec},$ so that $16 = \lambda\left(a^2 - 169\right)$ (2)

and $g = -\lambda x_2$.

Now,

(i) Let $x_2 > x_1, therefore\, u > v$.

From (1) and (3), we have:

$$U^2 - v^2 = \lambda\left(a^2 - x_1^2\right) - \lambda(a^2 - x_2^2)$$

$$= -\lambda\left(x_1^2 - x_2^2\right) = -\lambda\left(x_1 + x_2\right)\left(x_1 - x_2\right).$$

From (2) and (4), we have:

And f + g = $-\lambda\left(x_1 + x_2\right).$ (6)

Combining (5) and (6), we have:

$$x_1 - x_2 = \frac{u^2 - v^2}{f + g}.$$

(ii) If $x_2 > x_1$, then g > f. From (2), (4) and (7):

$$g - f = \lambda\left(x_1 - x_2\right) = \lambda\left(\frac{u^2 - v^2}{f + g}\right)$$

or $\lambda = \dfrac{g^2 - f^2}{u^2 - v^2}.$

Thus, the time period = $\dfrac{2\pi}{\sqrt{\lambda}} = 2\pi\sqrt{\dfrac{u^2 - v^2}{g^2 - f^2}}..$

(iii) From (2), $= -\dfrac{f}{\lambda},$ therefore (1) becomes

$$u^2 = \lambda\left(a^2 - \frac{f^2}{\lambda^2}\right)$$

or $\dfrac{u^2}{\lambda} = a^2 - \dfrac{f^2}{\lambda^2}$

or $a^2 = \dfrac{u^2}{\lambda} + \dfrac{f^2}{\lambda^2} = \dfrac{1}{\lambda}\left(u^2 + \dfrac{f^2}{\lambda}\right).$

Subtracting (2) from (1), we get,

$$\lambda = \frac{9}{25} \text{ or } \sqrt{\lambda} = \frac{3}{5}..$$

Hence, the time period = T = $\frac{2\pi}{\sqrt{\lambda}} = \frac{10\pi}{3} sec.$

Example

A particle describes a simple harmonic motion in such a way that the velocity and acceleration at point p are u and f respectively, and the corresponding quantities at another point Q are b and g. Show that

(i) Distance PQ = $\frac{u^2 - v^2}{f + g}$

(ii) Time period $= 2\pi\sqrt{\frac{u^2 - v^2}{g^2 - f^2}}$

(iii) Amplitude = $\sqrt{\frac{(u^2 - v^2)(u^2g^2 - v^2f^2)}{g^2 - f^2}}$.

Solution

Let O be the origin, OP $= x_1$ and $PQ = x_2$. we know that

$v^2 = \lambda(a^2 - x^2)$ and $a = -\lambda x$

$x = x_1$ $x + x_2$

O $\overrightarrow{P}$ Q

Acceleration = f acceleration = g

Velocity = u velocity = v

Therefore, at point P

$u^2 - \lambda(a^2 - x_1^2)$

and $f = -\lambda x_1$

and at point Q,

$v^2 = \lambda(a^2 - x_2^2)$

u sin (8),

we get

$$a^2 = \left(\frac{u^2 - v^2}{g^2 - f^2}\right)\left\{u^2 + f^2\left(\frac{u^2 - v^2}{g^2 - f^2}\right)\right\}$$

$$= \left(\frac{u^2 - v^2}{g^2 - f^2}\right)\left\{\frac{u^2 g^2 - u^2 f^2 + f^2 u^2 - f^2 v^2}{g^2 - f^2}\right\}.$$

Thus $a = \sqrt{\frac{(u^2 - v^2)\left(u^2 g^2 - v^2 f^2\right)}{g^2 - f^2}}$ is the anplitude.

Example

(a) Show that if the displacement of the particle at any time is given by

$x = a\cos(nt) + b\sin(nt)$,

then the particle executes a simple harmonic motion of amplitude, such that:

$\sqrt{a^2 + b^2}$ and period $\frac{2\pi}{n}$.

(b) If a = 3, b = 4 and n= 4, find the time period, amplitude, maximum velocity and maximum acceleration.

Solution

Given that

$x = a\cos(nt) + b\sin(nt)$,

We have:

$$\frac{dx}{dt} == -an\,sn(nt) + bn\cos(nt)$$

$$\frac{d^2x}{dt^2} = -an^2\cos(nt) + bn^2\sin(nt) = n^2 x.$$

This shows that acceleration is proportional to the distance. Thus, the motion is a simple harmonic motion.

Now $\lambda = n^2, \text{or} \sqrt{\lambda} = n, \text{then}:$

The time period = $\dfrac{2\pi}{\sqrt{\lambda}} = \dfrac{2\pi}{n}$

$x = a\cos(nt) + b\sin(nt)$, therefore

$$x = \sqrt{a^2+b^2}\left\{\frac{a}{\sqrt{a^2+b^2}}\cos(nt) + \frac{b}{\sqrt{a^2+b^2}}\sin(nt)\right\}.$$

Let $sin\theta = \dfrac{a}{\sqrt{a^2+b^2}}$ and $cos\theta = \dfrac{b}{\sqrt{a^2+b^2}}$.

Then $tan\theta = \dfrac{b}{a}$ or $\theta = \tan^{-1}\left(\dfrac{b}{a}\right)$. Thus,

$$x = \sqrt{a^2+b^2}\{\sin\theta\cos(nt) + \cos\theta\sin(nt)\}$$

$$= \sqrt{a^2+b^2}\sin(nt+\theta)$$

$$= \sqrt{a^2+b^2}\sin(nt + \tan^{-1}\frac{b}{a}).$$

The maximum value of $\sin\left(ny + tan^{-1}\dfrac{b}{a}\right) = 1, \text{thus}:$

Maximum displacement = $x = \sqrt{a^2+b^2}$

Or amplitude = $\sqrt{a^2+b^2}$.

(b) When a =3, b = 4, n = 2, we have:

the time period $T = \dfrac{2\pi}{n} = \pi$ and

amplitude = $\sqrt{a^2+b^2} = 5$.

Since $v = \sqrt{\lambda}\sqrt{a^2-x^2}$, v is maximum when $x = 0$, we have:

Maximum velocity $=\sqrt{\lambda}\sqrt{5^2-0}=2(5)=10.$

Acceleration $= -\lambda x$.

Acceleration is maximum when x = a, thus the maximum value of the magnitude of acceleration $= \lambda a=(4)(5)=20.$

Damped Harmonic Oscillator

A harmonic oscillator whose amplitude decreases gradually with time and ultimately vanishes is known as a damped harmonic oscillator.

Consider a free (no driving force) motion as a spring with no mass, with one end attached to an immovable wall and other end to a particle of mass m, as shown in the figure below.

The whole system is immersed in a fluid which causes a retarding force on m, proportional to the velocity of m. Let x be the displacement of m from the mean position, then the equation of motion can be written as:

$$mx=-c_1 x-c_2 x,$$

where c_1 is the force constant of the spring and c_2 is the constant of proportionality for the retarding force, derived from the equation above, then

$$x+2kx+\omega^2 x=0,$$

where 2k $=\frac{c^2}{m}$ and $\omega^2=\frac{c_1}{m}$. ω is know as the undamped angular frequency.

Eq.(1) can be written as

$$\frac{d^2x}{dt^2}+2k\frac{dx}{dt}+\omega^2 x=0.$$

Or $\left(D^2+2kD+\omega^2\right)x=0; d=\frac{d}{dt}$

The auxiliary equation is:

$D^2 + 2kD + \omega^2 = 0$.

Here are three cases;

(i) Underdamped case (light damping)

In this case, $k^2 < \omega^2$, therefore $k^2 - \omega^2$ is negative. Let $k^2 - \omega^2 = -n^2$, then $D = K \pm in.$

Thus, the solution of (1) is

$x = e^{-kt}\{a\sin(nt) + b\cos(nt)\}$,

where a and b are constants.

Letting a = d $\sin\alpha$ and $b = d\,cos\alpha$, we have

$x = de^{-ki}\cos(nt - \alpha)$.

This exhibits an oscillatory motion, with d = $\sqrt{a^2 + b^2}$ and $\alpha = \tan^{-1}\left(\frac{a}{b}\right)$.

If K is small, then the amplitude de^{-kt} gradually decrease with time and tends to zero as $t \to \infty$.

(ii) Critically damped case

In this case $k^2 = \omega^2$, therefore $k^2 - \omega^2$ is zero. Letting $k^2 - \omega^2 = 0$, then : $D = -\omega \pm 0$.

Thus, the solution of (1) is $x = e^{-\omega t}$ (A + Bt),

where a and b are constants.

For the limiting case where $t \to \infty$,

$x = LIM_{t\to\infty}\left(\frac{A + Bt}{e^{\omega t}}\right)$.

Using L. Hospital Rule:

$$x = LIM_{t \to \infty} \frac{B}{\omega e^{\omega t}} \to 0$$

(iii) Over damped case (heavy damping)

In this case $k^2 > \omega^2$, therefore $k^2 - \omega^2$ is positive. Letting

$k^2 - \omega^2 = \lambda^2$, then

$D = -K \pm \lambda$.

Thus, the solution of (1) is

$$x = Ce^{(-k+\lambda)t} + De^{(-k-\lambda)t}$$

$$= Ce^{(-k-\lambda)t} + De^{(-k+\lambda)t}.$$

This does not represent an oscillatory motion. Also, \ $t \to \infty$, $t \to 0$, which is dead a beat motion.

Damped Force Oscillator

Let a particle of mass m execute a simple harmonic motion, written as:

$$mx = -m\omega^2 x$$

$$x = -\omega^2 x.$$

Suppose that there is a damping force proportionality for damping force. Also, we consider an applied force F cos(pt) upon the particle, then equation (2) becomes:

$$x = -\omega^2 x - 2kx + F \cos pt$$

$$x = 2kx + \omega^2 x = F \cos pt.$$

For the complimentary function, we have

$$x = 2kx + \omega^2 x = 0,$$

whose auxiliary equation is

$$D^2 + 2kd + \omega^2 = 0.$$

This implies that

$$D = -K \pm \sqrt{k^2 - \omega^2}$$

For $k^2 < \omega^2, put\, k^2 - \omega^2 = -n^2$, then

$$D = k \pm in\,.$$

Therefore,

$$x_c = Ce^{-kt}\left\{A\sin(nt) + A\cos(nt)\right\}$$

$$x_c = Ce^{-ki}\sin(nt + \alpha),$$

where C $\sqrt{A^2 + B^2}$ and $\alpha = \tan^{-1}\left(\frac{A}{B}\right)$.

For the particular integral, we have:

$$x_p = \frac{F}{D^2 + 2KD + \omega^2}\cos pt$$

$$x_p = \frac{F}{-P^2 + 2KD + \omega^2}\cos pt$$

$$x_p = \frac{F}{2KD + (\omega^2 - p^2)}\cos pt$$

$$x_p = \frac{F\left[2kd - (\omega^2 - p^2)\right]}{4k^2D^2 - (\omega^2 - p^2)^2}\cos pt$$

$$x_p = \frac{F\left[2kd - (\omega^2 - p^2)\right]}{-4k^2D^2 - (\omega^2 - p^2)^2}\cos pt$$

$$x_p = \frac{-F\left[2kd - (\omega^2 - p^2)\right]}{4k^2D^2 + (\omega^2 - p^2)^2}\cos pt$$

$$x_p = \frac{-F\left[2kd\cos pt - (\omega^2 - p^2)\cos pt\right]}{4k^2D^2 + (\omega^2 - p^2)^2}$$

$$x_p = \frac{-F\left[2kp\sin pt - \left(\omega^2 - p^2\right)\cos pt\right]}{4k^2D^2 + \left(\omega^2 - p^2\right)^2}$$

$$x_p = \frac{F\left[2kp\sin pt + \left(\omega^2 - p^2\right)\cos pt\right]}{4k^2D^2 + \left(\omega^2 - p^2\right)^2}$$

$$x_p = \frac{FR\sin\left(pt+\alpha\right)}{R^2} = \frac{FR\sin\left(pt+\alpha\right)}{R},$$

where

$2kp = R\,cos\alpha\,.\,\omega^2 - p^2 = R\sin\alpha$

$$R^2 = 4k^2p^2 + \left(\omega^2 - P^2\right)^2.\alpha = \tan^{-1}\left(\frac{\omega^2 - p^2}{2kp}\right).$$

Thus, the complete solution is

$x = x_e + x_p$

$$= Ce^{-ki}\sin\left(nt+\alpha\right) + \frac{F\sin\left(pt+\alpha\right)}{R}.$$

Hence, x_c (complementary function) $\to 0$ as t $\to\infty$. Therefore,

$$x = \frac{F\sin\left(pt+\alpha\right)}{R}.$$

The amplitude of the particle integral is F / R, and therefore oscillates between F / R and has no tendency of becoming permanently zero. For this reason, the complimentary function x_c is called transient (temporary) and the particular integral x_p is called the steady state solution.

Example

A particle of mass m moving under the action of forces

$F_1 = -m\omega^2 x\,, F_2 = mF_o t, F_3 = -2m\mu x$

Assuming that damping is small, set up and solve the equation of motion.

Solution

The equation of motion for F_1 can be written as

$mx = m\omega^2 x$, where damping exists.

Vertical Motion with Air Resistance

When a body moves within the atmosphere, air slows down the motion. For velocities that are neither very small nor very large, this retardation is very nearly proportional to the square of the velocity.

(1) Downward Motion

A particle of mass m is released from rest at an initial height above the ground. The resistance of the air is assumed to be proportional to the square of the velocity.

SUMMARY AND DISCUSSION

Simple Harmonic Motion

The simple harmonic motion is the motion of a particle moving in a straight line with an acceleration that is always directed toward a fixed point in the line and is proportional to the distance of the particle from that point.

A	o	p	B
$x=-a$	$x=0$	$x=x$	$x=a$

$x = a\cos(\sqrt{\lambda}\mathrm{t})$, which gives the displacement at time t.

Nature of simple harmonic motion

Consider the equation

$x = a\cos\sqrt{\lambda}\,t$.

Since -1 $\le \cos\left(\sqrt{\lambda t}\right) \le 1$, we have $-a \le x \le a$, thus the maximum displacement from the fixed-point O is denoted as the center of the motion. The maximum displacement from the center is known as the amplitude of the motion.

A	o	p	B
$x=-a$	$x=0$	$x=x$	$x=a$

Therefore, the motion is repeated after time $t=\frac{2\pi}{\sqrt{\lambda}}$ and the particle oscillates between x = -a and

x = a.

$velocity = v = \sqrt{\lambda}\sqrt{a^2-x^2} = \sqrt{\lambda}\sqrt{a^2-x^2} = 0$.

Acceleration = a = - $\sqrt{\lambda}x = -\sqrt{\lambda}\,a,$ which has a maximum magnitude. Thus, at point B, x = a and its velocity is zero but its acceleration is maximum and directed towards O. Due to the acceleration the particle moves towards O. At point O, x = 0, thus, its acceleration is zero but its velocity is maximum and the particle moves towards point A.

At point A, x = -a, velocity is zero and acceleration is maximum. The particle moves toward O and finally comes to rest at A. Thus, the particle completes an oscillation or vibration. The motion is then repeated after time $2\pi/\lambda$. The time taken to complete one oscillation is called the time period and denoted by T, therefore, $T=\frac{2\pi}{\sqrt{\lambda}}$.

The number of oscillations completed in a unit time is called the, frequency and is denoted by υ, thus

$$\upsilon = \frac{1}{T} = \frac{\sqrt{\lambda}}{2\pi}.$$

Damped Harmonic Oscillator

A harmonic oscillator whose amplitude gradually decreases with time and ultimately vanishes is known as a damped harmonic oscillator.

(i) Underdamped case (light damping)

$$x = e^{-kt}\{a\sin(nt) + b\cos(nt)\}.$$

Where a and b are constants.

Letting a = d $\sin\alpha,\ b = d\,cos\alpha$,

$x = de^{-ki}\cos(nt-\alpha)$ exhibits an oscillatory motion,

where d = $\sqrt{a^2+b^2}$ and $\alpha = \tan^{-1}\left(\frac{a}{b}\right)$.

If K is quantity, then the amplitude de^{-kt} decrease gradually with time and tends to zero as $t \to \infty$

(ii) Critically damped case

$x = e^{-\omega t}$ (A + Bt), where A and B are constants.

For the limiting case, we have $t \to \infty$, and $x = LIM_{t\to\infty}\left(\frac{A+Bt}{e^{\omega t}}\right)$.

Using L. Hospital Rule. $x = LIM_{t\to\infty}\frac{B}{\omega e^{\omega t}} \to$ 0.

(iii) Over damped case (heavy damping)

$x = Ce^{(-k+\lambda)t} + De^{(-k-\lambda)t}$

$= Ce^{(-k-\lambda)t} + De^{(-k+\lambda)t}$.

This does not represent an oscillatory motion. Also \ $t \to \infty$ as $t \to 0$, which is a dead beat motion.

CHAPTER 9

CENTRAL FORCE

CENTRAL FORCE

A force acting on a particle P which always passes through a fixed point is called a central force. The fixed point is known as the center of force, and is usually taken as the origin. The central force is either attractive (toward the center of force) or repulsive (away from the center of force).

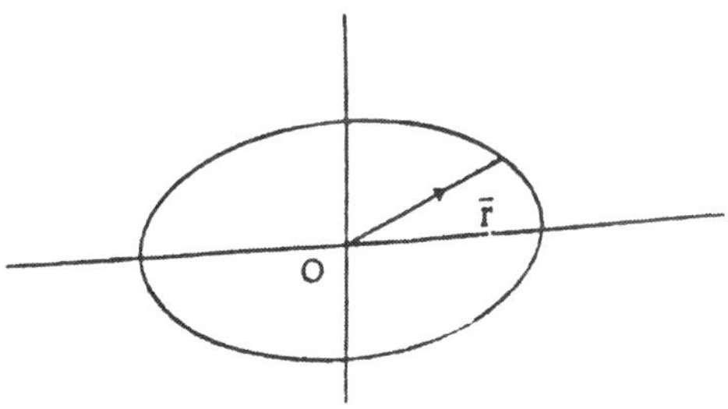

The central force at a point is a function of the distance of that point from the origin. This relation is known as the law of force. The path described by the particle moving under a central force is called the central orbit, thus, $\overline{F}$ is a central force if:

$\overline{F} = f(r)\hat{r}$.

Where $\hat{r}$ is a unit vector in the direction of $\hat{r}$ and r is the magnitude of the position vector $\hat{r}$.Also , $\overline{r}\, x \overline{F} = f(r)\overline{r}\, x \overline{r} = f(r) r\hat{r}\, x \hat{r} = 0,$

i.e $\overline{r}\, x \overline{F} = 0$.

Properties

(i) **Show that the orbit of a particle moving under a central force must be a plane curve.**

Proof:

Let $\overline{r}$ be the position vector of a particle moving under the action of the central force $\overline{F}$. Then:

$\overline{r}\, x \overline{F} = 0$

$\overline{r}\, x \dfrac{d\overline{v}}{dt} = 0$

$\dfrac{d}{dt}(\overline{r}\, x \overline{v}) = 0$.

By integrating

$\overline{r}\, x \overline{v} = \overline{c}; \overline{c}$ is a constant vector.

This shows that the normal to the plane containing $\overline{r}$ (posiyion vector) and $\overline{v}$ (tangent vector) has a constant direction. This is possible for the case of a plane curve i.e. the path of the particle lying on a plane.

(ii) Show that the angular momentum is constant and find its magnitude.

Proof

The angular momentum of a particle of mass m, moving with a velocity $\overline{v}$ is defined as

$\overline{J} = \overline{r}\, x m\overline{v}$

$\overline{j} = m\overline{r}\, x \overline{v}$.

m is a constant scalar and $\overline{r}\, x \overline{v}$ isa constant vector by property (i), therefore $\overline{j}$ ia a constant vector. Hence its magnitude is constant.

Take the velocity vector $\overline{v}$ in radial and transverse components as

$\overline{v} = \dot{r}\hat{r} + r\theta\,\hat{s}$

$$\bar{j} = m\bar{r}\, x\left(\overset{\text{ÿ}}{r\hat{r}}+r\theta\hat{s}\right)$$

$$= mr\hat{r}\, x\ (r\hat{r}+r\theta\hat{s}$$

$$= 0+mr^2\,\theta\left(\hat{r}\,x\,\hat{s}\right)$$

$$j=\left|\bar{j}\right| = mr^2\,\theta\left|\hat{r}x\,\hat{s}\right| = mr^2\theta$$

$$\frac{j}{m} = r^2\theta$$

Or

$$h = r^2\theta,$$

where h = $\frac{j}{m}$.

Differential Equation of the Orbit in Polar Form

If a particle of mass m is moving under the action of an attractive central force of magnitude $f(r)$ per unit mass, then the equation of motion can be written as:

$$m\ddot{\bar{r}} = mf(r)\hat{r}$$

$$\ddot{\bar{r}} = f(r)\hat{r}.$$

By using the radial and transverse components:

$$\left(\ddot{\bar{r}} - r\theta^2\right)\hat{r} + \frac{1}{r}\frac{d}{dt}\left(r^2\theta\right)\hat{s} = -f(r)\hat{r}$$

Equating the radial and transverse components, we get

$$\left(\ddot{\bar{r}} - r\theta^2\right) = -f(r)\hat{r} \qquad (1)$$

$$\frac{1}{r}\frac{d}{dt}\left(r^2\theta\right) = 0. \qquad (2)$$

Consider eq. (2),

$$\frac{d}{dt}\left(r^2\theta\right)=0, \text{this}$$ implies that

$$r^2\theta = h, \tag{3}$$

where h is the constant of integration and known as the angular momentum of a particular of unit mass.

Now consider equation (1):

$$\left(\ddot{r}-r\frac{h^2}{r^4}\right)=-f(r); \text{by using } (3)$$, we get:

$$\left(\ddot{r}-\frac{h^2}{r^3}\right)=-f(r)$$.

Let $r=\frac{1}{u}$, then,

$$r=\frac{1}{u^2}\frac{du}{d\theta}=-\frac{1}{u^2}\frac{du}{d\theta}\theta$$

$$r=-h\frac{du}{d\theta}$$ using (3)

$$r=-h\frac{d^2u}{d\theta}\theta$$

$$r=-h^2u^2\frac{d^2u}{d\theta^2}$$; using (3).

Inserting the value of r into the equation, we get:

$$-h^2u^2\frac{d^2u}{d\theta^2}-h^2u^3=-f(u)$$

$$h^2u^2\left(\left(\frac{d^2u}{d\theta^2}\right)+u\right)=f(u),$$ which is the differential equation of the orbit.

Differential Equation of the Orbit in Pedal From

From calculus, we know that:

$$\frac{1}{p^2}=\frac{1}{r^2}+\frac{1}{r^4}\left(\frac{dr}{d\theta}\right)^2$$

Letting $r=\frac{1}{u}$, we get $\frac{dr}{d\theta}=-\frac{1}{u^2}\frac{du}{d\theta}$.

Now eq. (1) becomes

$$\frac{1}{p^2}=u^2+\left(\frac{du}{d\theta}\right)^2.$$

Differentiate $w.r.t.r,$ we get

$$-\frac{2}{p^3}\frac{dp}{dr}=2u\frac{du}{d\theta}\frac{d\theta}{dr}+2\left(\frac{du}{d\theta}\right)\frac{d^2u}{d\theta^2}\frac{d\theta}{dr}$$

$$-\frac{2}{p^3}\frac{dp}{dr}=\frac{du}{d\theta}\frac{d\theta}{dr}\left(u\frac{d^2u}{d\theta^2}\right)$$

$$\frac{d^2u}{d\theta^2}+u=\frac{f(u)}{h^2u^2},$$

Therefore

$$-\frac{1}{p^3}\frac{dp}{dr}=\frac{f(u)}{h^2u^2}\frac{du}{dr}.$$

Since $u=\frac{1}{r}, therefore, \frac{du}{dr}=-\frac{1}{r^2}=-u^2$.

Thus $\frac{1}{p^3}\frac{dp}{dr}=\frac{f(r)}{h^2}$, and

$$f(r)=\frac{h^2}{p^3}\frac{dp}{dr}.$$

This is the pedal equation of the orbit.

Apse and Apsidal Distance

An apse is a point on an orbit where the radius vector drawn from the center of force exhibits an extremum (maximum or minimum) magnitude. Therefore, at an apse we have:

$$\frac{dr}{dt} = 0 \qquad (1)$$

i.e. the radial component of velocity is zero. From (1)

$$\frac{dr}{dt} = \frac{dr}{d\theta}\frac{d\theta}{dt} = 0 .$$

Since $\frac{d\theta}{dt} \neq 0,$ we have :

$$\frac{dr}{d\theta} = 0 .$$

Let r = 1 /u, then by differentiating:

$$\frac{dr}{d\theta} = \frac{1}{u^2}\frac{du}{d\theta} .$$

By using (2), we have

$$\frac{dr}{d\theta} = 0 \quad \text{or} \quad \frac{du}{d\theta} = 0 .$$

The magnitude of the radius vector at an apse is known as the apsidal distance.

The line joining an apse to the center of the force is an apse line.

The angle between two consecutive apse lines is an apse angle.

The radius vector is perpendicular to the tangent at an apse.

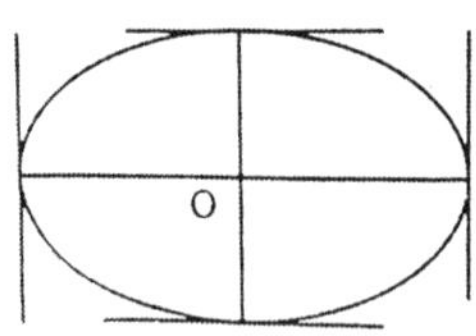

Let $p(r,\theta)$ be a point on orbit with position vector $\overline{r}$. If $\varnothing$ is the angle between the radius vector and the tangent at p, then from calculus we have:

$$\tan\varnothing = r\frac{d\theta}{dr},$$

where $\varnothing$ is the angle between the radius vector and the tangent.

Or $$\cot\varnothing = \frac{1}{r}\frac{dr}{d\theta}$$

But at an apse we have:

$$\frac{dr}{d\theta} = 0, \text{such that}$$

$\cot\varnothing = 0$ or $\varnothing = \frac{\pi}{2}, \frac{3\pi}{2}$.

Hence, the radius vector is perpendicular to the tangent at an apse.

Example

Find the apse of an ellipse.

Solution

The equation of the ellipse is:

$$\frac{x^2}{a^2} + \frac{y^2}{b^2} = 1,$$

where a is the semi major axis and b is the semi minor axis. To convert the equation into polar coordinates, we use:

$$x = r\cos\theta, \quad y = r\,sin\theta$$

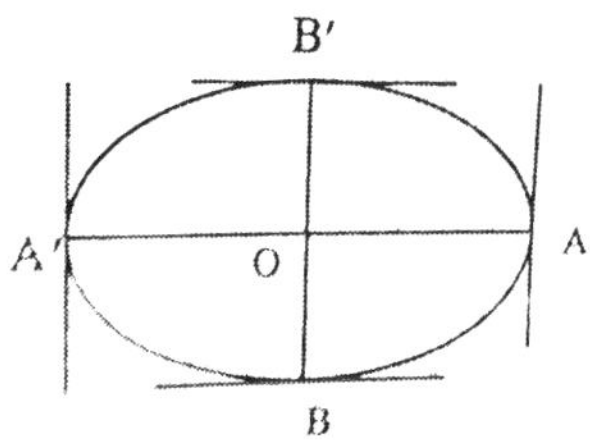

and the equation of an ellipse becomes:

$$r^2\left(\frac{\cos^2\theta}{a^2}+\frac{\sin^2\theta}{b^2}\right)=1.$$

Differentiating w.r.t θ, we get

$$2r\frac{dr}{d\theta}\left(\frac{\cos^2\theta}{a^2}+\frac{\sin^2\theta}{b^2}\right)+2r^2\sin\theta\cos\theta\left(-\frac{1}{a^2}+\frac{1}{b^2}\right)=0$$

For the apse, let $\frac{dr}{d\theta}=0,$ then

$$0+2r^2\sin\theta cos\theta\left(-\frac{1}{a^2}+\frac{1}{b^2}\right)=0.$$

Or $\sin\theta cos\theta=0$.

This implies that either $sin\theta=0$, i.e. $\theta=0,\pi$

or $\cos\theta=0$, i.e. $\theta=\frac{\pi}{2},\frac{3\pi}{2}.$

Hence, the apses of the ellipse are A, A', B and B'.

Areal Velocity

Let $\bar{r}$ *be* the position vector of a point P on the orbit, Q a neighboring point with position vector $\bar{r}+d\bar{r}$ and $d\theta$ the angle between $\bar{r}$ and $\bar{r}+d\bar{r}$, then from the figure below:

$$\left|\bar{r}+d\bar{r}\right|\sin(d\theta)=\left|d\bar{r}\right|.$$

Where $\left|d\bar{r}\right|=rd\theta.$

Area swept by the position vector $\bar{r}$ is

$$dA=\frac{1}{2}\left|\left\{\bar{r}x\left(\bar{r}+d\bar{r}\right)\right\}\right|=\frac{1}{3}\left|\bar{r}\right|\left|\bar{r}+d\bar{r}\right|\sin(d\theta).$$

However, $\left|\bar{r}+d\bar{r}\right|\sin(d\theta)=\left|d\underline{r}\right|$

$$\frac{1}{2}\left|\bar{r}\right|\left|d\bar{r}\right| = \frac{1}{2}r\,r d\theta = \frac{1}{2}r^2\,d\theta$$

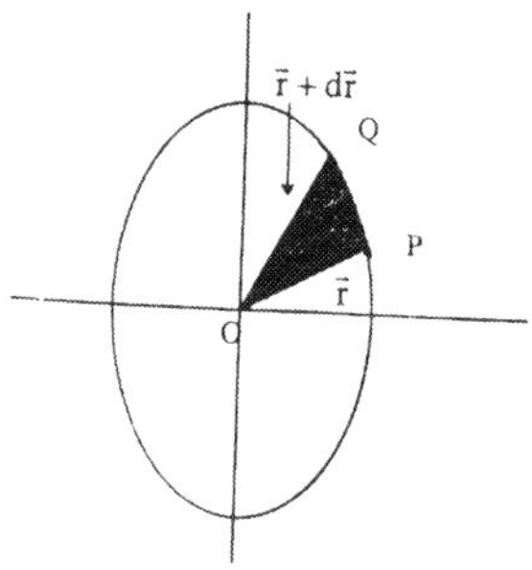

Areal Velocity = $\frac{dA}{dt} = \frac{1}{2}r^2\,\theta$,

for circular orbit $r^2\,\theta = h,\text{therefore}:$

Areal velocity = $\frac{dA}{dt} = \frac{h}{2}$.

The area described at any time t_1 from a fixed point where $t=0$ can be obtained by integrating the real velocity, therefore:

$$A = \frac{h}{2}\int_0^{t_1} dt = \frac{h}{2}t_1$$

Periodic Time

The time required to complete a rotation is known as the periodic time. Since h/2 is the area described in unit time, it follows that if T is the time that the particle takes to describe the whole area of the ellipse, then:

Area of the ellipse = $\frac{h}{2}T$

Or $\pi ab = \frac{h}{2}T$

Or $T = \dfrac{2\pi ab}{h}$.

Kepler's laws

In 1609 Johannes Kepler stated the first two laws of planetary motion, which are given below.

First Law

Every planet follows an ellipse with the sun at one focus.

Second Law

The radius vector drawn from the sun to a planet describes equal areas at equal intervals of time.

Ten years later, in 1619, Kepler announced the third law.

Third Law

The squares of the periods of the different planets are proportional to the cubes of their respective semi axes.

After half a century, Newton derived the gravitational law using Kepler's Law.

Derivation of Newton's Law of Gravitation from Kepler's laws

A particle moves in an ellipse under a central force directed towards its focus, whose polar equation is:

$$\frac{l}{r} = 1 + e\,cos\theta.$$

Let $\dfrac{l}{r} = u,$ then,

$$lu = 1 + e\,cos\theta. \qquad (1)$$

Therefore, by differentiating two times w.r.t.θ, we get:

$$l\frac{d^2u}{d\theta}=-e\cos\theta.$$

Using equation (1):

$$l\frac{d^2u}{d\theta}=1-lu$$

Or $\frac{d^2u}{d\theta^2}=\frac{1}{l}-u$. (2)

We know that the differential equation of the orbit is:

$$h^2u^2\left(\frac{d^2u}{d\theta^2}+u\right)=f(u).$$

Using (2), we have:

$$h^2u^2\left(\frac{1}{l}-u+u\right)=f(u)$$

$$f(u)=\frac{h^2}{l}u^2,$$

or $f(r)=\frac{h^2}{lr^2}=\frac{\mu}{r^2}$ where $\mu=\frac{h^2}{l}$.

This is Newton's law of gravitation.

DERIVATION OF KEPLER'S LAWS

First Law

The differential equation of circular orbit is

$$h^2u^2\left(\frac{d^2u}{d\theta^2}+u\right)=f(u).$$

By Newton's law of gravitation, we have:

$$f(u)=\frac{h^2}{l}u^2.$$

Therefore, (1) becomes

$$\frac{d^2u}{d\theta^2}+u=\frac{1}{l}$$

for the complimentary function

$$\frac{d^2u}{d\theta^2}+u=0$$

whose solution is $u_c = A\cos\theta + B\sin\theta = C\cos(\theta-\alpha)$

where $C=\sqrt{A^2+B^2}, A=C\cos\alpha$ and $B=C\sin\alpha$.

For the particular integral, we have:

$$u_p=\frac{1}{D^2+1}\left(\frac{1}{1}\right)=\frac{1}{l}.$$

Thus,

$$u=u_c+u_p=C\cos(\theta-\alpha)+\frac{1}{l}.$$

Rotating initial line through an angle α

$$u=C\cos(\theta)+\frac{1}{l}$$

$$u=lC\cos(\theta)+1$$

$$\frac{l}{r}=1+e\cos\theta.$$

This equation represents an ellipse where $e=lC$ and l is the semi latus rectum of an ellipse equal to $\frac{b^2}{a}$, with b as the semi minor axis and a is semi major axis.

Second Law

We know that the areal velocity of a particle equals $\frac{dA}{dt} = \frac{h}{2}$.

Suppose that $A_1 \text{and } A_2$ are the areas swept out by a particle in times $t_1 \text{ and } t_2$ respectively, then

$$A_1 = \frac{h}{2}\int_0^{t_1} dt = \frac{h}{2}t_1 \quad \text{and} \quad A_2 = \frac{h}{2}\int_0^{t_2} dt = \frac{h}{2}t_2.$$

Thus, we get: $A_2 - A_1 = \frac{h}{2}(t_2 - t_1)$.

This shows that the radius vector drawn from the sun to the planet sweeps out equal areas at equal times.

Third Law

We know that periodic time of a particle is $T = \frac{2\pi ab}{h}$,
where a is semi major axis and b is semi minor axis. Also

$$h = \sqrt{\mu l}$$

where l is the semi latus rectum and equal to $\frac{b^2}{a}$. Therefore,

$$h = \sqrt{\mu l} = \frac{\sqrt{\mu}}{\sqrt{a}}b.$$

Using this value of h in the periodic time equation, we get:

$$T^2 = \frac{4\pi^2}{\mu}a^3.$$

Hence, the squares of the periods of the planets are proportional to the cubes of the major axis of their orbits.

SUMMARY AND DISCUSSION

Central Force

A force acting on a particle P that always passes through a fixed point is called a central force. The fixed point is known as the center of force, and is usually taken as the origin. The central force is either attractive (towards the center of force) or repulsive (away from the center of force).

The central force at a point is a function of the distance of that point from the origin. This relationship is called the law of force. The path described by the particle moving under a central force is called the central orbit. Thus, $\overline{F}$ is a central force if

$$\overline{F} = f(r)\hat{r},$$

where $\hat{r}$ is a unit vector in the direction of $\hat{r}$ and r is the magnitude of the position vector $\hat{r}$. Also

$$\overline{r}\, x\, \overline{F} = f(r)\overline{r}\, x\, \overline{r} = f(r) r\hat{r}\, x\, \hat{r} = 0$$

i.e. $\overline{r}\, x\, \overline{F} = 0.$

Properties

(i) The orbit of a particle moving under a central force must be a plane curve.

(ii) The angular momentum is constant.

Differential Equation of the Orbit in Polar Form

$$h^2 u^2 \left(\left(\frac{d^2 u}{d\theta^2} \right) + u \right) = f(u).$$

Differential Equation of the Orbit in pedal From

$$f(r) = \frac{h^2}{p^3} \frac{dp}{dr}.$$

Apse and Apsidal Distance

An apse is the point on an orbit at which the radius vector drawn from the center of force has and extremum (maximum or minimum) magnitude.

Therefore, at an apse $\frac{dr}{dt}=0,$ i.e. the radial component of the velocity is zero.

i. The magnitude of the radius vector at an apse is known as the apsidal distance.

ii. The line joinsing an apse to the center of the force is known as an apse line.

iii. The angle between two consecutive apse lines is the apse angle.

iv. The radius vector is perpendicular to the tangent at an apse.

Areal Velocity

Let $\bar{r}$ be the position vector of a point P on the orbit, Q the neighboring point with position vector $\bar{r}+d\bar{r}$ and angle between $\bar{r}$ and $\bar{r}+d\bar{r}$ as $d\theta,$ then we have:

Areal Velocity = $\frac{dA}{dt}=\frac{1}{2}r^2\,\theta$

For circular orbit $r^2\,\theta=h,$ therefore :

Area velocity = $\frac{dA}{dt}=\frac{h}{2}$.

The area described at any time t_1 from a fixed point where $t=0$ can be obtained by integrating the real velocity, therefore:

$$A=\frac{h}{2}\int_0^{t_1}dt=\frac{h}{2}t_1$$.

Periodic Time

The time required to complete a rotation is known as the periodic time. Since h/2 is the area described in unit time, it follows that if T is the time that the particle takes to describe the whole area of the ellipse, we have:

Area of the ellipse $= \frac{h}{2}T$

Or $\pi ab = \frac{h}{2}T$

Or $T = \frac{2\pi ab}{h}$.

Kepler's Laws

In 1609 Johannes Kepler stated the first two laws of planetary motion, which are

First Law

Every planet follows an ellipse with the sun at one focus.

Second Law

The radius vector drawn from the sun to a planet describes an equal area at equal intervals of time.

Ten years later, in 1619, Kepler announced the third law.

Third Law

The squares of the periods of the different planets are proportional to the cubes of their respective semi axes.

After half a century, Newton derived the gravitational law from Kepler's Law.

Derivation of Newton's Law of Gravitation from Kepler's Laws.

A particle moves in an ellipse under a central force directed towards its focus, with a polar equation:

$$\frac{l}{r} = 1 + e\cos\theta$$

$$f(r) = \frac{h^2}{lr^2} = \frac{\mu}{r^2} \quad \text{where} \quad \mu = \frac{h^2}{l}$$

This is Newton's Law of Gravitation.

CHAPTER 10

COLLISIONS

THEORY OF COLLISION

A collision is said to have taken place when two bodies physically collide with each other.

Linear Momentum of a Particle

If a particle of mass m moves with velocity v, then the linear momentum of the particle, usually denoted by L, is the product of mass and velocity.

Therefore $l = mv$

Angular Momentum

If l Is the linear momentum of the particle, then the angular momentum $\underline{\grave{U}}$ is define as

$\underline{\Omega} = \underline{r} + \underline{l}$.

If the particle is moving in the $xy-$plane then

$\underline{\Omega} = (x\underline{i} + y\underline{j})x(L\underline{i} + L\underline{j})$

or $\underline{\Omega} = (x\underline{i} + y\underline{j})x(mv\ \underline{i} + mv\ \underline{j})$

where $\underline{L}\ m\,v = m(v\underline{i} + v\underline{j})$

$\underline{\Omega} = m(xv - yv)\underline{K}$

or $\underline{\Omega} = m(xy - yx)\ \underline{k}$.

Principal of Linear Momentum

The sudden change in the linear momentum of a system is equal to the total external impulsive force.

Proof. Let m be the mass of the ith particle and X, Y, the components of external force in the direction of the axes. Then:

$$\sum_i \frac{d}{dt}(m,x) = \sum X, \sum_I \frac{d}{dt}(m_I, y_I) = \sum_i y_I.$$

Integrating these equation with respect to t over the interval (t_1, t_2) and taking the limit as t_2 approaches t_1, we get:

$$Lt_{t_2 \to t_1} \left| \sum_I m_I x_I \right|_{t_1}^{t_2} = Lt_{t_2 \to t_1} \int_{t_1}^{t_2} \Sigma X_I \, dt.$$

Thus,

$$Lt_{t_2 \to t_1} \left| \sum_I m_I y_I \right|_{t_1}^{t_2} = Lt_{t_2 \to t_1} \int_{t_1}^{t_2} \Sigma Y_I \, dt.$$

The left hand sides of these equations denotes the instantaneous change in the momentum and the right hand sides represent the total impulsive forces in the directions of the coordinates axes. The above equations may be written in the vector from as:

$$\underline{M} = \hat{I},$$

where $\hat{I} = \hat{X}\,\underline{i} + \hat{Y}\,\underline{j}$

or $M\overline{v} = \hat{I} \; (\because \underline{M} = M\overline{v})$

where M is the mass of the system and $\overline{v}$ is the sudden change in the velocity of the mass center. If there are no external impulsive forces, we have

$$\underline{\hat{I}} = 0$$

And hence

$$\underline{\hat{v}} = 0.$$

Thus, when the impulsive forces are purely internal, there is no sudden change in the velocity of the mass center. This result is important for solving problems of collision and explosion.

Note. It should be remembered that a blow changes only the velocity or momentum of a particle, not its instantaneous position.

The Principle of Angular Momentum

Now we prove that the sudden change in angular momentum of a system about a fixed axis is equal to the moment of the external impulsive forces about the axis We will just prove the principal for a single particle, however, the proof can quite easily be extended to a system.

Taking the fixed axis as the z-axis we have, by the principal of angular momentum,

$$\frac{d}{dt}\left[m\left(xy-yx\right)\right]=xY-yX.$$

Where the external force on the particle is $\underline{F}=X\underline{i}+Y\underline{j}$.

Integrating the last equation over the interval (t_1,t_2) and proceeding to the limit of t_2 approaching t_1 we have:

$$Lt_{t_2\to t_1}\left|m\ (xu-yx\right|_{t_1}^{t_2}=Lt_{t_2\to t_1}\int_{t_1}^{t_2}\left(xY-yX\right)dt$$

$$x\left\{Lt_{t_2\to t_1}\int_{t_1}^{t_2}(Y)\,dt\right\}-y\left\{Lt_{t_2\to t_1}\int_{t_1}^{t_2}(X)\,dt\right\}$$

($\because$ x,y Remain unchanged)

$$x\hat{Y}-y\hat{X}=\hat{N}.$$

Thus, the left hand side represents h, the instantaneous change in the angular momentum and the right hand side denotes the moment of the external force about z- axis.

Similarly, considering the motion relative to the center of mass, we can prove that the sudden change in angular momentum relative to the mass center is equal to the moment of the external impulsive forces about the mass center.

Impulse

If a particle of mass m is moving under the action of a constant force $\underline{F}$ producing an acceleration $\underline{a}$, then by second law of Newton $\underline{F} = m\underline{a}$, an impulse of the force during an interval of time t is defined as the product of the force and t. Thus, impulse $\underline{i}$ is given by

$$\underline{i} = \underline{F}t.$$

Using Newton's second law:

$$\underline{I} = m\underline{a}t.$$

Also, using first law of motion we have

$$v_f - v_t = \underline{a}t$$

$$\underline{I} = m\left(v_f - v_t\right),$$

where v_f is the final velocity and v_t is the initial velocity of the particle. Hence, the impulse of the constant force is equal to change in momentum of the particle. When the force is variable in magnitude but constant in the direction, then impulse $\underline{I}$ during the interval from $t = t_1$ to $t = t_2$ is defined as

$$\underline{\text{I}} = \int_{t_2}^{t_1} \underline{f}dt$$

$$\underline{\text{I}} = \int_{t_2}^{t_1} m\frac{dv}{dt} : F = m\underline{\text{a}} = m\frac{dv}{dt}$$

$$= \int_{t_2}^{t_1} m\underline{dv},$$

where $v_1 \text{ and } v_2$ are terminal velocities at $t = t_1 \; and\, t = t_2$ respectively. Hence, whether the force is constant or variable momentum in the interval.

Impulse Force

When a bat strikes a cricket ball, the beginning and the end of the collision can be determined fairly precisely. The bat is in contact with the ball for an interval that is relatively short compared to the time during which we are watching the ball. During the collision, the bat exerts a large force on the ball. This force varies with time and is difficult to measure.

Both the bat and the ball are deformed during the collision. A force that acts for a time that is short compared with the time of observation of the system is known as an impulsive force.

Consider the integral such that the force $\underline{F} = X\underline{i} + Y\underline{j}$ increases without limit and the time interval $t_2 - t_1 = t$ tends to zero in such way that the integral remains constant or approaches a finite limit. In such a case, the value $\underline{I}$ of the integral is known as the impulsive force and its components $\hat{X}, \hat{Y}$ are given as:

$$\hat{X} = Lt_{t_2 \to t_1} \int_{t_1}^{t_2} X dt \text{ and } \hat{Y} = Lt_{t_2 \to t_1} \int_{t_1}^{t_2} Y dt$$

A force that remains finite during the interval (t_1, t_2) contributes nothing to the impulsive force, as if $\underline{F}$ is the magnitude of such a force, the corresponding impulsive force:

$$= Lt_{t_2 \to t_1} \int_{t_1}^{t_2} \underline{F}\, dt = \underline{F}.0 = 0$$

Collision of Two Spheres and the Coefficient of Restitution

Newton's law of restitution states that when two bodies impinge, their relative velocity along the common normal after impact bears a constant ratio, denoted as e, to that before the impact. Here, e ($0 \le e \le 1$) is a constant depending on the nature of the material of the bodies. This constant e is called the coefficient of restitution. For an elastic impact, e = 1 and for a perfectly inelastic impact, e = 0. If 0 < e < 1, the collision is said to be inelastic.

Now we consider the problem of collision of two spheres which are moving along the line joining their centers. We take ox along the line of the center, and use the following notation:

Masses of center = m_1 ,m_2

Velocities o center before collision u_1 ,u_2

Velocities of center after collision = v_1, v_2.

Let $\hat{P}$ be the magnitude of the impulsive reaction. Then we have:

$$m_1(v_1 - u_1) = -\hat{P}$$

And $m_2(v_2 - u_2) = \hat{P}$

And so $m_1(v_1 - u_1) = -- m_2(v_2 - u_2)$

or $m_1v_1 + m_2v_2 = m_1u_1 + m_2u_2 + m_2u_2$.

The last equation is evident otherwise also, since impulsive action and reaction are only internal to system and therefore the total linear momentum of the system should remain unaffected by the collision By Newton's law of restitution.

$$v_2 - v_1 = -e(u_2 - u_1)$$

$$= e(u_1 - u_2).$$

Solving equation (1) and (2) for $v_1$and v_2. we get

$$v_1 = \frac{m_1 - em_2}{m_1 + m_2}u_1 + \frac{m_2}{m_1 + m_2}u_2$$

$$v_2 = (1+e)\frac{m_1}{m_1 + m_2}u_1 + \frac{m_2 - em_1}{m_1 + m_2}u_2.$$

For a perfectly inelastic sphere, e = o and we obtained

Thus, there is no rebound and after the collision the spheres move with the same velocity.

It can be proved that if spheres are elastic (e =1) and have the same mass, they exchange velocities after the impact.

Poisson Hypothesis

This is an alternative to Newton's law of restitution. According to this hypothesis, when two bodies come into contact, there is a short interval in which they undergo compression. During this interval, the centers of the two bodies get closer and closer as a result of the deformation due to the impact. This is followed by another short interval in which the original shape of the bodies move with a common velocity and the distance between the center of the two bodies is at a minimum. During the interval of restitution (the interval in which the bodies regain their shape), the impulsive pressure between the bodies.

SUMMARY AND DISCUSSION

Collision

A collision takes place when two bodies physically collide with each other.

Linear Momentum of a Particle

If a particle of mass m moves with velocity v, then the linear momentum of the particle is the product of mass and velocity, and is usually denoted by l:

Therefore $l = mv$.

Angular Momentum

If l is the linear momentum of the particle, then the angular momentum $\underline{\grave{U}}$ is define as

$$\underline{\grave{U}} = \underline{r} + \underline{1}.$$

Principal of Linear Momentum

The sudden change in linear momentum of a system is equal to the total external impulsive force.

The Principle of Angular Momentum

The sudden change in angular momentum of a system on a fixed axis is equal to the moment of the external impulsive forces about the axis.

Impulse

Ii a particle of mass m is moving under the action of a constant force $\underline{F}$ producing an acceleration $\underline{a}$, then by the second law of Newton $\underline{F} = m\underline{a}$ and the impulse of the force during an interval of time t is defined as the product of the force and the time t. Thus impulse $\underline{i}$ is given by:

$$\underline{i} = \underline{F}t.$$

Impulse Force

When a bat strikes a cricket ball, the beginning and end of the collision can be determined fairly precisely. The bat is in contact with the ball for an interval that is relatively short compared with the time during which we are watching the ball. During the collision, the bat exerts a large force on the ball. This force varies with time and is difficult to measure.

Both the bat and ball are deformed during the collision. A force that acts for a time that is short compared with the time of observation of the system is known as an impulsive force.

Collision of Two Spheres and the Coefficient of Restitution

Newton's law of restitution states that when two bodies impinge, their relative velocity along the common normal after impact bears a constant ratio, denoted as e, to that before the impact. Here, e ($0 \le e \le 1$) is a constant depending on the nature of the material of the bodies. This constant e is called the coefficient of restitution. For an elastic impact, e = 1 and for a perfectly inelastic impact, e = 0. If 0 < e < 1, the collision is said to be inelastic.

Poisson Hypothesis

This is an alternative to Newton's law of restitution. According to this hypothesis, when two bodies come into contact, there is a short interval in which they undergo compression. During this interval, the centers of the two bodies get closer and closer as a result of the deformation due to the impact. This is followed by another short interval in which the original shape of the bodies move with a common velocity and the distance between the center of the two bodies is at a minimum. During the interval of restitution (the interval in which the bodies regain their shape), the impulsive pressure between the bodies.

CHAPTER 11

SOLUTION OF ALGEBRAIC AND TRANSCENDENTAL EQUATIONS

INTRODUCTION

One of the basic problems in science and engineering is the computation of roots of an equation of the form $f(x) = 0$. The equation $f(x)=0$ is denoted as an algebraic equation if it is purely a polynomial in x and it is called a transcendental equation if $f(x)$ contains trigonometric, exponential of logarithmic functions. For example.

$x^3 + 5x^2 - 6x + 3 = 0$

is an algebraic equation, whereas

$M = E - e \sin E$ and $ax^2 + \log(x-3) + e^x \sin x = 0$ are transcendental equations.

To find the solution of an equation $f(x) = 0$, we find those values of x for which $f(x) = 0$ is satisfied. Such values of x are called the roots of $f(x) = 0$. Thus a is a root of an equation $f(x) = 0$, if and only if, $f(a) = 0$

Before we evaluate various numerical methods, we shall list below some of the basic properties of algebraic equations.

(i) Every algebraic equation of nth degree, when n is a positive integer, has n and only n roots.

(ii) Complex roots occur in pairs. That is, if $(a+ib)$ is a root of $f(x)=0$, then $(a-ib)$ is also a root of this equation.

(iii) If $x=a$ isa root of $f(x)=0$, a polynomial of degree n, then $(x-a)$ is factor of $f(x)$. By dividing $f(x)$ by (x-a) we obtain a polynomial of degree $(n-1)$.

(iv) Descartes rule of sign: The number of positive roots of an algebraic equation $f(x)=0$ with real coefficients cannot exceed the number of changes in sign of the coefficients in the polynomial $f(x)=0$. Similarly, the number of negative roots $f(x)=0$ cannot exceed the number of changes in the sign of the coefficient of $f(-x)=0$. For example, consider an equation:

$x^3-3x^2+4x-5=0$.

There are three changes in the sign, also the degree of the equation is three, and hence the given equation will have three positive roots.

(v) Intermediate value property: if $f(x)$ is a real valued continuous function in the closed interval $a \le x \le b$ and if $f(a)$ and $f(b)$ have opposite signs, then the graph of the function $y=f(x)$ crosses the $x-axis$ at least once; that is $f(x)=0$ has at least one root ξ.

Broadly speaking, all the known numerical methods for solving either a transcendental equation or an algebraic equation can be classified into two group: direct methods and iterative methods. Direct methods require no knowledge of initial approximate of a root of the equation $f(x)=0$, while interactive methods require first approximation to initiate iteration. How do we determine the first approximation? We can find the approximation value of the root of $f(x)=0$, either by a graphical method or by an analytical method, as explained below:

Graphical Method

Often, the equation $f(x)=0$ can be rewritten as $f_1(x)=f_2(x)$, and the first approximation to a root of $f(x)=0$ can be taken as the abscissa of the point of intersection of the graphs of $y=f_1(x)$ and $y=f_2(x)$. For example:

$$f(x)=x-x-1=0.$$

can be written as x- 1 = sin x. Now, we shall draw the graph of

$y=x-1$ and $y=\sin x$.

As shows in fig, the approximate value of the root is found to be 1.9.

Analytical Method

This method is based on an 'intermediate value property'. We shall illustrate it through an example. Let

$$f(x)=3x-\sqrt{1+sinx}=0$$

We can easily verify

$$f(0)=-1$$

$$f(1)=3=\sqrt{1+\sin(1x\frac{180}{\pi})}=3-\sqrt{1+0.84147}=1.64299.$$

We observe that $f(0)$ and $f(1)$ are of opposite signs. Therefore, using the intermediate value property we infer that there is at least one root between $x=0$ and $x=1$. This method is often used to find the first approximation to a root of either a transcendental equation or an algebraic equation. Hence, for the analytical method, we must always start with an initial interval (a,b), so that $f(a)$ and $f(b)$ have opposite signs.

Bisection Method

Suppose we wish to locate the root of an equation $f(x)=0$ in an interval, say (x_o,x_1). Let $f(x_o)$ and $f(x_o)$and $f(x_1)$ have opposite signs. Such that $f(x_o)f(x_1)<0$.

Then the graph of the function crosses the x-axis between x_o and x_1, which guarantees the existence of at least one root in the interval (x_o,x_1). The desired root is approximately defined by the midpoint:

$$x_2=\frac{x_o+x_i}{2}\ .$$

If $f(x_2)=0$, then x_2 is the desired root of f(x) = 0. However, if $f(x_2)\neq 0$, then the root may be between x_o and x_2 and x_1.Now, we defined the next approximation by

$$x_3=\frac{x_o+x_2}{2}$$

Provided $f(x_0)f(x_2)<0$, then the root may be found between x_o and x_2 or by:

$$x_3=\frac{x_1+x_2}{2}.$$

Provided $f(x_1)f(x_2)<0$, then the root lies between x_1 and x_2.

Thus, at each iteration, we either find the desired root to the required accuracy or narrow the range to half the previous interval. This process of halving the intervals is continued to determine a smaller and smaller interval in which the desired root lies. Continuation of this process eventually gives us the desired root. This method is illustrated in the following example.

Example

Solve $x^3 - 9x + 1 = 0$ for the root between $x = 2$ and $x = 4$ by the bisection method.

Solution

Given $f(x) = x^3 - 9x + 1$. we can verify f(2) = -9, f(4) = 29. Therefore, $f(2)f(4) < 0$ and hence the root lies between 2 and 4. Let $x_o = 2$ $x_1 = 4$. Now we define

$$x_2 = \frac{x_o + x_1}{2} = \frac{2+4}{2} = 3$$

as a first approximation to a root of $f(x) = 0$ and note that $f(3) = 1$, so that $f(2)f(3) < 0$. Thus, the root lies between 2 and 3. We further define,

$$x_3 = \frac{x_o + x_2}{2} = \frac{2+3}{2} = 2.5$$

and note that $f(x_3) = f(2.5) < 0$, so that $f(2.5)f(3) < 0$. Therefore, we define the mid-point,

$$x_4 = \frac{x_3 + x_2}{2} = \frac{2.5+3}{2} = 2.75.$$

Similarly, we find that:

$x_5 = 2.875$ and $x_6 = 2.9375$

and the process can be continued until the root is obtained to the desired accuracy. These results are presented in the table below

n	x_n	$f(x_n)$
2	3	1.0
3	2.5	-5.875
4	2.75	-2.9531
5	2.875	-1. 1113
6	2.9375	-0.0901

REGULA–FALSI METHOD

This method is also known as the method of false position. In the method, we choose two points x_n and x_{n-1} such that $f(x_n)$ and $(f(x_{n-1})$ are of opposite signs. The intermediate value property suggest that the graph of $y = f(x)$ crosses the x-axis between these two points and therefore, a root say $x = \xi$ lies between these two points. Thus, to find a real root of f(x) = 0 using the regula- falsi method, we replace the part of the curve between the points $A[x_n, f(x_n)]$ and $B[x_{n-1}, f(x_{n-1})$ by a chord in that interval and we take the point of intersection of this chord with the x – axis as a first approximation to the root. Now, the equation of the chord joining the point A and B is:

$$\frac{y - f(x_n)}{f(x_{n-1}) - f(x_n)} = \frac{x - x_n}{x_{n-1} - x_n}.$$

Setting $y = 0$ in Eq. we get

$$x = x_n - \frac{x_{n-1} - x_n}{f(x_{n-1}) - f(x_n)} f(x_n).$$

Hence, the first approximation to the root of $f(x) = 0$ is given by

$$X_{n+1} = x_n - \frac{x_n - x_{n-1}}{f(x_n) - f(x_{n-1})} f(x_n).$$

We observe that $f(x_{n-1})$ and $f(x_{n+1})$ are of opposite signs. Thus, it is possible to apply the above procedure to determine the line through B and A_1 and so on. Hence, the successive approximations to the root of $f(x)=0$ is given by. This method can be better understood through the following examples.

Example

Use the Regula-Falsi method to compute a real root of the equation $x^3-9x+1=0$

(i) if the root lies between 2 and 2,

(ii) if the root lies between 2 and 3.

Comment on the results.

Solution

Let $f(x)=x^3-9x+1$.

(i) $f(2)=-9$ and $f(4)=29$. Since $f(2)$ and $f(4)$ are opposite signs, the root of $f(x)=0$ lies between 2 and 4. Taking $x_1=2, x_2=4$ and using the Regula-Falsi method, the first approximation is given by:

$$x_3 = x_2 - \frac{x_2 - x_3}{f(x_2)-f(x_1)} f(x_2) = 4 - \frac{2x29}{38} = 2.47368$$

and $f(x_3)=-6.12644$. Since $f(x_2)$ and $f(x_3)$ are of opposite signs, and the root lies between x_2 and x_3. The second approximations to the root is given as

$$x_4 = x_3 - \frac{x_3 - x_2}{f(x_3)-f(x_2)} f(x_3) = 2.73989$$

,

and $f(x_4)=-3.090707$. Note that since $f(x_2)$ and $f(x_4)$ are of opposite signs, the third approximation is obtained from

$$x_5 = x_4 - \frac{x_4 - x_2}{f(x_4) - f(x_2)} f(x_4) = 2.86125$$

and $f(x_5) = -1.32686$. This procedure can be continued until we get the desired result. The first three iterations are shown in the table:

n	x_{n+1}	$f(x_{n+1})$
2	2.47368	6.12644
3	2.73989	3.090707
4	2.86125	1.32686

(ii) $f(2) = -9$ and $f(3) = 1$. Since $f(2)$ and $f(3)$ are opposite signs, the root of $f(x) = 0$ lies between 2 and 3. Taking $x_1 = 2,\ x_2 = 3$ and using the Regula-Falsi method, the first approximation is given by:

$$x_3 = x_2 - \frac{x_3 - x_1}{f(x_2) - f(x_1)} f(x_2) = 3 - \frac{1}{10} = 2.9$$

and $f(x_3) = -0.711$. Since $f(x_2)$ and (x_3) are of opposite signs, the root lies between x_2 and x_3. The second approximation to the root is given as

$$x_4 = x_3 - \frac{x_3 - x_2}{f(x_3) - f(x_2)} f(x_3) = 2.94156$$

and $f(x_4) = -0.0207$. We observe that $f(x_2)$ and $f(x_4)$ are of opposite signs. The third approximation is obtained from

$$x_5 = x_4 - \frac{X_4 - X_2}{f(x_4) - f(x_2)} f(x_4) = 2.94275$$

and $f(x_5) = -0.0011896$. This procedure can be continued until we get the desired result. The first three iterations are shown in the table:

n	x_{n+1}	$f(x_{n+1})$
2	2.9	-0.711
3	2.94156	-0.0207
4	2.944275	-0.0011896

From the above computations, we observe that the value of the root as a third approximation is evidentially different in both the cases, while the value of x_5 when the interval considered is (2,3), is closer to the root. Hence, an important observation in this method is that the interval (x_1, x_2) chosen initially in which the root of the equation lies must be sufficiently small.

Example

Use the Regula-Falsi method to find a real root of the equation

$\log x - \cos x = 0$

accurate to four decimal places after three successive approximations

Solution

Given $f(x) = \log x - \cos x$. We observe that

$f(1) = 0 - 0.5403 = -0.5403$

and

$f(2) = 0.69315 + 0.41615 = 1.1093$

Since $f(1)$ and $f(2)$ are of opposite signs, the root lies between $x_1 = 1,\ x_2 = 2$ This first approximation is obtained from

$$x_3 = x_2 - \frac{x_2 - x_1}{f(x_2) - f(x_1)} f(x_2) = 2 - \frac{1.1093}{1.6496} = 13275$$

and

$f(x_3) = 0.2833 - 0.2409 = 0.0424$

Now, since $f(x_1)$and $f(x_3)$ are opposite signs, the second approximation is obtained as

$$x_4 = 1.3375 - \frac{(.3275)(.0424)}{0.0424 + 0.5403} = 1.3037$$

and

$f(x_4) = 1.24816x10^{-3}$

Similarly, we observe that $f(x_1)$ and $f(x_4)$ are of opposite signs, so the third approximation is given by:

$$x_5 = 1.3037 - \frac{(0.3037)(0.001248)}{0.001248 + 0.5403} = 1.3030$$

and

$f(x_5) = 0.62045x10^{-4}$.

Hence, the require real root is 1.3030.

Example

Using the Regula-Falsi method, find the real root of the following equation correct to three decimal places:

$x\log_{10} x = 1.2$.

Solution

Let $f(x) = x\log_{10} x - 12$. We observe that $f(2) = -0.5979, f(3) = 0.2314$.

Since $f(2)$and $f(3)$ are of opposite signs, the real root lies between $x_1 = 2, x_2 = 3$. The first approximation is obtained from

$$x_3 = x_2 - \frac{x_2 - x_1}{f(x_2) - f(x_1)} f(x_2) = 3 - \frac{0.2314}{0.8293} = 2.72097$$

and $f(x_3) = -0.01713$. Since $f(x_2)$and $f(x_3)$ are of opposite signs, the root of $f(x)=0$ lies between x_2 and x_3. Now, the second approximation is given by

$$x_4 = x_3 - \frac{x_3 - x_2}{f(x_3) - f(x_2)} f(x_3) = 2.7402$$

and $f(x_3) = -3.8905x10^{-4}$. Thus, the root of the given equation correct to three decimal places is 2.740

Method of Iteration

The method of iteration can be applied to find a real root of the equation

$f(x)=0$ by rewriting in the form,

$x = \varnothing(x)$.

For example, $f(x) = \cos x - 2x + 3 = 0$. it can be rewritten as

$$x = \frac{1}{2}(\cos x + 3) = \varnothing(x).$$

Let $x = \xi$ be the desired root. Suppose x_0 is its initial approximation.

The first and successive approximations to the root can be obtained as

$$\left.\begin{aligned} x_1 &= \varnothing(x_o) \\ x_2 &= \varnothing(x_1) \\ &\;\;. \\ &\;\;. \\ x_{n-1} &= \varnothing(x_n) \end{aligned}\right\}.$$

Definition

Let $\{x_i\}$ be the sequence obtained by a given method and let $x = \xi$ denote the root of the equation $f(x)=0$. Then the method is said to be convergent, if an only if

$Lt_{t\to\infty}\left|x_n-\xi\right|=0$.

Theorem

Let $x=\xi$ be a root of the equation $f(x)=0$, which can be rewritten as $x=\varnothing(x)$, contained in an interval I. Then, if $\left|\varnothing'(x)\right|<1$ for all x in I, the iterative process defined by $x_{n+1}=\varnothing(x_n)$ converges to the root $x=\xi$, if the initially chosen approximation x_o ò I.

This method is shown through the following examples:

Example

Use the method of iteration to determine the real root of the equation $e^{-x}=10x$ correct to four decimal places.

Solution

Let $f(x)=e^{-x}-10x=0$, we observe that $f(0)=1$ and $f(1)=-9.6321$. Since $f(0)<f(1)$ numerically, the root is near to $x=0$. Now, we shall rewrite the given equation in the form

$$x=\frac{1}{10}e^{-x}=\varnothing(x).$$

Therefore,

$$\varnothing'(x)=\frac{1}{10}e^{-x}$$

And

$$\left|\varnothing'(x)\right|=\frac{1}{10}e^{-x}=\frac{1}{10e^{x}}<1.$$

For all x in (0,1). Hence, the method of iteration can be applied. Thus, we start with the initial value $x_o=0$, then

$$x_1 = \varnothing(x_o) = \frac{1}{10} = 0.1, \quad f(x_t) = -0.09516.$$

Similarly, the successive approximations are given as:

$$x_2 = \varnothing(x_1) = \frac{1}{10}e^{-0.r} = \frac{0.904837}{10} = 0.09048 \; f(x_2) = -0.00869$$

$$x_3 = \varnothing(x_2) = 0.091349, \quad f(x_3) = -7.90877\,x10^{-4}$$

$$x_4 = \varnothing(x_3) = 0.091274\varnothing, \quad f(x_4) = 2.75784\,x10^{-5}.$$

Hence, the required root is 0.0913.

Example Find a real root of the equation

$f(x) = x^3 + x^2 - 1 = 0$ by the method of iteration.

Solution

We observe that $f(0) = -1, f(1) = 1,$ which shows that there is a real root between x = 0 and x = 1. To find the real root, we rewrite the equation in the form

$$x^2(x+1) = 1 \; or \; x = \frac{1}{\sqrt{x+1}} = \varnothing(x).$$

Therefore,

$$\varnothing(x) = -\frac{1}{2(x+1)^{\frac{3}{2}}}.$$

We note that $|\varnothing'(x)| < 1,$ for all x in (0,1). Hence, the method of iteration is applicable here.

Taking the initial value $x_o = 1,$ we successively obtain the following values.

$x_1 = \varnothing(x_0) = 1/\sqrt{2} = 0.70711, f(x_1) = -0.14644$

$x_2 = \varnothing(x_1) = 0.76537, \qquad f(x_2) = 0.03414$

$x_3 = \varnothing(x_2) = 0.75263, \qquad f(x_3) = 7.2213x10^{-3}$

$x_4 = \varnothing(x_3) = 0.75536, \qquad f(x_4) = 1.55658x10^{-3}$

$x_5 = \varnothing(x_4) = 0.754777, \qquad f(x_5) = -3.44323x10^{-4}$

$x_6 = \varnothing(x_5) = 0.7549, \qquad f(x_6) = 7.38295x10^{-5}.$

Hence, the required root is 0.7549.

Note: The given equation can be rewritten in many ways. Suppose, we rewrite

$$x^2 = 1 - x^3 \quad or \quad x = \left(1 - x^3\right)^{1/2} = \varnothing(x),$$

then we have

$$\left|\varnothing'(x)\right| = \frac{3x^2}{2\left(1 - x^2\right)^{1/2}}.$$

If we take $x = 1$ in the interval (0,1), $\left|\varnothing'(x)\right| = \infty$, then the conditional $\left|\varnothing'(x)\right| < 1$ is violated.

Newton-Raphson Method

This is a very powerful method for finding the real root of an equation in the form, $f(x) = 0$. Suppose x_0 is an approximate root of f(x) = 0. Let $x_1 = x_0 + h$, where h is small, be the exact root of f(x) = 0, then $f(x_1) = 0$. Now, expanding $f(x_0 + h)$ by Taylor's theorem, we get

$$f(x_0+h)=f(x_0)+hf'(x_0)+\frac{h^2}{2}f'(x_0)+....=0.$$

Since h is small, we neglect terms containing h^2 and its higher powers, then

$$f(x_0)+hf'(x_0)=0 \quad or\, h=\frac{-f(x_0)}{f'(x_0)}.$$

Improved approximations $x_2, x_3 ...., x_n$ to the root can obviously be obtained from the iteration formula, and

$$x_{n+1}=x_n-\frac{f(x_n)}{f'(x_n)}$$

is known as the Newton – Raphson iteration formula, and has the following geometrical interpretation:

Suppose, the graph of the function $y=f(x)$ crosses the $x-axis$ at a, then $x=a$ is the root of the equation $f(x)=0$.

Let x_0 be a point closer to the root α, then the equation of the tangent at $P_0(x_0, f(x_0))$ is

$$y-f(x_0)=f'(x_0)(x-x_0).$$

This tangent cuts the x axis at $R_0(x_1,0)$. Therefore,

$$x_1=x_0-\frac{f(x_0)}{f'(x_0)}$$

Which is a first approximation to the root α. If p_1 is a point on the curves corresponding to x_i then the tangent p_i cut the x-axis at $r_1(x_2,0)$, which is still closer to α, than x_i. Therefore, x_2 is a second approximation to the root. Continuing this process, we rapidly arrive at the root α, which is evident from . Thus, in this method, we have replaced the part of the curve between the point p_0 and the $x-axis$ by a tangent to the curve at p_0 and so on. In order to illustrate this method, we shall consider the following example.

Example

Find the real root of the equation $xe^x - 2 = 0$ correct two decimal places, using the Newton-Raphson method.

Solution

Given $f(x) = xe^x + e^x$ and $f''(x) = xe^x + 2e^x$

Clearly, we have

$f(0) = -2$ and $f(1) = e - 2 = 0.71828$.

Hence, the required root lies in the interval (0,1) and is nearer to 1.

Also $f'(x)$and $f'(x)$ do not vanish in (0,1) and $f(x)$and $f''(x)$ will have the same sign at x =1. Therefore, we take the first approximation $x_0 = 1$, and using the Newton- Raphson method, we get

$$x_1 = x_0 - \frac{f(x_0)}{f(x_0)} = \frac{e+2}{2e} = 0.867879$$

And

$f(x_1) = 6.71607x10^{-2}$.

The second approximation is

$$x_2 = x_1 - \frac{f(x_1)}{f(x_1)} = 0.867879$$

and

$f(x_2) = 7.655x10^{-4}$.

Thus, the required root is 0.853.

Example

Find a real root of the equation $x^3 - x - 1 = 0$ using the Newton-Raphson method, correct to four decimal places.

Solution

Let $f(x) = x^3 - x - 1,$ then we observe that $f(1) = -1, f(2) = 5.$ Therefore, the root lies in the interval $(1, 2).$ We also observe

$f(x) = -1,\ f''(1) = 6,\ f(2) = 5,\ f''(2) = 12$.

Since $f(2)$ and $f''(2)$ are of the same sign, we chose $x_0 = 2$ as the first approximation to the root. The second approximation is computed using the Newton-Raphson method as

$$x_1 = x_0 - \frac{f(x_0)}{f'(x_0)} = 2 - \frac{5}{11} = 1.54545$$ and $f(x_1) = 1.14573$.

The successive approximation are

$$x_2 = 1.54545 - \frac{1.14573}{6.16525} = 1.35961,\ f(x_2) = 0.15369$$

$$x_3 = 1.35961 - \frac{0.15369}{4.54562} = 1.32579,\ f(x_3) = 4.60959\,X10^{-3}$$

$$x_4 = 1.32579 - \frac{4.60959x10^{-3}}{6.16525} = 1.32471,\ f(x_4) = 3.39345x10^{-5}$$

$$x_5 = 1.32471 + \frac{3.39345x10^{-5}}{4.26457} = 1.324718, f(x_5) = 1.823x10^{-7}$$

Hence, the required root is 1.3247.

Convergence of the Newton-Raphson Method

To determine the convergence of the Newton-Raphson formula,

$$x_{n+1} = x_n - \frac{f(x_n)}{f'(x_n)},$$

we compare it with the general iteration formula $x_{n+1} = \varnothing(x_n)$, and thus obtain

$$\varnothing(x_n) = x_n - \frac{f(x_n)}{f'(x_n)}.$$

In general, we write this as

$$\varnothing(x) = -x - \frac{f(x)}{f'(x)}.$$

We have already noted that the iteration method converges if $|\varnothing'(x)| < 1$. Therefore, the Newton-Raphson formula converges, provided that

$$|f(x)f''(x)| < |f'(x)|^2$$

in the interval considered. The Newton-Raphson formula therefore converges, provided the initial approximation x_0 is chosen sufficiently close to root and $f(x), f'(x)$ and $f''(x)$ are continuous and bounded in any small interval containing the root.

Definition

Let

$$x_n = a + \varepsilon_n,\ x_{n+1} = a + \varepsilon_{n+1},$$

where a is a root of $f(x) = 0$ and ε_n is the error involved at the nth step. Finding the root by Newton-Raphson formula

$$\alpha + \varepsilon_{n+1} = \alpha - \frac{f(\alpha + \varepsilon_n)}{f'(\alpha + \varepsilon_n)}$$

i.e. $$\varepsilon_{n+1} = \varepsilon_n - \frac{f(\alpha + \varepsilon_n)}{f'(\alpha + \varepsilon_n)} = \frac{\varepsilon_n f'(\alpha + \varepsilon_n) - f(\alpha + \varepsilon_n)}{f'(\alpha + \varepsilon_n)}$$

Using Taylor expansion, we get,

$$\varepsilon_{n+1} = \frac{1}{f(a)+\varepsilon_n f''(a)+.....}\{\varepsilon_n\left[f'(a)+\varepsilon_n f''(a)+...\right]$$

$$-\left[f(a)+\varepsilon_n f'(a)+\frac{\varepsilon_n^2}{2}f''(\alpha)+....\right]\}.$$

Since α is a root, $f(\alpha)=0.$ Therefore, the above expression is simplified to

$$\varepsilon_{n+1} = \frac{\varepsilon_n^2}{2}f''(\alpha)\frac{1}{f'(\alpha)+\varepsilon_n f''(\alpha)}$$

$$=\frac{\varepsilon_n^2}{2}\frac{f''(\alpha)}{f'(\alpha)}\left[1+\varepsilon_n\frac{f''(\alpha)}{f'(\alpha)}\right]^{-1}$$

$$=\frac{\varepsilon_n^2}{2}\frac{f''(\alpha)}{f'(\alpha)}\left[1-\varepsilon_n\frac{f''(\alpha)}{f'(\alpha)}\right]^{-1}$$

Or

$$\varepsilon_{n+1} = \frac{\varepsilon_n^2}{2}\frac{f''(\alpha)}{f'(\alpha)}+O\left(\varepsilon_n^3\right).$$

On neighboring terms of order ε_n^3 and higher power, we obtain

$$\varepsilon_{n+1} = K\varepsilon_n^2$$

where

$$k = \frac{f''(\alpha)}{2f'(\alpha)}.$$

This shows that the Newton-Raphson method converges quadratically

Example

Set up Newton's scheme of iteration to find the square root of a positive number N.

Solution

The square root of N can be carried out as a root of the equation $x^2 - N = 0$.$Let\ f(x) = x^2 - N$ .By Newton's method , we have

$$x_{n+1} = x_n \frac{f(x_n)}{f''(x_n)}.$$

In this process, $f(x) = x^2 - N, f'(x) = 2x.$ Therefore,

$$x_{n+1} = x_n - \frac{x_n^2 - N}{2x_n} = \frac{1}{2}\left(x_n + \frac{N}{x_n}\right).$$

Example 10

Evaluate $\sqrt{12}$ by Newton's formula.

Solution

Since $\sqrt{9} = 3, \sqrt{16} = 4,$ we take $x_0 = (3+4)/2 = 3.5$ and we have

$$x_1 = \frac{1}{2}\left(x_0 + \frac{N}{x_0}\right) = \frac{1}{2}\left(3.5 + \frac{12}{3.5}\right) = 3.4643$$

$$x_2 = \frac{1}{2}\left(3.4643 + \frac{12}{3.4643}\right) = 3.4641$$

$$x_3 = \frac{1}{2}\left(3.4641 + \frac{12}{3.4641}\right) = 3.4641,$$

Hence, $\sqrt{12} = 3.4641$.

Example

Obtain the Newton-Raphson extended formula

$$x_1 = x_0 - \frac{f(x_0)}{f'(x_0)} - \frac{1}{2}\frac{\left[f(x_0)\right]^2}{2\left[f'(x_0)\right]^3} f''(x_0)$$

for finding the root of the equation $f(x)=0$.

Solution

Expanding $f(x)$ using Taylor series, in the neighborhood of x_0, we obtain after retaining the first order term only:

$$0\,f(x) = f(x_0) + (x - x_0) f'(x_0) + ..$$

which gives

$$x = x_0 \frac{f(x_0)}{f'(x_0)}.$$

This is the first approximation to the root. Therefore,

$$x_1 = x_0 - \frac{f(x_0)}{f'(x_0)}.$$

Again, expanding $f(x)$by the Taylor series and retaining up to the second order term, we have

$$0 = f(x) = f(x_0) + (x - x_0) f'(x_0) + \frac{x_1 - x_0}{2} f''(x_0).$$

Therefore,

$$f(x_1) = f(x_0) + (x_1 - x_0) f'(x_0) + \frac{x_1 - x_0}{2} f''(x_0) = 0.$$

Hence, we have

$$x_1 = x_0 - \frac{f(x_0)}{f'(x_0)} - \frac{1}{2}\frac{[f(x_0)]^2}{[f'(x_0)]^3} f''(x_0)$$
.

This is also known as Chebyshev's formula of third order.

Muller's Method

In Muller's method, $f(x)=0$ is approximated by a second degree polynomial; that is by a quadratic equation that fits through three points in the vicinity of a root. The roots of this quadratic equation are then assumed to be approximated by the roots of the equation $f(x)=0.$ This method is iterative in nature and does not require the evaluation of the derivates of the real and complex roots of $f(x)=0$.

Suppose that the quadratic passes through the point (x_{i-2}, f_{i-2}), (x_{i-1}, f_{i-1}) and (x_i, f_i), then the following equations will be satisfied:

$$ax_{i-2}^2 + bx_{i-2} + c = f_{i-2}$$

$$ax_{i-1}^2 + bx_{i-1} + c = f_{i-1}$$

$$ax_i^2 + bx_i + c = f_i$$
.

Eliminating a, b, c in the above Eqs.

$$\begin{vmatrix} x^2 & x & 1 & f \\ x_{i-2}^2 & x_{i-2} & 1 & f_{i-2} \\ x_{i-1}^2 & x_{i-1} & 1 & f_i \\ x_i^2 & x_i & 1 & f_i \end{vmatrix} = 0$$

Which can be written conveniently as

$$f = \frac{(x - x_{i-1})(x - x_i)}{(x_{i-2} - x_{i-1})(x_{i-2} - x_i)} f_{i-2} + \frac{(x - x_{i-2})(x - x_i)}{(x_{i-1} - x_{i-2})(x_{i-i} - x_i)} f_{i-1}$$

$$+ \frac{(x - x_{i-2})(x - x_{i=1})}{(x_i - x_{i-2})(x_i - x_{i-1})} f_i$$

This is obviously is a second-degree polynomial. Now, introducing the notation

$$h = x - x_i, \qquad h_i = x_i - x_{i-1}, \quad h_{i-1} = x_{i-1} - x_{i-2}$$

The above equation can be written as

$$f = \frac{(h - h_i)h}{-h_{i-1}(-h_{i-1} - h_i)} f_{i-2} + \frac{(h - h_i + h_{i-1})h}{(h_{i-1}) - (h_i)} f_{i-1}$$

$$+ \frac{(h - h_i + h_{i-1})(h + h_i)}{(h_i + h_{i-1})h_i} f_i$$

We further define

$$\lambda = \frac{h}{h_i} = \frac{x - x_i}{x_i - x_{i-1}}$$

$$\lambda_i = \frac{h_i}{h_{i-1}}$$

And $\delta_i = 1 + \lambda_i$

Thus, Eq can be further simplified to

$$f = \frac{1}{\delta_i}\left[\lambda(\lambda+1)\lambda_i^2 f_{i-2} - \lambda(\lambda + 1 + \lambda_i^{-1})\lambda_i \delta_i f_{i-1} + (\lambda+1)(\lambda + 1 + \lambda_i^{-1})\lambda_i f_i\right]$$

Or

$$f = \lambda^2 (f_{i-2}\lambda_i^2 - f_{i-1}\lambda_i \delta_i + f_i \lambda_i)\delta_i^{-1}$$

$$+\lambda[f_{i-2}\lambda_i^2 - f_{i-1}\delta_i^2$$

$$+f_i(\lambda_i+\delta_i)]\delta_i^{-1}+f_i.$$

Now, to compute λ, set $f=0\,Eq$ *uations* and obtain the quadratic equation as

$$\lambda_i(f_{i-2}\lambda_i - f_{i-1}\delta_i + f_i)\lambda^2 + g_i\lambda + \delta_i f_i = 0$$

Where

$$g_i = f_{i=2}\lambda_i^2 - f_{i-1}\delta_1^2 + fi(\lambda_i+\delta).$$

A direct solution of $Eq.$ leads to a loss of accuracy and therefore to obtain the maximum accuracy we rewrite, as follow:

$$\frac{f_i\delta_i}{\lambda^2} = \frac{g_i}{\lambda} + \lambda_i(f_{i-2}\lambda_i - f_{i-1}\delta_i + f_i).$$

So that,

$$\frac{1}{\lambda} = \frac{-g_i \pm \left[g_i^2 - 4f_{i-2}\delta_i(f_{i-2}\lambda_i - f_{i-1}\delta_i + f_i)\right]^{1/2}}{2f_i\delta_i}$$

$$= \lambda = \frac{-2f_i\delta_i}{g_i \pm \left[g_i^2 - 4f_i\delta_i\lambda_i(f_{1-2}\lambda_i - f_{i-1}\delta_i + f_i)\right]^{\frac{1}{2}}}.$$

Here, the positive sign must be so chosen so that the denominator becomes largest in magnitude. Using $x_{i+1} = x_i + h_i\lambda$ we can get a better approximation to the root.

Example

Find the root of the equation $x^3 - x - 1 = 0$ using Muller's method.

Solution

Let $f(x)=x^3-x-1, \text{then } f(0)=-1, f(1)=-1, f(2)=5.$ Therefore, a root lies between 1 and 2 and is close to 1. Muller's method can be conveniently started by taking

$$x_{i\_2}=0,\ x_{i-1}=1 \text{ and } x_i=2.$$

Correspondingly, we get

$$f_{i-2}=-1,\quad f_{i-1}=-1 \text{ and } f_i=5.$$

We define

$$h_{i-1}=x_{i-1}-x_{i-2}=1-0=1, \lambda=\frac{h}{h_i}=x-2$$

$$\lambda_i=\frac{h_i}{h_{i-1}}=1, \delta_i=1+\lambda_i=2.$$

We note that

$$g_i=(-1)(1)-(-1)(2)^2+(5)(3)=18.$$

Now, the quadratic equation for $1/\lambda$ is given by:

$$\frac{f_i\delta_i}{\lambda^2}+\frac{g_i}{\lambda}+\lambda_i\left(f_{i-2}\lambda_i-f_{i-1}\delta_i+f_i\right)=0.$$

That is,

$$\frac{5}{\lambda^2}+\frac{9}{\lambda}+3=0,$$

which gives

$$\frac{1}{\lambda}=\frac{-9\pm\sqrt{21}}{10}.$$

Taking a negative sign, so that the numerator is the largest value in magnitude, we get $\lambda = -0.736236$. Therefore, the first approximation to the required root is given by

$$x_{i+1} = x_i + \lambda h_i = 2 - 0.736236 = 1.26376.$$

Graeffe's Root Squaring Method

This method is particularly attractive for finding all the roots of a polynomial equation. We shall illustrate this method by considering a polynomial of third degree. However, it may be noted that this method is also applicable for higher degree polynomials.

Consider a polynomial of third degree:

$$f(x) a_0 + a_1 x + a_2 x^2 + a_3 x_3.$$

On putting x = - x, we get

$$f(-x) = a_0 + a_1 x + a_2 x^2 + a_3 x_{3.}$$

On multiplying the above equation, we get

$$f(x) f(-x) = a_3^2 t^3 - \left(a_2^2 - 2a_1 a_3\right) t^2 \left(a_1^2 - 2a_0 a_2\right) t - a_0^2 \quad .$$

Here, we have replaced x^2 by t. It may be observed that the roots of the equation are indeed the squares or $2^i (i = 1)$ powers of the original roots. Here i = 1 indicates that squaring is performed once.

Now, we can square again. This squaring process is repeated as many times as required. After each squaring, the coefficient become large and an overflow is possible as i increases.

Suppose, we have squared the given polynomial *i* times, then we can estimate the value of the roots by evaluating the 2^i root of

$$\left|\frac{a_i}{a_{i-1}}\right|, \; i = 1, 2, \ldots\ldots, n$$

,

where n is the degree of the given polynomial. The proper sign of each root can be determined by recalling the original equation. This method of course fails when the roots of the given polynomial are repeated. This technique can be understood better from the following example.

Example

Using the Graeffe root squaring method, find all the roots of the equation

$x^3 - 6x^2 + 11x - 6 = 0$.

Solution

Using the Graeffe root squaring method, the first three squared polynomials are:

For $i = 1,$ the polynomial is

$$x^3 - (36 - 22)x^2 + (121 - 72)x - 36 = x^3 - 14x^2 + 49x - 36. \tag{1}$$

For $i = 2,$ the polynomial is

$$x^3 - (196 - 98)x^2 + (2402 - 1008)x - 1296 = x^3 - 98x^2 + 1393x - 1296. \tag{2}$$

For $i = 3,$ the polynomial is

$$x^3 - (9604 - 2786)x^2 + (1940449 - 254016)x - 1679616.$$

The roots of the polynomial (1) are

$$\sqrt{\frac{36}{49}} = 0.85714, \sqrt{\frac{49}{14}} = 1.8708 \sqrt{\frac{14}{1}} = 3.7417.$$

Similarly, the roots of polynomial (2) are

$$\sqrt[4]{\frac{1296}{1393}} = 0.9821, \sqrt[4]{\frac{1393}{98}} = 1.9417, \sqrt[4]{\frac{98}{1}} = 3.1464.$$

The improved estimates of the roots obtained from polynomial (3) are

$$\sqrt[8]{\frac{1679616}{1686433}}=0.99949\ ,\ \sqrt[8]{\frac{1686433}{6818}}=1.99143\ ,\sqrt[8]{\frac{6818}{1}}=3.0144.$$

It may be observed that the exact values of the given polynomial are 1, 2 and 3.

SUMMARY AND DISCUSSION

One of the basic problems in science and engineering is the computation of roots of an equation of the form $f(x)=0$. The equation $f(x)=0$ is denoted as an algebraic equation if it is purely a polynomial in x and it is called a transcendental equation if $f(x)$ contains trigonometric, exponential of logarithmic functions.

To find the solution of an equation $f(x)=0$, we find those values of x for which $f(x)=0$ is satisfied. Such values of x are called the roots of $f(x)=0$. Thus a is a root of an equation $f(x)=0,$ if and only if, $f(a)=0$

Basic Properties of an Algebraic Equation

(i) Every algebraic equation of nth degree, when n is a positive integer, has n and only n roots.

(ii) Complex roots occur in pairs. That is, if $(a+ib)$ is a root of $f(x)=0,$ then $(a-ib)$ is also a root of this equation.

(iii) If $x=a$ isa root of $f(x)=0$, a polynomial of degree n, then $(x-a)$ is factor of $f(x)$. Bydividing $f(x)$ by (x-a) we obtain a polynomial of degree $(n-1)$.

(iv) Descartes rule of sign: The number of positive roots of an algebraic equation $f(x)=0$ with real coefficients cannot exceed the number of changes in sign of the coefficients in the polynomial $f(x)=0.$ Similarly, the number of negative roots $f(x)=0$ cannot exceed the number of changes in the sign of the coefficient of $f(-x)=0$.

(v) Intermediate value property: if $f(x)$ is a real valued continuous function in the closed interval $a \le x \le b$ and if $f(a)$ and $f(b)$ have opposite signs, then the graph of the function $y=f(x)$ crosses the $x-axis$ at least once; that is $f(x)=0$ has at least one root ξ.

Numerical Methods Introduction

Numerical methods used to solve either a transcendental equation or an algebraic equation can be classified into two group: direct methods and iterative methods. Direct methods require no knowledge of an initial approximate of a root of the equation $f(x)=0$, while interactive methods do require a first approximation of an initial iteration.

Bisection Method

Suppose we wish to locate the root of an equation $f(x)=0$ in an interval, say (x_o, x_1). Let $f(x_o)$ and $f(x_o)$ and $f(x_1)$ have opposite signs. Such that $f(x_o)f(x_1)<0.$

Then the graph of the function crosses the x-axis between x_o and x_1, which guarantees the existence of at least one root in the interval (x_o, x_1). The desired root is approximately defined by the midpoint:

$$x_2 = \frac{x_o + x_i}{2}.$$

If $f(x_2)=0,$ then x_2 is the desired root of f(x) = 0. However, if $f(x_2) \neq 0,$ then the root may be between x_o and x_2 and x_1.Now, we defined the next approximation by

$$x_3 = \frac{x_o + x_2}{2}$$

Provided $f(x_0)f(x_2)<0,$ then the root may be found between x_o and x_2 or by:

$$x_3 = \frac{x_1 + x_2}{2}.$$

Provided $f(x_1)f(x_2)< 0,$ then the root lies between x_1 and x_2.

REGULA–FALSI METHOD

This method is also known as the method of false position. In the method, we choose two points x_n and x_{n-1} such that $f(x_n)$ and $(f(x_{n-1})$ are of opposite signs. The intermediate value property suggests that the graph of $y=f(x)$ crosses the x-axis between these two points and therefore, a root say $x=\xi$ lies between these two points. Thus, to find a real root of f(x) = 0 using the regula- falsi method, we replace the part of the curve between the points $A\left[x_n, f(x_n)\right]$ and $B[x_{n-1}, f(x_{n-1})$ by a chord in that interval and we take the point of intersection of this chord with the x – axis as a first approximation to the root. Now, the equation of the chord joining the point A and B is:

$$\frac{y-f(x_n)}{f(x_{n-1})-f(x_n)} = \frac{x-x_n}{x_{n-1}-x_n}.$$

Setting $y=0$ in Eq. we get

$$x = x_n - \frac{x_{n-1}-x_n}{f(x_{n-1})-f(x_n)} f(x_n).$$

Hence, the first approximation to the root of $f(x)=0$ is given by

$$X_{n+1} = x_n - \frac{x_n - x_{n-1}}{f(x_n) - f(x_{n-1})} f(x_n)$$

We observe that $f(x_{n-1})$ and $f(x_{n+1})$ are of opposite signs. Thus, it is possible to apply the above procedure to determine the line through B and A_1 and so on.

Method of Iteration

The method of iteration can be applied to find a real root of the equation $f(x) = 0$ by rewriting in the form,

$$x = \varnothing(x).$$

For example, $f(x) = \cos x - 2x + 3 = 0.$ it can be rewritten as

$$x = \frac{1}{2}(\cos x + 3) = \varnothing(x).$$

Let $x = \xi$ be the desired root. Suppose x_0 is its initial approximation.

The first and successive approximations to the root can be obtained as

$$\left.\begin{aligned} x_1 &= \varnothing(x_o) \\ x_2 &= \varnothing(x_1) \\ &\cdot \\ &\cdot \\ x_{n-1} &= \varnothing(x_n) \end{aligned}\right\}$$

Definition

Let $\{x_i\}$ be the sequence obtained by a given method and let $x = \xi$ denote the root of the equation $f(x) = 0.$ Then the method is said to be convergent, if an only if

$Lt_{t\to\infty}|x_n - \xi| = 0$.

Newton-Raphson Method

This is a very powerful method for finding the real root of an equation in the form, $f(x)=0.$ Suppose x_0 is an approximate root of f(x) = 0. Let $x_1 = x_0 + h,$ where h is small, be the exact root of f(x) = 0, then $f(x_1)=0$. Now, expanding $f(x_0+h)$ by Taylor's theorem, we get

$$f(x_0+h) = f(x_0) + h f'(x_0) + \frac{h^2}{2} f'(x_0) + ... = 0$$

Since h is small, we neglect terms containing h^2 and its higher powers, then

$$f(x_0) + h f'(x_0) = 0 \quad \text{or}\, h = \frac{-f(x_0)}{f'(x_0)}$$

Improved approximations $x_2, x_3 ...., x_n$ to the root can obviously be obtained from the iteration formula, and

$$x_{n+1} = x_n - \frac{f(x_n)}{f'(x_n)}$$

is known as the Newton – Raphson iteration formula.

Suppose, the graph of the function $y=f(x)$ crosses the $x-axis$ at $a,$ then $x=a$ is the root of the equation $f(x)=0$.

Let x_0 be a point closer to the root $\alpha,$ then the equation of the tangent at $P_0(x_0, f(x_0))$ is

$$y - f(x_0) = f'(x_0)(x - x_0).$$

This tangent cuts the x axis at $R_0(x_1,0)$. Therefore,

$$x_1 = x_0 - \frac{f(x_0)}{f'(x_0)}$$

Which is a first approximation to the root α.If p_1 is a point on the curves corresponding to x_i then the tangent p_i cut the x-axis at $r_1(x_2,0)$, which is still closer to α, than x_i. Therefore, x_2 is a second approximation to the root. Continuing this process, we rapidly arrive at the root α, which is evident from . Thus, in this method, we have replaced the part of the curve between the point p_0 and the $x-axis$ by a tangent to the curve at p_0 and so on.

Muller's Method

In Muller's method, $f(x)=0$ is approximated by a second-degree polynomial; that is by a quadratic equation that fits through three points in the vicinity of a root. The roots of this quadratic equation are then assumed to be approximated by the roots of the equation $f(x)=0.$ This method is iterative in nature and does not require the evaluation of the derivates of the real and complex roots of $f(x)=0$.

Graeffe's Root Squaring Method

This method is particularly attractive for finding all the roots of a polynomial equation. We shall illustrate this method by considering a polynomial of third degree. However, it may be noted that this method is also applicable for higher degree polynomials.

CHAPTER 12

SOLUTION OF LINEAR SYSTEMS OF EQUATIONS AND MATRIX INVERSION

INTRODUCTION

Many real-life problems in engineering give rise to a system of linear equations. For example, such systems occur in certain applications of statistical analysis and in finding the numerical solution of partial differential equations. It is therefore natural to seek efficient methods for solving these equations numerically.

The generally from of a system of m linear equations in n unknowns $x_1, x_2, x_3 \ldots,$ can be represented in matrix from:

$$\begin{bmatrix} a_{11}\, a_{12}\, a_{13} \ldots\ldots a_{1n} \\ a_{21}\, a_{22}\, a_{23} \ldots\ldots a_{2n} \\ \cdot \\ \cdot \\ \cdot \\ a_{m1}\quad a_{m2}\quad a_{m3} \ldots a_{mn} \\ \cdot \end{bmatrix} \begin{pmatrix} x_1 \\ x_2 \\ \cdot \\ \cdot \\ \cdot \\ x_n \end{pmatrix} = \begin{pmatrix} b_1 \\ b_2 \\ \cdot \\ \cdot \\ \cdot \\ b_m \end{pmatrix}$$

Using matrix notation, the above system can be written in compact form as:

$$[A](X)=(B).$$

The solution of the system of equation gives n unknown values $x_1, x_2 \ldots, x_m$, which satisfy the system simultaneously. If $m>n$, we may not be able to find a solution, in principle, which satisfies all the equations. If $m>n$, the system will usually have an infinite number of solutions. However, in this chapter, we shall restrict to the case to $m=n$. In this case, if $|A| \neq 0$, then the system will have a unique solution, while if $|A|=0$, then there exist no solution.

Various numerical methods are available for finding the solution of the system of equations and are classified as direct and iterative methods. For the direct methods, we get the solution of the system after performing all the steps involved in the procedure. The direct methods consist of elimination methods and decomposition methods. In this chapter, we will consider the Gaussian and Gauss-Jordan elimination methods. Crout's reduction, also known as Cholesky's reduction, is a decomposition method. Out of the iterative methods, the initial approximate solution is assumed to an improved method working towards the exact solution. All these methods are easily adoptable as computational procedures and can be used to solve even hundreds or more simultaneous linear equations.

GAUSSIAN ELIMINATION METHOD

In the Gaussian elimination method, the solution to a system of equations is obtained in two stages. In the first stage, the given system of equations is reduced to an equivalent upper triangle system solved using the back substitution procedure, by which we obtain the solution in the order $x_n, x_{n-1}, x_{n-2}, \ldots\ldots x_2, x_1$.

This method is explained by considering a system of n equations in n unknowns in the form as follows:

$$\left.\begin{array}{l} a_{11}x_1 + a_{12}x_2 + \ldots + a_{1n}x_n = b_1 \\ a_{21x_1} + a_{22}x_2 + \ldots\ldots a_{2n}x_n = b_2 \\ \qquad\ldots \\ \qquad\ldots \\ \qquad\ldots \\ a_n x_1 + a_{n2}x_2 + \ldots\ldots a_{nn}x_n = b_n \end{array}\right\}$$

Stage1: We divide the first equation by a_{11} and then subtract the product of this equation and $a_{21}, a_{31}, \ldots\ldots, a_{n1}$. from the 2nd, 3rd, ….., n^{th} the equation:

$$\left.\begin{array}{l} x_1 + a_{12}x_2 + \ldots + a_{1n}x_n = b_1 \\ \quad a_{22}x_2 + \ldots\ldots a_{2n}x_n = b_2 \\ \qquad\ldots \\ \qquad\ldots \\ \qquad\ldots \\ \quad a_{n2}x_2 + \ldots\ldots a_{nn}x_n = b_n \end{array}\right\}$$

re, we can observe that the last $(n-1)$ equations are independent of x_1, that is, x_1 is eliminated from the last $(n-1)$ equations.

This procedure is repeated with the second equation. That is, we divide the second equation by a_{22} and the x_2 is eliminated from 3rd, 4th, ….., n^{th} equations. The same procedure is repeated again and again until the given system assumes the following upper triangular form:

$$\left.\begin{array}{l} c_{11}x_1 + c_{12}x_2 + \ldots + c_{1n}x_n = d_1 \\ \quad c_{22}x_2 + \ldots\ldots c_{2n}x_n = d_2 \\ \qquad\ldots \\ \qquad\ldots \\ \qquad\ldots \\ \qquad c_{nn}x_n = d_n \end{array}\right\}$$

Stage II: The values of the unknowns are determined by the back substitution procedure, in which we obtain x_n from the last equation and then substituting this value of x_{n-1}. Continuing this way, we can find the values of all other unknowns in the order $x_n, x_{n-1}, \ldots x_2, x_1$.

In this method, we observe that the determinant of the coefficient matrix is obtained as a by-product, that is:

$$|A| c_{11} c_{22} \ldots c_{nn}.$$

To familiarize ourselves with the method, we consider the following example:

Example

Solve the following system of equations using the Gaussian elimination method.

$$2x + 3y - z = 5$$

$$4x + 4y - 3z = 3$$

$$-2x + 3y - z = 1.$$

Solution

The given system of equations is solved in two stages

Stage 1 (reduction to upper-triangular from): We divide the first equation by 2 and then subtract the resulting equation (multiplied by 4 and -2) from the second and third equations, respectively. Thus, we eliminate x from the 2nd and 3rd equations. The resulting new system given by

$$\left.\begin{aligned} x + \frac{3}{2}y - \frac{z}{2} = \frac{5}{2} \\ -2y - z = -7 \\ 6y - 2z = 6 \end{aligned}\right\}$$

Now, we divide the second equation of (1) by -2 and eliminate from the last equation. The modified system is given by:

$$\left.\begin{aligned} x+\frac{3}{2}y-\frac{z}{2}=\frac{5}{2} \\ y+\frac{z}{2}=\frac{7}{2} \\ -5z=-15 \end{aligned}\right\}$$

Stage II (back substitution): From the last equation of (2), we immediately get

$Z=3$

Using this value of z, the second equation of (2) gives us:

$$y=\frac{7}{2}-\frac{3}{2}=2.$$

Using this value of y and z in the first equation of (2), we get

$$x=\frac{5}{2}+\frac{3}{2}-3=1.$$

Thus, the solution of the given system is given by the equations:

Partial and Full Pivoting

The Gaussian elimination method fails if any one of the pivot elements becomes zero. In such a situation, we rewrite the equations in a different order to avoid zero pivots. Changing the order of the equations is called pivoting.

We now introduce the concept of partial pivoting. In this technique, if the pivot a_{ii} happens to be zero, then as the numerically largest elements are searched for in the ith column elements. Let the jth row (j> i) contain this element, then we interchange the ith equation with the jth equation and proceed for the elimination. For example, let us examine the solution of the following simple system

$$10^{-5}x_1+x_2=2$$

$$x_1+x_2=2.$$

We will use the Gaussian elimination method with and without partial pivoting, and assume that we require the solution accurate to only four decimal places. The solution by Gaussian elimination gives $x_1 = 0, x_2 = 1$. If we use partial pivoting, the system takes the form

$$x_1 + x_2 = 2$$

$$10^{-5} x_1 + x_2 = 1$$.

Using the Gaussian elimination method, the solution is found to $x_1 = 1, x_2 = 1$.

For full pivoting, which is also known as complete pivoting, we interchange rows as columns, such that the largest element in the matrix of the system becomes the pivot element. In this process, the position of the unknown variable also get changed. Full pivoting, in fact, is more complicated than partial pivoting. Partial pivoting is preferred for hand computation.

Example

Solve the system of equations

$$x + y + z = 7$$

$$3x + 3y + 4z = 24$$

$$2x + y + 3z = 16$$

using the Gaussian elimination method with partial pivoting.

Solution

In matrix notation, the given system can be written as

$$\begin{bmatrix} 111 \\ 334 \\ 213 \end{bmatrix} \begin{pmatrix} x \\ y \\ z \end{pmatrix} = \begin{pmatrix} 7 \\ 24 \\ 16 \end{pmatrix}$$

To start with, we observe that the pivot element $a_{11} = 1(\neq 0)$. However, a glance at the first column reveals that the numerically largest element is 3, which is in the second row. Hence, we interchange the first row with the second row and then proceed for elimination, after partial pivoting:

$$\begin{bmatrix} 334 \\ 111 \\ 213 \end{bmatrix} \begin{pmatrix} x \\ y \\ z \end{pmatrix} = \begin{pmatrix} 24 \\ 7 \\ 16 \end{pmatrix}$$

Stage1(reduction to upper triangular from): By dividing the first row of the system and then subtracting the resulting row multiplied by 1 and 2 from the second and third rows of the system (2), we get

$$\begin{bmatrix} 1\,1\frac{4}{3} \\ -\frac{1}{3} \\ -1\;\frac{1}{3} \end{bmatrix} \begin{pmatrix} x \\ y \\ z \end{pmatrix} = \begin{pmatrix} 8 \\ -1 \\ 0 \end{pmatrix}$$

The second row cannot be used as the pivot row, as $a_{22} = 0$. Interchanging the second and third rows, we obtain

$$\begin{bmatrix} 1\,1\frac{4}{3} \\ -1\;\frac{1}{3} \\ -\frac{1}{3} \end{bmatrix} \begin{pmatrix} x \\ y \\ z \end{pmatrix} = \begin{pmatrix} 8 \\ 0 \\ -1 \end{pmatrix}$$

Which is in the upper triangular from.

Stage II: From the last row we get

$z = 3$

The second row with this value of z gives us

$y+1=0 \ or \ y=1$.

Using these values of y and z, the first row gives us:

$x+1+4=8 \ or \ x=3$

Thus ^{Eqs}. constitute the solution to the given system of equations.

Example

Solve by Gaussian elimination method with partial pivoting applied to the following system of equations:

$$0x_1+4x^2+2x_3+8x_4=24$$

$$4x_1+10x_2+5x_3+4x_4=32$$

$$4x_1+5x_2+6.5x_3+2x_4=26$$

$$9x_1+4x_2+4x_3+)x_4=21$$

Solution

In matrix notation, the given system can be written as

$$\begin{bmatrix} 0 & 4 & 2 & 8 \\ 4 & 10 & 5 & 4 \\ 4 & 5 & 6.5 & 2 \\ 9 & 4 & 4 & 0 \end{bmatrix} \begin{pmatrix} x_1 \\ x_2 \\ x_3 \\ x_4 \end{pmatrix} = \begin{pmatrix} 21 \\ 32 \\ 26 \\ 21 \end{pmatrix}$$

To start with, we observe that the first row has a zero-pivot element $(a_{11}=0)$. This row should be interchanged with any proceeding row, which on becoming a pivot row should not have a zero-pivot element. While interchanging rows, it is better to interchange with a row having the largest pivotal element. Thus, we interchange the first and fourth row, known as partial pivoting and get,

$$\begin{bmatrix} 9 & 4 & 4 & 0 \\ 4 & 10 & 5 & 4 \\ 4 & 5 & 6.5 & 2 \\ 0 & 4 & 2 & 8 \end{bmatrix} \begin{pmatrix} x_1 \\ x_2 \\ x_3 \\ x_4 \end{pmatrix} = \begin{pmatrix} 21 \\ 32 \\ 26 \\ 24 \end{pmatrix}$$

We observe that, in partial pivoting, the unknown vector remains unaltered, while the right-hand side vector changes.

Now, we shall carry out the Gaussian elimination process in two stages.

Stage 1 (reduction to upper-triangular from): In this stage, by dividing the first row of system (2) by 9 and subtracting this resulting row, multiplied by 4 and 4, from the second and third rows of Eq we get:

$$\begin{bmatrix} 1 & \frac{4}{9} & \frac{4}{9} & 0 \\ 0 & 8.2222 & 3.2222 & 4 \\ 0 & 3.2222 & 4.7222 & 2 \\ 0 & 4 & 4 & 0 \end{bmatrix} \begin{pmatrix} x_1 \\ x_2 \\ x_3 \\ x_4 \end{pmatrix} = \begin{pmatrix} 2.3333 \\ 22.6666 \\ 16.6666 \\ 24 \end{pmatrix}$$

Now we divide the second pivot row by 8.2222 and subtract the resultant row multiplied by 3.2222 and 4 from the third and fourth rows of Eq. to get

$$\begin{bmatrix} 9 & \frac{4}{9} & \frac{4}{9} & 0 \\ 4 & 1 & 3.2222 & 0.4865 \\ 4 & 0 & 3.4594 & 0.4324 \\ 0 & 0 & 0.4324 & 6.0540 \end{bmatrix} \begin{pmatrix} x_1 \\ x_2 \\ x_3 \\ x_4 \end{pmatrix} = \begin{pmatrix} 2.3333 \\ 2.7568 \\ 7.7836 \\ 12.9728 \end{pmatrix}$$

Finally, we divide the third pivot row by 3.4594 and subtract the resultant row multiplied by 0.4324 from fourth row of Eq thus getting the upper triangular form.

$$\begin{bmatrix} 9 & \frac{4}{9} & \frac{4}{9} & 0 \\ 4 & 1 & 3.2222 & 0.4865 \\ 4 & 0 & 1 & 0.1250 \\ 0 & 0 & 0 & 5.9999 \end{bmatrix} \begin{pmatrix} x_1 \\ x_2 \\ x_3 \\ x_4 \end{pmatrix} = \begin{pmatrix} 2.3333 \\ . \\ 2.7568 \\ 2.2500 \\ 11.9999 \end{pmatrix}$$

Stage II (Black Substitution): From the last row of $Eq.$ we immediately get $x_4 = 2.0000.$ Using this value of x_4 for the third row of $Eq.$ we obtain

$$x_3 + 0.25 = 2.25 \; or \; x_3 = 2.0000$$

Similarly, we get

$$x_2 = 1.0000, \;\; x_1 = 1.0000 \; .$$

Thus, the solution of the given system is given by

$$x_1 = 1.0, \;\; x_2 = 1.0. \;\; x_3 = 2.0, \;\; x_4 = 2.0 \; .$$

Gauss-Jordan Elimination Method

This method is a variation of the Gaussian elimination method. In this method, the elements above and below the diagonal are simultaneously made zero and thereby the given system is reduced to an equivalent diagonal using elementary transformations. The solution of the resulting diagonal system can then be readily obtained.

Sometimes, we normalize the pivot row with respect to the pivot element, before elimination. Partial pivoting is also used whenever the pivot element becomes zero.

This method is illustrated through the following examples

Example

Solve the system of equations

$$\left.\begin{array}{r} x+2y+z=8 \\ 2x+3y+4z=20 \\ 4x+3y+2z=16 \end{array}\right\}$$

using Gauss-Jordan elimination method.

Solution

In matrix notation, the given system (1) can be written as

$$\begin{bmatrix} 1 & 2 & 1 \\ 2 & 3 & 4 \\ 4 & 3 & 2 \end{bmatrix}\begin{pmatrix} x \\ y \\ z \end{pmatrix}=\begin{pmatrix} 8 \\ 20 \\ 16 \end{pmatrix}$$

We subtract the first row multiplied by 2 and 4 from the second and third rows respectively and eliminate x.

$$\begin{bmatrix} 1 & 2 & 1 \\ 0 & -1 & 2 \\ 0 & -5 & -2 \end{bmatrix}\begin{pmatrix} x \\ y \\ z \end{pmatrix}\begin{pmatrix} 8 \\ 4 \\ -16 \end{pmatrix}$$

Now, we eliminate y from the first and third rows using the second row. Thus, we get

$$\begin{bmatrix} 1 & 0 & 1 \\ 0 & -1 & 2 \\ 0 & 5 & -12 \end{bmatrix}\begin{pmatrix} x \\ y \\ z \end{pmatrix}\begin{pmatrix} 16 \\ 4 \\ -36 \end{pmatrix}$$

Before eliminating z from the first and second row, we normalize the third row with respect to the pivot element, and get

$$\begin{bmatrix} 1 & 0 & 5 \\ 0 & -1 & 2 \\ 0 & 0 & 1 \end{bmatrix} \begin{pmatrix} x \\ y \\ z \end{pmatrix} \begin{pmatrix} 16 \\ 4 \\ 3 \end{pmatrix}$$

Using the third row of $Eq.$ eliminating z from the first and second rows of $Eq.$

$$\begin{bmatrix} 1 & 0 & 0 \\ 0 & -1 & 0 \\ 0 & 0 & 1 \end{bmatrix} \begin{pmatrix} x \\ y \\ z \end{pmatrix} \begin{pmatrix} 1 \\ -2 \\ 3 \end{pmatrix}$$

From $Eq.$, we get the solution directly as $x = 1, y = 2, z = 3.$

Crout's Reduction Method

This method is based on the fact that the coefficient matrix $[A]$ of the system of equations can be decomposed into the product of two matrices [L] and [U], where [L] is a lower-triangular matrix and [U] is an upper triangular matrix with 1 in its main diagonal. The rules for getting [L] and [U] can be obtained from the following:

$$[L][U]=[A].$$

For the purpose of illustration, let us consider a (3×3) general matrix in the form

$$\begin{bmatrix} l_{11} & 0 & 0 \\ l_{12} & l_{22} & 0 \\ l_{31} & l_{32} & l_{33} \end{bmatrix} \begin{bmatrix} 1\frac{11}{12}\frac{11}{13} \\ 01\frac{11}{23} \\ 001 \end{bmatrix} = \begin{bmatrix} a_{11} & a_{12} & a_{13} \\ a_{21} & a_{22} & a_{23} \\ a_{31} & a_{32} & a_{33} \end{bmatrix}.$$

The sequence of steps for getting [L] and [U] are given below:

*Step*1: Multiply all rows of [L] is same as the first column of [U], we get

$l_{11} = a_{11}, l_{21} = a_{21}, l_{31} = a_{31}$

Thus, we observe that the first column of [L] is same as the first column of [A].

Step II. Now, multiplying the first row of [L] by the second column of [U], we obtain

$l_{11 ll_{12}} = a_{12}, \qquad l_{11 ll_{13}} = a_{13}$

or

$ll_{12} = \frac{a_{12}}{l_{11}}, \quad u_{13} = \frac{a_{13}}{l_{11}}.$

Thus, the first row of [U] is obtained. Now, we continue this process, alternately deriving a column of [L] and a row of [U].

Step III. Multiply the second and third rows of [L] by the second column of [U] to get

$l_{21}u_{12} + l_{22n} = a_{22}, \qquad l_{31}u_{12} + l_{32} = a_{32}.$

Which gives

$l_{22} = a_{22} - l_{21}u_{12}, \qquad l_{32} = a_{32} - l_{31}u_{12}.$

Thus, the second column of [L] is obtained.

Step Iv: Now, multiply the second row of [L] by the third column of [U], which yields

$L_{12}u_{13} + l_{32}u_{23} = a_{23} \qquad \text{or} \quad u_{23} = \frac{a_{23} - l_{21}u_{13}}{l_{22}}.$

Step V: Lastly, we multiply the third row of [L] by the third columns of [U] and get

$l_{31}u_{13} + l_{32}u_{23} + l_{33} = a_{33}.$

Which gives

$l_{33} = a_{33} - l_{31}u_{13} - l_{32}u_{23}.$

Thus, the above five steps determine [L] and [U]. This algorithm can be generalized to any linear system of order n.

Now, we shall obtain the solution of the linear system

$$\left.\begin{aligned} a_{11}x_1 + a_{12}x_2 + a_{13}x_3 &= b_1 \\ a_{21}x_1 + a_{22}x_2 + a_{23}x_3 &= b_2 \\ a_{31}x_1 + a_{32}x_2 + a_{33}x_3 &= b_3 \end{aligned}\right\}$$

in matrix notation as $[A](X) = (B)$. Let $[A]=[L][U]$, then we get

$[L][U](X)=(B).$

Substituting $[U](X)=(Z)$ in Eq we obtain

$[L](Z)=(B)$.

Now, Eq is equivalent to

$$\left.\begin{aligned} l_{11}z_1 &= b_1 \\ l_{21}z_1 + l_{22}z_2 &= b_2 \\ l_{31}z_1 + l_{32}z_2 + l_{32}z_3 &= b_3 \end{aligned}\right\}$$

The first of these equations gives us z_1 By knowing z_1, we use the second equation for z_2; then the third equation can be used to compute x_1, x_2 and x_3 from equation [U] (x) = (z) or from

$$\begin{bmatrix} 1 & l_{12} & u_{13} \\ 0 & 1 & u_{23} \\ 0 & 0 & 1 \end{bmatrix} \begin{pmatrix} x_1 \\ x_2 \\ x_3 \end{pmatrix} = \begin{pmatrix} z_1 \\ z_2 \\ z_3 \end{pmatrix}$$

This method is also known as Cholesky's reduction method. This technique is widely used in the numerical solutions of partial differential equations.

This method is very popular from a computer programming point of view, since the storage space reserved for matrix [A] can be used to store the elements of [L] and [U] at the end of computation.

It may be noted that this method fails if any $a_{ii} = 0.$ In this case, the system in singular. In order to familiarize ourselves with Crout's reduction method, we consider the following examples.

Example

Solve the following system of equations:

$$5x_1 - 2x_2 + x_3 = 4$$

$$7x_1 + x_2 - 5x_3 = 8$$

$$3x_1 + 7x_2 + 4x_3 = 10$$

using Crout's reduction method and hand computation.

Solution

Let the coefficient matrix [A] be written as [L] [U]. Thus,

$$\begin{bmatrix} l_{11} & 0 & 0 \\ l_{21} & l_{22} & 0 \\ l_{31} & L_{32} & L_{33} \end{bmatrix} \begin{bmatrix} 1 & u_{12} & u_{13} \\ 0 & 1 & u_{23} \\ 0 & 0 & 1 \end{bmatrix} = \begin{bmatrix} 5 & -2 & 1 \\ 7 & 1 & -5 \\ 3 & 7 & 4 \end{bmatrix}$$

Step 1:

Multiply all the rows of [L] by the first column of [U] to get

$$l_{11} = 5,\ l_{21} = 7,\ l_{31} = 3 .$$

Step II:

Multiply the first row of [L] by the second and third columns of [U]:

$$l_{11}u_{12} = -2\ ,\ l_{11}u_{13} = 1$$

Using Eq. we get

$$u_{12} = -\frac{2}{5},\ u_{13} = \frac{1}{5}.$$

Step III:

Multiply the second and third rows of [L] by the second column of [U]. Using *Eqs.* We get

$$\left.\begin{aligned} l_{21}u_{12} + l_{22} = 1 \quad or\ l_{22} = 1 + \frac{14}{5} = \frac{19}{5} \\ l_{31}`u_{12} + l_{32} = 7 \quad or\ l_{32} = 7 + \frac{6}{5} = \frac{41}{5} \end{aligned}\right\}$$

Step IV:

Multiply the second row of [L] by the third column of [U], which yields $l_{21}u_{13} + l_{22}i_{23}$ = - 5. Using *Eqs.*, we get

$$\frac{19}{5}u_{23} = -5 - \frac{7}{5}$$

Therefore,

$$u_{23} = -\frac{32}{19}.$$

Step V:

Finally, multiply the third row of [L] with the third column of [U], to obtain

$$l_{31}u_{13} + l_{32}\,u_{23} + l_{33} = 4.$$

Using *Eqs.*we get

$$l_{32} = \frac{327}{19}.$$

Thus, the given system of equations takes the form $[L][U][X]=(B)$. That is,

$$\begin{bmatrix} 5 & 0 & 0 \\ 7 & \frac{19}{5} & 0 \\ 3 & \frac{41}{5} & \frac{327}{19} \end{bmatrix} \begin{bmatrix} 1 & -\frac{2}{5} & \frac{1}{5} \\ 0 & 1 & -\frac{32}{19} \\ 0 & 0 & 1 \end{bmatrix} \begin{pmatrix} x_1 \\ x_2 \\ x_3 \end{pmatrix} = \begin{pmatrix} 4 \\ 8 \\ 10 \end{pmatrix}$$

Let [U] (x) = (z), then

[L](Z) = $(4810)^T$

or

$$\begin{bmatrix} 5 & 0 & 0 \\ 7 & \frac{19}{5} & 0 \\ 3 & \frac{41}{5} & \frac{327}{19} \end{bmatrix} \begin{pmatrix} z_1 \\ z_2 \\ z_3 \end{pmatrix} = \begin{pmatrix} 4 \\ 8 \\ 10 \end{pmatrix}$$

Which gives

$$z_1 = \frac{4}{5}, \quad z_2 = \frac{12}{19}, \; z_3 = \frac{46}{327}.$$

Utilizing these values of z, Eq becomes

$$\begin{bmatrix} 1 - \frac{2}{5}\frac{1}{5} \\ 0\ 1 - \frac{32}{19} \\ 0\ \ 0\ \ 1 \end{bmatrix} \begin{pmatrix} x_1 \\ x_2 \\ x_3 \end{pmatrix} = \begin{pmatrix} \frac{4}{5} \\ \frac{12}{19} \\ \frac{46}{327} \end{pmatrix}$$

By back substitution method, we obtain

$$x_3 = \frac{46}{327}, \ x_2 = \frac{284}{327}, \ x_1 = \frac{366}{327}.$$

This is the required solution.

Jacobi's Method

Jacobi's method is an iterative method, where the initial approximate solution to a given system of equations is estimated and improved towards the exact solution in an iterative way. In general, when the coefficient matrix of the system of equations is a sparse matrix (many elements are zero), iterative methods have a definite advantage over direct methods in respect to computer memory. Such sparse matrices arise in computing the numerical solution of partial differential equations.

To illustrate Jacobi's method, let us consider a linear system given by

$$\left.\begin{array}{l} a_{11}x_1 + a_{12}x_2 + \ldots + a_{1n}\,x_n = b_1 \\ \qquad\qquad\quad . \\ a_{21}x_1 + a_{22}x_2 + \ldots + 1_{2n}\,x_n = b_2 \\ \qquad\qquad\quad . \\ a_{n1}x_1 + a_{n2}x_2 + \ldots + a_{nn}\,x_n = b_n \end{array}\right\}$$

In this method, we assume that for the coefficient matrix [A], the modulus of the diagonal element exceeds the sum of the off-diagonal elements. We also assume that the diagonal element vanishes, and that the equations can always be rearranged to satisfy this condition. Now the system can be written as:

$$\left.\begin{aligned} x_1 &= \frac{b_1}{a_1} x_2 - \ldots - \frac{a_{in}}{a_{11}} x_n \\ x &= \frac{b_2}{a_{22}} - \frac{a_{21}}{a_{22}} x_1 - \ldots - \frac{a_{2n}}{a_{22}} x_n \\ x_n &= \frac{b_n}{a_{nn}} - \frac{a_{n1}}{a_{nn}} x_1 - \ldots - \frac{a_n(n-1)}{a_{nn}} x_{n-1} \end{aligned}\right\}$$

We shall take this solution vector $(x_1, x_2, \ldots, x_n)^T$ as a first approximation to the exact solution of a system. For convenience, let us denote the first approximation vector by $(x_1^1, x_n^1, \ldots, x_n^1)$ after taking (0, 0, …, 0) as an initial starting vector. Substituting this first approximation in the right-hand side of system, we obtain the second approximation to the given system in the form

$$\left.\begin{aligned} x_1^2 &= \frac{b_1}{a_{11}} - \frac{a_{12}}{a_{11}} x_2^1 - \ldots - \frac{a_{in}}{a_{11}} x_n^1 \\ x_2^2 &= \frac{b_2}{a_{22}} - \frac{a_{21}}{a_{22}} x_1^1 - \ldots - \frac{a_{2n}}{a_{22}} x_n^1 \\ x_n^2 &= \frac{b_n}{a_{nn}} - \frac{a_{n1}}{a_{nn}} x_1^1 - \ldots - a_{\frac{n(n-1)}{a_{nn}} x_{n-1}^1} \end{aligned}\right\}$$

The second approximation is substituted into the right-hand side of *Eqs.* to obtain the third approximation and so on. This process is repeated and $(r+1)$, the approximation, is calculated from

$$\left.\begin{aligned} x_1^{(r+1)} &= \frac{b_1}{a_{11}} - \frac{a_{12}}{a_{11}} x_2^{(r)} - \ldots - \frac{a_{in}}{a_{11}} x_n^{(r)} \\ x_2^{(r+1)} &= \frac{b_2}{a_{22}} - \frac{a_{21}}{a_{22}} x_1^{(r)} - \ldots - \frac{a_{2n}}{a_{22}} x_n^{(r)} \\ x_n^{(r+1)} &= \frac{b_n}{a_{nn}} - \frac{a_{n1}}{a_{nn}} x_1^{(r)} - \ldots - a_{\frac{n(n-1)}{a_{nn}} x_{n-1}^{(r)}} \end{aligned}\right\}$$

Briefly, we can rewrite $Eqs.$ as

$$x_i^{r+1} = \frac{b_i}{a_{ii}} - \sum_{\substack{j=1 \\ j \neq i}}^{n} \frac{a_{ii}}{a_{ii}} x_j^r, \quad r = 1,2,\ldots, i = 1\ ,2,\ldots,n$$

This method is called Jacobi's iterative method. It is also known as the method of simultaneous displacement, since no element of x_i^{r+1} is used in this iteration until every element is computed.

A sufficient condition for convergence of the iterative solution to the exact solution is

$$|a_{ii}| > \sum_{\substack{j=1 \\ j \neq i}}^{n} |a_{ij}|, \quad i = 1\ ,2,\ldots,n$$

When this condition (diagonal dominance) is true, the Jacobi method converges.

Example

Find the solution to the following system of equations

$83x + 11y - 4z = 95$

$7x + 52y + 13z = 104$

$3x + 8y + 29z = 71$

using Jacobi iterative method for the first five iterations.

Solution

A t first, we rewrite the given system in the form

$$\left.\begin{aligned} x &= \frac{95}{83} - \frac{11}{83}y + \frac{4}{83}z \\ y &= \frac{104}{52} - \frac{7}{52}x - \frac{13}{52}z \\ z &= \frac{71}{29} - \frac{3}{29}x - \frac{8}{29}y \end{aligned}\right\}$$

Taking the initial starting of solution vector as $(0,0,0)^T$, from $Eq.$ we have the first approximation as:

$$\begin{pmatrix} x^{(1)} \\ y^{(1)} \\ z^{(1)} \end{pmatrix} = \begin{pmatrix} 1.1446 \\ 2.0000 \\ 2.4483 \end{pmatrix}$$

Now, using Eq., the second approximation is compute from the equations

$$\left.\begin{aligned} x^2 &= 1.1446 - 0.1325y^1 + 0.0482_z^1 \\ y^z &= 2.0 - 0.1346_x^1 - 0.25_z^1 \\ z^2 &= 2.4483 - 0.1035x^1 - 0.2759y^1 \end{aligned}\right\}$$

Substituting Eq. into Eq we get the second approximation as:

$$\begin{pmatrix} x^{(1)} \\ y^{(1)} \\ z^{(1)} \end{pmatrix} = \begin{pmatrix} 0.9976 \\ 1.2339 \\ 1.7424 \end{pmatrix}$$

Gauss-Seidal Iteration Method

This is another well-known iterative method for solving a system of linear equations.

For the Jacobi method, the $(r+1)$ th approximation of the system is given by $Eqs.$ from which we can observe that no element of x_i^{r+1} replaces

x_i^r entirely for the next cycle of computation.

However, for the Gauss-Seidel method, the corresponding elements of x_i^{r+1} replace those of x_i^r as soon as they become available. Hence, it is called the method of successive displacement. The $(r+1)_{\text{th}}$ approximation or iteration for a system using the Gauss-Seidel method is computed from:

$$\left.\begin{aligned} x_1^{(r+1)} &= \frac{b_1}{a_{11}} - \frac{a_{12}}{a_{11}} x_2^{(r)} - \ldots\ldots - \frac{a_{in}}{a_{11}} x_n^{(r)} \\ x_2^{(r+1)} &= \frac{b_2}{a_{22}} - \frac{a_{21}}{a_{22}} x_1^{(r)} - \ldots\ldots - \frac{a_{2n}}{a_{22}} x_n^{(r)} \\ x_n^{(r+1)} &= \frac{b_n}{a_{nn}} - \frac{a_{n1}}{a_{nn}} x_1^{(r)} - \ldots\ldots - a_{\frac{n(n-1)}{a_{nn}} x_{n-1}^{(r)}} \end{aligned}\right\}$$

Thus, the general procedure can be written in the following compact from

$$x_i^{r+1} = \frac{b_i}{a_{ii}} - \sum_{j=1}^{i-1} \frac{a_{ii}}{a_{ii}} x_j^{r+1} - \sum_{j=i+1}^{n} \frac{a_{ii}}{a_{ii}} x_j^r.$$

For all $i=1,2,\ldots,n$ and $r=1,2,\ldots$

To describe the system; in the first equation, we substitute the rth approximation into the right-hand side and denote the result by x_1^{r+1}. In the second equation, we substitute $x_1^{r+1}, x_3^r, \ldots, x_n^r$ and denote the result by x_2^{r+1}.In the third equation, we substitute $(x_i^{r+1}, x_2^{r+1}, x_4^r, \ \ldots \ x_n^r)$ and denote the result x_3^{r+1}. For illustration purposes, we consider the following example:

Example

Find the solution of the following system

$$x_1 - \frac{1}{4}x_2 - \frac{1}{4}x_3 = \frac{1}{2}$$

$$-\frac{1}{4}x_1 + x_2 - \frac{1}{4}x_4 = \frac{1}{2}$$

$$\frac{1}{4}x_1 + x_3 - \frac{1}{4}x_4 = \frac{1}{4}$$

$$-\frac{1}{4}x_2 - \frac{1}{4}x_3 + x_4 = \frac{1}{4}$$

Using the Gauss-Seidel method and performing the first five iterations:

$$\left.\begin{aligned} x_1 &= 0.5 + 0.25x_2 + 0.25x_3 \\ x_2 &= 0.5 + 0.25x_1 + 0.25x_4 \\ x_3 &= 0.25 + 0.25x_1 + 0.25x_4 \\ x_4 &= 0.25 + 0.25x_2 + 0.25x_3 \end{aligned}\right\}$$

Taking $x_2 = x_3 = x_4 = 0$ on the right-hand side of the first equation of system (1), we get $x_1^1 = 0.5$. Taking $x_3 = x_4 = 0$ and the current value of x_i, we get $x_2^1 = 0.5 + (0.25)(0.5) + 0 = 0.625$

from the second equation of system (1). Furthermore, we take $x_4 = 0$ and the current value of x_i:

$$x_3^1 = 0.25 + (0.25)(0.5) + 0 = 0.375$$

from the third equation of the system. Now, using the current values of x_2 and x_3, the fourth equation of system (1) gives us

$$x_4^1 = 0.25 + (0.25)(0.625) + (0.25)(0.375) = 0.5$$

The Gauss-Seidel iterations for the given set of equations can be written as

$$x_1^{r+1} = 0.5 + 0.25x_2^r + 0.25x_3^r$$

$$x_2^{r+1} = 0.5 + 0.25x_1^{r+1} + 0.25x_4^r$$

$$x_3^{r+1} = 0.25 + 0.25x_1^{r+1} + 0.25x_4^r$$

$$x_4^{r+1} = 0.25 + 0.25x_2^{r+1} + 0.25x_3^r$$

Now, by the Gauss-Seidel procedure, the second and subsequent approximations can be obtained.

The Relaxation Method

To explain this iterative method, consider again the system of equations. Let

$$X^p = \left(x_1^p, x_2^p, \ldots, x_n^p\right)^T$$

be the solution vector obtained iteratively after the pth. Let R_i^p denote the residual of the ith equation of system:

$$a_{i1} + a_{i2}x_2 + \ldots + a_{in}x_n = b_i$$

defined as

$$R_i^p = b_i - a_{il}x_1^p - a_{i2}x_2^p - \ldots - a_{in}x_n^p$$

Then, we can improve the solution vector successively by reducing the largest residual to zero at that iteration. This is the basic idea of the relaxation method.

To achieve fast convergence of the procedure, we take all terms to one side and then reorder the equations so that the largest negative coefficient in the equations appear on the diagonal. Now, if at any iteration, R_i is the largest residual in magnitude, then we give denote the increment

$$dx_i = \frac{R_i}{a_{ii}}$$

for x_i, with a_{ii} being the coefficient of x_i. In other words, we change $x_i to \left(x_i + dx_i\right)$ to relax R_i, that is to reduce R_i to zero. We shall illustrate this procedure through an example.

Example

Solve the system of equations:

$6x_1 - 3x_2 + x_3 = 11$

$2x_1 + x_2 - 8x_3 = -15$

$x_1 - 7x_2 + x_3 = 10$

using the relaxation method, starting with the vector (0, 0, 0).

Solution

At first, we transfer all the terms to the right-hand side and re-order the equations, so that the largest coefficient in the equation appears in the diagonal. Thus, we get

$$\left.\begin{aligned} 0 &= 11 - 6x_1 + 3x_2 - x_3 \\ 0 &= 10 - x_1 + 7x_2 - x_3 \\ 0 &= -15 - 2x_1 - x_2 + 8x_3 \end{aligned}\right\}$$

after interchanging the second and third rows. Starting with the initial solution vector (0, 0, 0), that is, taking $x_1 = x_2 = x_3 = 0,$ we find the residuals $R_1 = 11, R_2 = 10, R_3 = -15,$ with R_3 having the largest magnitude. This indicates that the third equation has a greater error and needs immediate attentions for improvement. Thus, we introduce the change of dx_3 in x_3 which is obtained from the formula:

$$dx_3 = -\frac{R_3}{a_{33}} = \frac{15}{8} = 1.875$$

Similarly, we find the new residuals of large magnitude and relax them to zero, and so on. We continue this process, until all the residuals are zero, or very small.

At this stage, we observe that all the residual $R_1\ R_2$ and R_3 are small enough and therefore we may take the corresponding values of x_1 at this iteration as the solution. Hence, the numerical solution to the given system is given by

$$x_1 = 1.0017, \quad x_2 = -1.0, \quad x_3 = 2.0.$$

Matrix Inversion

Consider a system of equations in the form

$$[A](X) = (B)$$

One way of writing its solution is in the form

$$(X) = [A]^{-1}(B)$$

Thus, the solution to the system can also be obtained if the inverse of the coefficient matrix [A] is known. Alternatively, if the product of two square matrices is an identity matrix, that is, if

[A] [B] = [*I*]

then, [B] $= [A]^{-1}$ and $[A] = [B]^{-1}$.

Every square non-singular matrix will have an inverse. Gauss elimination and Gauss-Jordan are popular among many methods available for finding the inverse of a matrix.

Gaussian Elimination Method

For this method, consider the matrix *A*. To find its inverse, we first place an identity matrix, with an order equal to *A*, adjacent to A. This identity matrix is known as the augmented matrix. The inverse of A is then computed in two stages. In the first stage, A is converted into an upper triangular from, using the Gaussian elimination method as discussed in this section. In the second stage, the above upper triangular matrix is reduced to an identity matrix by row transformations. These operations are also performed on the adjacently placed identity matrix. Finally, when A is transformed into an identity matrix, the adjacent matrix gives the inverse of *A*. In order to increase the accuracy of the result, it is essential to employ partial pivoting. To understand the sequence of the steps involved, we consider an example.

Example

Use the Gaussian elimination method to find the inverse of the matrix:

$$A = \begin{bmatrix} 1 & 1 & 1 \\ 4 & 3 & -1 \\ 3 & 5 & 3 \end{bmatrix}$$

Solution

Firstly, we place an identity matrix of the same order adjacent to the given matrix. Thus, the augmented matrix can be written as:

$$\left[\begin{array}{ccc|ccc} 1 & 1 & 1 & 1 & 0 & 0 \\ 4 & 3 & -1 & 0 & 1 & 0 \\ 3 & 5 & 3 & 0 & 0 & 1 \end{array}\right]$$

Stage 1: (reduction to upper triangular form): Let , and denote the first, second and third rows of a matrix. In the first column of Eq. (1), 4 is the largest element, thus interchanging to bring the pivot element 4 to the place of we have the augmented matrix in the form:

$$\left[\begin{array}{ccc|ccc} 4 & 3 & -1 & 0 & 1 & 0 \\ 1 & 1 & 1 & 1 & 0 & 0 \\ 3 & 5 & 3 & 0 & 0 & 1 \end{array}\right]$$

Divide

$$\left[\begin{array}{ccc|ccc} 1 & \frac{3}{4} & -\frac{1}{4} & 0 & \frac{1}{4} & 0 \\ 1 & 1 & 1 & 1 & 0 & 0 \\ 3 & 5 & 3 & 0 & 0 & 1 \end{array}\right]$$

Performing $R_2 - R_1$ ::

$$\left[\begin{array}{ccc|ccc} 1 & \frac{3}{4} & -\frac{1}{4} & 0 & \frac{1}{4} & 0 \\ 0 & \frac{1}{4} & \frac{5}{4} & 1 & -\frac{1}{4} & 0 \\ 3 & 5 & 3 & 0 & 0 & 1 \end{array}\right]$$

Performing $R_3 - 3R_1 \to R_3$ in $Eq.$ which yields:

$$\left[\begin{array}{ccc|ccc} 1 & \frac{3}{4} & -\frac{1}{4} & 0 & \frac{1}{4} & 0 \\ 0 & \frac{1}{4} & \frac{5}{4} & 1 & -\frac{1}{4} & 0 \\ 3 & \frac{11}{4} & \frac{15}{4} & 0 & -\frac{3}{4} & 1 \end{array}\right]$$

Now, looking at the second column for the pivot, the max of $(1/4, 11/4)$ is $11/4$. Therefore, we interchange R_2 and R_3 in Eqs and get:

$$\left[\begin{array}{ccc|ccc} 1 & \frac{3}{4} & -\frac{1}{4} & 0 & \frac{1}{4} & 0 \\ 0 & \frac{11}{4} & \frac{15}{4} & 0 & -\frac{3}{4} & 1 \\ 3 & \frac{1}{4} & \frac{5}{4} & 1 & -\frac{1}{4} & 0 \end{array}\right]$$

Now, divide R_2 by the pivot $a_{22} = 11/4$, to obtain

$$\left[\begin{array}{ccc|ccc} 1 & \frac{3}{4} & -\frac{1}{4} & 0 & \frac{1}{4} & 0 \\ 0 & 1 & \frac{15}{4} & 0 & -\frac{3}{4} & \frac{4}{11} \\ 3 & \frac{1}{4} & \frac{5}{4} & 1 & -\frac{1}{4} & 0 \end{array}\right]$$

Performing $R_3 - (1/4)R_2 \rightarrow R_3$:

$$\left[\begin{array}{ccc|ccc} 1 & \frac{3}{4} & -\frac{1}{4} & 0 & \frac{1}{4} & 0 \\ 0 & 1 & \frac{15}{11} & 0 & -\frac{3}{11} & \frac{4}{11} \\ 0 & 0 & \frac{10}{11} & 1 & -\frac{2}{11} & -\frac{1}{11} \end{array}\right]$$

Finally, we divide R_3 by (10 / 11), thus getting an upper triangular form

$$\left[\begin{array}{ccc|ccc} 1 & \frac{3}{4} & -\frac{1}{4} & 0 & \frac{1}{4} & 0 \\ 0 & 1 & \frac{15}{11} & 0 & -\frac{3}{11} & \frac{4}{11} \\ 0 & 0 & 1 & \frac{1}{10} & -\frac{1}{5} & -\frac{1}{10} \end{array}\right]$$

Stage II (Reduction to an identity matrix): Multiply R_3 $by -$ ¼ and 15/11 respectively and subtract from R_1 and R_2 we get:

$$\left[\begin{array}{ccc|ccc} 1 & \frac{3}{4} & 0 & \frac{11}{40} & \frac{1}{5} & -\frac{1}{40} \\ 0 & 1 & 0 & -\frac{3}{2} & 0 & \frac{1}{2} \\ 0 & 0 & 1 & \frac{11}{10} & -\frac{1}{5} & -\frac{1}{10} \end{array}\right]$$

Finally, performing $R_1 - (3/4)\, R_2 \rightarrow R_1$ in $Eq.$ we obtain:

$$\left[\begin{array}{ccc|ccc} 1 & 0 & 0 & \frac{7}{5} & \frac{1}{5} & -\frac{2}{5} \\ 0 & 1 & 0 & -\frac{3}{2} & 0 & \frac{1}{2} \\ 0 & 0 & 1 & \frac{11}{10} & -\frac{1}{5} & -\frac{1}{10} \end{array}\right]$$

Thus, we have:

$$A^{-1} \begin{bmatrix} \frac{7}{5} & \frac{1}{5} & -\frac{2}{5} \\ -\frac{3}{2} & 0 & \frac{1}{2} \\ \frac{11}{10} & -\frac{1}{5} & -\frac{1}{10} \end{bmatrix}$$

We can easily check this, i.e., $[A]\left[A^{-1}\right]=[I]$.

Gauss-Jordan Method

This method is similar to the Gaussian elimination method, with the essential difference that the stage of reducing the given matrix to upper triangular form is not needed. However, the given matrix can be directly reduced to an identity matrix using elementary row transformations. This technique is illustrated in the following example.

Example

Find the inverse of

$$A = \begin{bmatrix} 1 & 1 & 1 \\ 4 & 3 & -1 \\ 3 & 5 & 3 \end{bmatrix}$$

using the Gauss-Jordan method.

Solution

Let R_1, R_2 and R_3 denote the first, second and third rows of the matrix. We place an identity matrix adjacent to the given matrix as the first step and the resulting augmented is given by

$$\left[\begin{array}{ccc|ccc} 1 & 1 & 1 & 1 & 0 & 0 \\ 4 & 3 & -1 & 0 & 1 & 0 \\ 3 & 5 & 3 & 0 & 0 & 1 \end{array}\right].$$

Performing $R_2 - 4R_1 \rightarrow R_2$, we get:

$$\left[\begin{array}{ccc|ccc} 1 & 1 & 1 & 1 & 0 & 0 \\ 4 & -1 & -5 & -4 & 1 & 0 \\ 3 & 5 & 3 & -3 & 0 & 1 \end{array}\right]$$

Carrying out the further operations of $R_2 + R_1 \rightarrow R_1$ and $R_3 + 2R_2 \rightarrow R_3$, we arrive at

$$\left[\begin{array}{ccc|ccc} 1 & 0 & -4 & -3 & 1 & 0 \\ 4 & -1 & -5 & -4 & 1 & 0 \\ 0 & 0 & -10 & 11 & 2 & 1 \end{array}\right]$$

Now, dividing the third row by – 10, we get

$$\left[\begin{array}{ccc|ccc} 1 & 0 & -4 & -3 & 1 & 0 \\ 0 & -1 & -5 & -4 & 1 & 0 \\ 0 & 0 & 1 & \frac{11}{10} & -\frac{1}{5} & -\frac{1}{10} \end{array}\right]$$

Further, we perform $R_1 + 4R_3 \to R_1$, and $R_2 + 5R_3 \to R_2$ to get

$$\left[\begin{array}{ccc|ccc} 1 & 0 & 0 & \frac{7}{5} & \frac{1}{5} & \frac{2}{5} \\ 0 & -1 & 0 & \frac{3}{2} & 1 & -\frac{1}{2} \\ 0 & 0 & 1 & \frac{11}{10} & -\frac{1}{5} & -\frac{1}{10} \end{array}\right]$$

Finally, multiplying $R_2\ by -1,$ we obtain

$$\left[\begin{array}{ccc|ccc} 1 & 0 & 0 & \frac{7}{5} & \frac{1}{5} & -\frac{2}{5} \\ 0 & 1 & 0 & -\frac{3}{2} & 0 & -\frac{1}{2} \\ 0 & 0 & 1 & \frac{11}{10} & -\frac{1}{5} & -\frac{1}{10} \end{array}\right]$$

Hence, we have:

$$A^{-1} = \begin{bmatrix} \frac{7}{5} & \frac{1}{5} & -\frac{2}{5} \\ -\frac{3}{2} & 0 & \frac{1}{2} \\ \frac{11}{10} & -\frac{1}{5} & \frac{1}{10} \end{bmatrix}$$

It can be easily verified that [A] $\left[A^{-1}\right]=[I]$.

CURVE FITTING

Introduction

In science and engineering, we often come across experimental observations involving two variables, such as concentration and time, pressure and temperature and load and deflection. The important question is: can we find a law or a relation connecting them. Suppose we wish to find a relation connecting load p (Kg) and the deflection of a beam y (cm), given the maximum deflection y for various loads p as shown in fig. Typically, a linear relation of the form $y=mx+c$ is used. In another situation, the graph may indicate a parabolic or an exponential curve. In fact, Kepler formulated the third law of planetary motion, T = $cx^{3/2}$, from observed data pairs (x, t) of various planets, where T is the period, and x is the mean distance of the planet from the sun. In general, the problem of finding an equation of an approximating curve that passes through as many points as possible, is called curve fitting. Hence, the basic problem is to find an equation of the curve of best fit.

The method of group averages, the least square method and method of moments are some of the well-known methods based on the principles of elementary statistics. Amongst them, the least square method gives a unique best fit and is highly recommended.

Method of Group Averages

Suppose we have n sets of observations in an experiment $(x_1, y_1),(x_2,y_2),\ldots\ldots(x_n,y_n)$. Consider the straight line $y = mx + c$.

Since the equation contains two constants, m and c, we require two equations in m and c to determine the value of these constants.

From the given data, when $x = x_1$, we note that the observed value of y is y_1. But, from the assumed empirical law, the expected value

y is $mx_1 + c$. Now, we define the residual, e, as the difference between the observed value of y and its expected value from the same x_1, and write

$$e_1 = y_1 - (mx_1 + c).$$

Similarly, for other sets of data, we construct the residuals as:

$$e_2 = y_2 - (mx_2 + c)$$

$$e_n = y_n - (mx_n + c).$$

These expressions clearly indicate that some of the residuals may be positive and some of them may be negative . However, the method of group averages state that the sum of all the residuals is zero. That is:

$$\sum_{i=1}^{n} e_1 = [y_1 - (mx_1 + c) + \left[y_2 - (mx_2 + c)\right] + \left[y_j - (mx_j + c)\right] = 0$$

and

$$\sum_{i=1}^{n} e_1 = \left[y_{j+1} - (mx_{j+1} + c)\right] + \left[y_{j+2} - (mx_{j+2} + c)\right] + \left[y_n - (mx_n + c)\right] = 0.$$

On simplification, the equations become:

$$(y_1 + y_2 + \ldots + y_n) - jc - m(x_1 + x_2 + \ldots\ldots + x_j) = 0$$

$$(y_{j+1} + y_{j+2} + \ldots\ldots + y_n) - (n-j)c - m(x_{j+1} + x_{j+2} + \ldots\ldots + x_n) = 0,$$

which can be rewritten as

$$\frac{y_1 + y_2 + \ldots . y_j}{j} = \frac{m(x_1 + x_2 + \ldots\ldots + x_j)}{j} + c$$

And

$$\frac{y_{j+1}+y_{j+2}+....y_n}{n-j}=\frac{m\left(x_{j+1}+x_{j+2}+.....+x_n\right)}{n-j}+c.$$

These equations are sufficient to determine the unknown constants m and c. Obviously, there is no unique way of dividing the observations into two group and hence, different choices will give different values for m and c. This is the major drawback of the method of group averages. We shall illustrate this method through the following examples.

Example

Use the method of group averages and find a curve of the form that fits the following data:

x	10	20	30	40	50	60	70	80
y	1.06	1.33	1.52	1.68	1.81	1.91	2.01	2.11

Solution

The required curve is of the form

$$y = mx^n.$$

Taking logarithms on both sides, we get

$\log_{10} y = Y,$ $\log_{10} m = c$ and $\log_{10} x = X,$ then Eq. becomes

$$Y = nX + c.$$

Now, we take the logarithm of the given pairs of data and divide them in two group, such that the first group contains the first four values and the remaining constitutes the second group, as follows:

Group 1

x	y	$y = \log_{10} y$	x= $\log_{10} x$
10	1.06	0.0253	1.0000
20	1.33	0.1239	1.3010
30	1.52	0.1818	1.4771
40	1.68	0.2253	1.6021
$\sum y = 0.5563$		$\sum x = 5.3802$	

Group 2

x	y	$y = \log_{10} y$	x= $\log_{10} x$
50	2.81	0.02577	1.6990
60	1.91	0.2810	1.7782
70	2.01	0.3032	1.8451
80	2.11	0.3243	1.9031
$\sum y = 1.1662$		$\sum x = 7.2254$	

Now, using the method of group averages, we determine the constants n and c from ^{eqs} and as follows:

$$\frac{1}{4}\left(y_1 + y_2 + y_3 + y_4\right) = \frac{n}{4}\left(x_1 + x_2 + x_3 + x_4\right) + c$$

$$\frac{1}{4}\left(y_5 + y_6 + y_7 + y_8\right) = \frac{n}{4}\left(x_5 + x_6 + x_7 + x_8\right) + c$$

Substituting the values from the above tables, we get:

$$\frac{0.5563}{4} = n\frac{53802}{4} + c$$

$$\frac{1.1662}{4} = n\frac{7.2254}{4} + c$$

On simplification, we have

4c + 5.3802n – 0.5563 = 0

4c + 7.2254n - 1.1662 = 0.

Solving Eqs. and by the method of cross multiplication we obtain

$c = -0.3055, \quad n = 0.3305.$

Therefore,

$m = antilog\, c = 0.4949$.

Hence, the required curve fit is:

$y = 0.4949x^{0.3305}$.

Example

Using the method of group averages, find a curve of the form $y = a + bc^2$ that fits the following data:

x	20	30	35	40	45	50	
y	10.0	11.0	11.8	12.4	13.5	14.4	

Solution

Let the required equation of fit be

$y = a + bx^2$,

Setting $x^2 = X$, the above equation becomes:

$y = bX + a$.

Now, the given data can be grouped and tabulated as shown:

Group 1

y	x	$x = x^2$
10.0	20	400
11.0	30	900
11.8	35	1225
$\sum y$ = 32.8		$\sum x$ = 2525

Group 2

y	x	$x = x^2$
12.4	40	1600
13.5	45	2025
14.4	50	2500
$\sum y = 40.3$		$\sum x = 6125$

To determine the unknown constants a and b using the method of group averages, we have equations:

$$\frac{32.8}{3} = \frac{2525b}{3} + a$$

$$\frac{40.3}{3} = \frac{6125b}{3} + a.$$

On simplification, we have

$$2525b + 3a - 32.8 = 0$$

$$6125b + 3a - 40.3 = 0.$$

Solving Eqs, we get

$$b = 2.083x10^{-3}, \ a = 9.1799.$$

Hence, the required curve is:

$$y = 9.1799 + 2.083x10^{-3}x^2.$$

Example

Using the method of group averages, fit a curve of the form

$$y = \frac{x}{a + bx}$$

for the following data:

x	8	10	15	20	30	40
y	13	14	15.4	16.3	17.2	17.8

Solution

Let the equation of the required fit be:

$$y = \frac{x}{a + bx}$$

Rewriting, we get:

$$\frac{1}{y} = \frac{a + bx}{x} = \frac{a}{x} + b$$

Setting

$$\frac{1}{y} = y, \frac{1}{x} = X,$$

the above equation becomes

$$Y = aX + b.$$

Which is linear in X and Y. Now, we shall tabulate the given data into two groups:

Group 1

x	$x = 1/x$	y	$y = 1/y$
8	0.1230	13	0.769
10	0.1000	14	0.0714
15	0.0667	15.4	0.0649
$\Sigma X = 0.2917$		$\Sigma y = 0.2132$	

Group 2

x	$x = 1/x$	y	$y = 1/y$
20	0.0500	16.3	0.0614
30	0.0333	17.2	0.0581
40	0.0250	17.8	0.0562
$\Sigma X = 0.1083$		$\Sigma y = 0.1757$	

Using the method of group averages, we have

$$\frac{02132}{3} = a\frac{0.2917}{3} + b$$

And

$$\frac{01757}{3} = a\frac{01083}{3} + \text{ b.}$$

On simplification, we get

$$0.2917a + 3b - 0.2132 = 0$$

$$0.1083a + 3b - 0.1757 = 0 .$$

We obtain

$$a = 0.2045, \quad b = 0.0512 .$$

Hence, the required curve is

$$y = \frac{x}{0.2045 + 0.0512x} .$$

The Least Squares Method

When fitting a curve to a given set of data, the method of least squares is considered to give the best fit, with prior knowledge about the shape of the curve. Suppose $(x_1, y_1), (x_2, y_2) \ldots\ldots (x_n, y_n)$ is a set of n observations in an experiment. Let us assume that:

$$Y = f(x)$$

is a relationship between x and y. When $x = x_1$, the observed value of y is y_1 and the expected value of y from Eqs is $f(x_1)$. Then the residual is defined by

$$e_1 = y_1 - f(x_1).$$

Similarly, we can define all other residuals $e_2, e_3, \ldots\ldots, e_n$ as:

$$e_2 = y_2 - f(x_2)$$

$$e_n = y_n - f(x_n).$$

It is clear from the difference of the residuals that some may be positive and some may be negative. In order to give equal importance to both positive and negative residuals, it is better to consider the sum of the square of the residuals, rather than sum of the residuals, as in the method of group averages. Thus, we consider:

$$E = e_1^2 + e_2^2 + \ldots . + e_n^2$$

$$= [y_1 - f(x_1)]^2 + [y_2 = f(x_2)]^2 + \ldots\ldots\ldots [y_n - f(x_n)]^2.$$

Here, the quantity E is a measure of how well the curve $y = f(x)$ fits the given set of observations as a whole. Therefore, E will be zero if and only if all the points of the observations lie on the curve given by Eq. The value of E decreases depending on the closeness of the observed data to the curve assumed. Hence, the best representative curve to the given set of observations is one for which E, the sum of the squares of the residuals, is minimum This concept is known as the principle of least squares. In general, we consider straight lines, parabolas or an exponential curves for fitting the given data.

Fitting a Straight Line

Suppose $(x_1, y_1), (x_2, y_2), ..., (x_n y_n)$ are a set of n observations in an experiment and we wish to fit a straight line

$$y = ax + b$$

to these observations. That is, we have to determine the constants a and b, using the principal of least squares.

For any $x_i (i = 1, 2, ..., n)$, the expected value of y is $ax_i + b$, while the observed value of y is y_i Hence, the residuals are:

$$e_1 = Y_I - (ax_i + b), \quad for \ \ i = 1, 2, ..., n.$$

Now, the sum of the squares of the residuals is given as:

$$E = \sum_{i=1}^{n} \left[y_i - (ax_i + b) \right]^2,$$

where E is a function of the parameters a and b. The necessary conditions for E to be minimum give:

$$\frac{\partial E}{\partial a} = \frac{\partial E}{\partial b} = 0.$$

In other words, the first condition yields:

$$2x_1(y_1 - ax_1 - b) + 2x_2(y_2 - ax_2 - b) + ... + 2x_n(y_n - ax_n - b) = 0.$$

Or:

$$a\sum_{i=1}^{n} x_i^2 + b\sum_{i=1}^{n} x_i = \sum_{i=1}^{n} x_i y_i.$$

Similarly, the second condition

$$\frac{\partial E}{\partial b} = 0$$

gives us:

$$(y_1 - ax_1 - b) + (y_2 - ax_2 - b) + \ldots\ldots + (y_n - ax_n - b) = 0.$$

That is,

$$a\sum_{i=1}^{n} x_i + nb = \sum_{i=1}^{n} y_i.$$

Equations and are called normal equations, which when solved, give the values of the constants a and b.

To compute *E,* the sum of the squares of the residuals, $Eq.$ can be rewritten as

$$\sum_{i=1}^{n} (y_1 - ax_1 - b)^2 = \sum_{i=1}^{n} \left[y_i^2 - 2ax_i y_i - 2by_i + (ax_i + b)^2 \right].$$

Now, using Eq., we obtain a convenient formula to compute E as:

$$E = \sum_{i}^{2} - \text{a} \sum x_i y_i - b \sum y_i$$

.

Example

Fit a parabola to the following data using the method of least squares.

x	1.0	1.2	1.4	1.6	1.8	2.0
y	0.98	1.40	1.86	2.55	2.28	3.20

Solution

Let the equation of the curve of parabolic fit be:

$$y = ax^2 + bx + c$$

The corresponding normal equations are

$$\left.\begin{aligned} a\sum x_i^2 + b\sum x_i^3 + c\sum x_i^2 &= \sum x_i^2 y_i \\ a\sum x_i^3 + b\sum x_i^2 + c\sum x_i &= \sum x_i y_i \\ a\sum x_i^2 + b\sum x_i + nc &= \sum y_i \end{aligned}\right\}$$

From the given data, we substitute these summations into the normal equations to get

$$39.9664a + 23.4b + 14.2c = 33.3568$$

$$23.4a + 14.2b + 9.0c = 19.848$$

$$14.2a + 9.0b + 6c = 12.27.$$

Solving Eqs. We obtain:

$$a = 0.1875, \quad b = 2.6239, \quad c = -1.4471$$

$$\log_{10} y = \log_{10} a + b\log_{10} x.$$

Let

$$\log_{10} y = Y, \qquad \log_{10} a = A, \quad \log_{10} \mathrm{x} = \mathrm{X}.$$

Then the above equations reduce to:

$$Y = A + bX.$$

Which is linear in $Y \, and \, X$. To illustrate this method, we consider an example.

Example

By using the method of least squares, find a relation of the form $y = ax^b$, that fits the data:

x	2	3	4	5
y	27.8	62.1	110	161

Solution

Let the equation of fit be

$y = ax^b$.

Taking the logarithm on both sides, we get

$\log_{10} y = \log_{10} a + \text{b} \log_{10} x$,

which is of form

$Y = A + bX = bX + A$

where,

$\log_{10} y = Y, \quad \log_{10} a = \text{A}, \quad \log_{10} x = X$

Now, the data for the modified variables X and Y is:

x	0.3010	0.4771	0.6021	0.6990
y	1.4440	1.7931	2.0414	2.2068

The corresponding normal equations are

$b\sum X_i^2 + A\sum X_i = \sum X_i Y_i$

$b\sum X_i + nA = \sum Y_i$.

From the above data, we have:

X	Y	XY	X^2
0.3010	1.4440	0.4346	0.0906
0.4771	1.7931	0.8555	0.2276
0.6021	1.7931	1.2291	0.3625
0.6990	2.2068	1.5426	0.4886
$\sum X_i = 2.0792$	$\sum y_i = 7.4853$	$\sum X_i y_i = 4.0618$	$\sum X_i^2 = 1.1693$

Substituting these summations into the normal equations we get

Solving these equations, we obtain $b = 1.9311, A = 0.8678$. Therefore, $a = antilog\, 0.8678 = 7.3575$. Hence, the required fit for the given data is:

$y = 7.375 x^{1.9311}$

Fitting an Exponential Curve

Suppose we have a set of n observations $(x_i, y_i), i = 1, 2, \ldots\ldots, n$ in an experiment and we wish to fit an exponential curve of the form

$$y = ae^{bx}$$

to these observation. At first, we take the logarithms on both sides to get

$$\log_{10} y = \log_{10} a + bx \log_{10} e.$$

Let the above equation become

$$Y = Bx + A.$$

Thus, it is equivalent to fitting a straight line. The procedure is illustrated through the following example.

Example

Using the principal of least squares, fit an equation of the form $y = ae^{bx}$ to the data:

x	1	2	3	4
y	1.65	2.70	4.50	7.35

Solution

Let

$$y = ae^{bx}$$

be the desired fit to the given data. Taking logarithms of both sides, we have

$$\log_{10} y = \log_{10} a + bx \log_{10} e.$$

Suppose that

$$\log_{10} y = Y, \qquad \log_{10} a = A, \qquad b \log_{10} e = B.$$

Then the above equations assume the form:

$Y = BX + A.$

The least square normal equations are:

$$B\sum x_i^2 + A\sum x_i = \sum x_i y_i$$

$$B\sum x_i + 4A = \sum y_i \, .$$

From the given data, we have:

X	Y	XY	X^2
0.3010	1.4440	0.4346	0.0906
0.4771	1.7931	0.8555	0.2276
0.6021	1.7931	1.2291	0.3625
0.6990	2.2068	1.5426	0.4886
$\sum X_i = 2.0792$	$\sum y_i = 7.4853$	$\sum X_i y_i = 4.0618$	$\sum X_i^2 = 1.1693$

Substituting these summations into the normal equations we get:

$$30B + 10A = 6.5051$$

$$10B + 4A = 2.1684 \, .$$

Solving these equations, we obtain B = 0.2168, A = 0. Now

B = b $\log_{10} e = 0.4343b = 0.2168,$

Therefore,

$$B = 0.4992, \quad a = 1.$$

Hence, the required equation of fit is

$$y = e^{0.4992x} \, .$$

Method of Moments

Consider a set of n observations $(x_i, y_i),\ i = 1, 2, \ldots, n$ in an experiment, such each x_i is equally spaced. That is:

$$x_2 - x_1 = x_3 - x_2 = \ldots.. = x_n - x_{n-1} = \Delta x$$

For such a set of observations, we shall define the observed moments as follow:

$$\left.\begin{aligned} \mu_1 &= \textit{The first moment} = \sum y_i\, \Delta x = \Delta x \sum y_i \\ \mu_2 &= \textit{The second moment} = \sum x_i y_i\, \Delta x = \Delta x \sum x_i y_i \\ \mu_3 &= \textit{The third moment} = \sum x_i^2\, y_i\, \Delta x = \Delta x \sum x_i^2\, y_i \end{aligned}\right\}$$

and so on. These moments are known as moments of the observed values of y. We shall also define the expected moments of the computed values of y, assuming that $y = f(x)$ is the equation of the curve fitting the given data, as:

$$\left.\begin{aligned} \gamma_1 &= \text{The first moment of} = \int y\, dx = \int f(x)\, dx \\ \gamma_2 &= \text{The second moment} = \int xy\, dx = \int xf(x)\, dx \\ \gamma_3 &= \text{The third moment} = \sum x_i^2\, y_i \Delta x = \Delta x \sum x_i^2\, y_i \end{aligned}\right\}$$

The method of moments states that the moments of the observed y's are respectively equal to the expected moments. That is,

$$\mu_1 = \gamma_1,\ \mu_2 = \gamma_2\ \mu_3 = \gamma_3$$

and so on. The observed moments can be easily calculated from the given data, while the moments of the expected y's can be computed as follows.

$$\left.\begin{aligned} \gamma_1 &= \int y\, dx = \int_{x_1+\left(\frac{\Delta x}{2}\right)}^{x_n+\left(\frac{\Delta x}{2}\right)} f(x)\, dx \\ \gamma_2 &= \int xy\, dx = \int_{x_1+\left(\frac{\Delta x}{2}\right)}^{x_n+\left(\frac{\Delta x}{2}\right)} xf(x)\, dx \\ \gamma_3 &= \int x^2\, y\, dx = \int_{x_n+(\Delta x/2)}^{x_n+(\Delta x/2)} x^2 f(x)\, dx \end{aligned}\right\}$$

From the principal of the method of moments, we have the following set of equations:

$$\left.\begin{aligned} \Delta x \sum y_1 &= \int_{x_1+\left(\frac{\Delta x}{2}\right)}^{x_n+\left(\frac{\Delta x}{2}\right)} xf(x)\,dx \\ \Delta x \sum x_i\, y_i &= \int_{x_1+\left(\frac{\Delta x}{2}\right)}^{x_n+\left(\frac{\Delta x}{2}\right)} xf(x)\,dx \\ \Delta x \sum x_i^2 y_i &= \int_{x_1+\left(\frac{\Delta x}{2}\right)}^{x_n+\left(\frac{\Delta x}{2}\right)} x^2 f(x)\,dx \end{aligned}\right\}$$

Given the form of $f(x)$, we can use Eqs to determine the unknown constant in $f(x)$. This method is illustrated through the following examples

Example

Fit a straight line of the form $y = ax + b,$ to the following data by the method of moments:

x	2	3	4	5
y	27	40	55	68

Solution

Since we have to determine two constants for the required fit, it is enough to compute the first-two moments. In this example, we note that $\Delta x = 1$. Thus the observed moments are:

$$\mu_1 = \Delta x \sum_{i=1}^{4} y_i = 1(27 + 40 + 55 + 68) = 190$$

$$\mu_2 = \Delta x \sum_{i=1}^{4} x_i y_i = 1(54 + 120 + 220 + 340) = 734$$

Now, we shall compute the expected moments for the given y. Therefore,

$$\gamma_1 = \int_{1.5}^{5.5} (ax + b)\, dx = \left[a\frac{x^2}{2} + bx \right]_{1.5}^{5.5} = 14a + 14b$$

.

The second moment of the expected value of y is

$$\gamma_2 = \int xy\, dx \int_{1.5}^{5.5} x(ax + b)\, dx = \left[a\frac{x^2}{2} + b\frac{x^2}{2} \right]_{1.5}^{5.5} = \frac{163}{3}a + 14b$$

.

Using the method of moments, we have:

$$14a + 4b = 190, \; \frac{163}{3}a + 14b = 734.$$

Solving these equations, we get $a = 12.9375, b = 22188$ hence, the required straight line fit is

$$y = 12.9375x + 2.2188$$

.

Example

Fit a parabola of the form $y = a + bx + cx^2$ to the data given below, using the method of moments.

x	3	4	5	6	7
y	31.9	34.6	33.8	27.0	31.6

Solution

Since we have to fit a parabola of the form $y = a + bx + cx^2$ to the given data, we have to determine three constants. Therefore, it necessary to compute the first-three moments. In this problem, the x's are equally spaced and we also note that $\Delta x = 1$.

Now, we compute the first moment as:

$\mu_1 = The\ first\ moment\ of\ the\ observed\ y's$:

$\Delta = x\sum y_i$

$= 1[31.9 + 34.6 + 33.8 + 27.0 + 31.6] = 158.9$

$\mu_2 = The\ second\ moment\ of\ the\ observed\ y's$:

$\Delta = x\sum x_i\ y_i$

$= [95.7 + 138.4 + 169.0 + 162 + 221.2] = 786.3$

$\mu_3 = The\ Third\ moment\ of\ the\ observed\ y's$:

$\Delta = x\sum x_i^2\ y_i$

$= [9x31.9 + 16x34.6 + 25x33.8 + 36x27 + 49x31.6] = 4206.1$

We also need to compute the moment of the expected values of y's, given by $y = a + bx + cx^2$. Thus, we have:

$\gamma_1 = The\ first\ moment\ of\ the\ expected\ values\ of\ y$

$$= \int y\,dx = \int_{2.5}^{7.5} \left(a + bx + cx^2\right) dx$$

$$= \left[ax + b\frac{x^2}{2} + c\frac{x^3}{3}\right]_{2.5}^{7.5}$$

$= 5a + 25b + 135.4167c$

Similarly,

$$\gamma_2 = \int xy\,dx \int_{2.5}^{7.5} x\left(a + bx + cx^2\right) dx$$

$$\left[a\frac{x^2}{2}+b\frac{x^3}{3}+c\frac{x^4}{4}\right]_{2.5}^{7.5}$$

$$= 25a + 135.42b + 781.25c$$

and

$$\gamma_3 = \int x^2 y\, dy \int_{2.5}^{7.5} x^2\left(a + bx + cx^2\right)dx$$

$$=\left[a\frac{x^3}{3}+b\frac{x^4}{4}+c\frac{x^5}{5}\right]_{2.5}^{7.5}$$

$$=135.42a + 781.25b + 4726.562c\,.$$

From the method of moments, we have:

$5a+25b+135.42c=158.9,$

25a + 135.42b + 781.25c = 786.3,

135.42a + 781.25b + 4726.562c = 42016.1.

Solving these equations, we obtain

$A=15.7217,\ \ b=7.9427,\ \ c=-0.8734\,.$

Hence, the required parabolic fit to the given data is:

$y=15.7217+7.9427x-0.8734x^2$.

SUMMARY AND DISCUSSION

Many real-life problems in engineering give rise to a system of linear equations. For example, such systems occur in certain applications of statistical analysis and in finding the numerical solution of partial differential equations. It is therefore natural to seek efficient methods for solving these equations numerically. The solution of the system of equations gives n unknown values $x_1, x_2 \ldots, x_m,$ which satisfy the system simultaneously. If $m > n,$ we may not be able to find a solution, in principle, which satisfies all the equations. If $m > n,$ the system will usually have an infinite number of solutions. However, in this chapter, we shall restrict to the case $m = n.$ In this case, $\text{if}|A| \neq 0,$ then the system will have a unique solution, while if $|A| = 0,$ then no solution exists.

Various numerical methods are available for finding the solution of the system of equations and are classified as direct and iterative methods. For the direct methods, we get the solution of the system after performing all the steps involved in the procedure. The direct methods consist of elimination methods and decomposition methods. In this chapter, we will consider the Gaussian and Gauss-Jordan elimination methods. Crout's reduction, also known as Cholesky's reduction, is a decomposition method. Out of the iterative methods, the initial approximate solution is assumed to an improved method working towards the exact solution. All these methods are easily adoptable as computational procedures and can be used to solve even hundreds or more simultaneous linear equations.

Gaussian Elimination Method

For the Gaussian elimination method, the solution to a system of equations is obtained in two stages. In the first stage, the given system of equations is reduced to an equivalent upper triangle system solved using the back substitution procedure, by which we obtain the solution in the order $x_n, x_{n-1}, x_{n-2}, \ldots\ldots x_2, x_1$.

Partial and Full Pivoting

The Gaussian elimination method fails if any one of the pivot elements becomes zero. In such a situation, we rewrite the equations in a different order to avoid zero pivots. Changing the order of equations is called pivoting.

We now introduce the concept of partial pivoting. In this technique, if the pivot happens to be zero, then as the numerically largest elements are searched for in the ith column elements. Let the jth row (j> i) contains this element, then, we interchange the ith equation with the jth equation and proceed for the elimination.

In full pivoting, also known as complete pivoting, we interchange rows as columns, such that the largest element in the matrix of the system becomes the pivot element. In this process, the position of the unknown variable also get changed. Full pivoting, in fact, is more complicated than partial pivoting. Partial pivoting is preferred for hand computation.

Gauss-Jordan Elimination Method

This method is a variation of the Gaussian elimination method. In this method, the elements above and below the diagonal are simultaneously made zero, and thereby the given system is reduced to an equivalent diagonal form using elementary transformations. The solution of the resulting diagonal system can then be readily obtained.

Sometimes, we normalize the pivot row with respect to the pivot element, before elimination. Partial pivoting is also used whenever the pivot element becomes zero.

Crout's Reduction Method

This method is based on the fact that the coefficient matrix $[A]$ of the system of equations can be decomposed into the product of two matrices [L] and [U], where [L] is a lower-triangular matrix and [U] is an upper triangular matrix with 1' s on its main diagonal. The rules for getting [L] and [U] can be obtained from the fact that:

$$[L][U]=[A].$$

This method is also known as Cholesky's reduction method. This technique is widely used for the numerical solution of partial differential equations.

This method is very popular from a computer programming point of view, since the storage space reserved for matrix [A] can be used to store the elements of [L] and [U] at the end of computation.

It may be noted that this method fails if any $a_{ii} = 0.$ In this case, the system in singular.

Jacobi's Method

Jacobi's method is an iterative method, where the initial approximate solution to a given system of equations is estimated and improved towards the exact solution in an iterative way. In general, when the coefficient matrix of the system of equations is a sparse matrix (many elements are zero), iterative methods have a definite advantage over direct methods with respect to computer memory. Such sparse matrices arise in computing the numerical solution of partial differential equations.

This method is called Jacobi's iterative method. It is also known as the method of simultaneous displacement, since no element of x_i^{r+1} is used in this iteration until every element is computed.

A sufficient condition for convergence of the iterative solution to the exact solution is:

$$|a_{ii}| > \sum_{\substack{j=1 \\ j \neq i}}^{n} |a_{ij}|, \quad i = 1, 2, \ldots, n$$

When this condition (diagonal dominance) is true, Jacobi method converges.

Gauss-Seidal Iteration Method

This is another well-known iterative method for solving a system of linear equations. For the Jacobi method, the $(r+1)$ th approximation of the system is given by $Eqs.$ from which we can observe that no element of x_i^{r+1} replaces x_i^r entirely for the next cycle of computation.

However, in the Gauss-Seidel method, the corresponding elements of x_i^{r+1} replaces those of x_i^r as soon as they become available. Hence, it is called the method of successive displacement.

Matrix Inversion

Consider a system of equations in the form:

[A] (X) = (B)

One way of writing its solution is in the form:

$$(X) = [A]^{-1}(B).$$

Thus, the solution to the system can also be obtained if the inverse of the coefficient matrix [A] is known. Alternatively, if the product of two square matrices is an identity matrix, that is, if

[A] [B] = [*I*],

then, [B] $= [A]^{-1}$ $and [A] = [B]^{-1}$.

Every square non-singular matrix will have an inverse. Gauss elimination and Gauss-Jordan are popular among many methods available for finding the inverse of a matrix.

Gaussian Elimination Method

For this method, consider the matrix *A*. To find its inverse, we first place an identity matrix, with an order equal to *A*, adjacent to A. This identity matrix is known as the augmented matrix. The inverse of A is then computed in two stages. In the first stage, A is converted into an upper triangular from, using the Gaussian elimination method as discussed in this section. In the second stage, the above upper triangular matrix is reduced to an identity matrix by row transformations. These operations are also performed on the adjacently placed identity matrix. Finally, when A is transformed into an identity matrix, the adjacent matrix gives the inverse of *A*. In order to increase the accuracy of the result, it is essential to employ partial pivoting.

Gauss-Jordan Method

This method is similar to the Gaussian elimination method, with the essential difference that the stage of reducing the given matrix to upper triangular form is not needed. However, the given matrix can be directly reduced to an identity matrix using elementary row transformations.

Curve Fitting

In science and engineering, we often come across experimental observations involving two variables, such as concentration and time, pressure and temperature and load and deflection. The important question is, can we find a law or a relation connecting them? Suppose that we wish to find a relationship connecting load p (Kg) and the deflection of a beam y (cm), given the maximum deflection y for various loads p. Typically, a linear relation of the form $y=mx+c$ can be used. In another situation, the graph may indicate a parabolic or an exponential curve. In fact, Kepler formulated the third law of planetary motion, T = $cx^{3/2}$, from observed data pairs (x, t) of various planets, where T is the period, and x is the mean distance of the planet from the sun. In general, the problem of finding an equation of an approximating curve that passes through as many points as possible, is called curve fitting. Hence. The basic problem is to find an equation of the curve of the best fit.

The method of group averages, the least square method and the method of moments are some of the well-known methods based on the principles of elementary statistics. Amongst them, the least square method gives a unique best fit and is highly recommended.

The Least Squares Method

In curve fitting for a given data set, the method of least squares is considered to give the best fit, with prior knowledge about the shape of the curve. Suppose that $(x_1, y_1),(x_2, y_2)\dots\dots(x_n, y_n)$ are a set of observations in an experiment. Let us assume that:

Y = f(x)

For x q a relationship between x and y, when $x = x_1$, the observed value of y is y_1 and the expected value of y from Eqs is $f(x_1)$. The residual is then defined by

$$e_1 = y_1 - f(x_1).$$

Similarly, we can define all other residuals $e_2, e_3\dots\dots, e_n$ as:

$$e_2 = y_2 - f(x_2)$$

$e_n = y_n - f(x_n)$.

It is clear from the difference of the residuals that some of them may be positive and some may be negative. In order to give equal importance to both positive and negative residuals, it is better to consider the sum of the square of the residuals, rather than the sum of the residuals as in the method of group averages.

Fitting a Straight Line

Let $(x_1, y_1), (x_2, y_2), \ldots\ldots, (x_n y_n)$ be a set of n observations in an experiment. We wish to fit a straight line:

$$y = ax + b$$

to these observations. That is, we have to determine the constants a and b, using the principal of least squares.

For any $x_i\ (i = 1, 2, \ldots\ldots, n)$, the expected value of y is $ax_i + b$, while the observed value of y is y_i. Hence, the residuals are:

$$e_1 = Y_I - (ax_i + b), \quad for\ \ i = 1, 2, \ldots\ldots, n.$$

Fitting an Exponential Curve

Suppose we have a set of n observations $(x_i, y_i), i = 1, 2, \ldots\ldots, n$ in an experiment and we wish to fit an exponential curve of the form,

$$y = ae^{bx}$$

to these observations. At first, we take logarithms of both sides to get

$$\log_{10} y = \log_{10} a + bx \log_{10} e.$$

The above equation then becomes:

$$Y = Bx + A.$$

Thus, it is equivalent to fitting a straight line.

Method of Moments

Consider n set of observations $(x_i, y_i),\ i = 1, 2, \ldots, n$ in an experiment, such that the $x_i\text{'s}$ are equally spaced. That is:

$$x_2 - x_1 = x_3 - x_2 = \ldots.. = x_n - x_{n-1} = \Delta x$$

For such a set of observations, we shall define the observed moments as follows:

$$\left.\begin{aligned} \mu_1 &= \text{The first moment} = \sum y_i\, \Delta x = \Delta x \sum y_i \\ \mu_2 &= \text{The second moment} = \sum x_i y_i\, \Delta x = \Delta x \sum x_i y_i \\ \mu_3 &= \text{The third moment} = \sum x_i^2\, y_i\, \Delta x = \Delta x \sum x_i^2\, y_i \end{aligned}\right\}$$

and so on. These moments are known as moments of the observed values of y. We shall also define the expected moments of the computed values of y, assuming that $y = f(x)$ is the equation of the curve fitting the given data:

$$\left.\begin{aligned} \gamma_1 &= \textit{The first moment of} = \int y\, dx = \int f(x)\, dx \\ \gamma_2 &= \textit{The second moment} = \int xy\, dx = \int xf(x)\, dx \\ \gamma_3 &= \textit{The third moment} = \sum x_i^2\, y_i \Delta x = \Delta x \sum x_i^2\, y_i \end{aligned}\right\}$$

The method of moments states that the moments of the observed y's are respectively equal to the expected moments. That is,

$$\mu_1 = \gamma_1,\ \mu_2 = \gamma_2\ \mu_3 = \gamma_3$$

and so on. The observed moments can easily be calculated from given data, while the moments of the expected y's can be computed as follows:

$$\left.\begin{aligned}
\gamma_1 &= \int y\,dx = \int_{x_1+\left(\frac{\Delta x}{2}\right)}^{x_n+\left(\frac{\Delta x}{2}\right)} f(x)\,dx \\
\gamma_2 &= \int xy\,dx = \int_{x_1+\left(\frac{\Delta x}{2}\right)}^{x_n+\left(\frac{\Delta x}{2}\right)} xf(x)\,dx \\
\gamma_3 &= \int x^2\,y\,dx = \int_{x_n+(\Delta x/2)}^{x_n+(\Delta x/2)} x^2 f(x)\,dx
\end{aligned}\right\}$$

From the principal of the method of moments, we have the following set of equations:

$$\left.\begin{aligned}
\Delta x \sum y_1 &= \int_{x_1+\left(\frac{\Delta x}{2}\right)}^{x_n+\left(\frac{\Delta x}{2}\right)} xf(x)\,dx \\
\Delta x \sum x_i\,y_i &= \int_{x_1+\left(\frac{\Delta x}{2}\right)}^{x_n+\left(\frac{\Delta x}{2}\right)} xf(x)\,dx \\
\Delta x \sum x_i^2\,y_i &= \int_{x_1+\left(\frac{\Delta x}{2}\right)}^{x_n+\left(\frac{\Delta x}{2}\right)} x^2 f(x)\,dx
\end{aligned}\right\}$$

Given the form of $f(x)$ we can use Eqs to determine the unknown constant in $f(x)$.

CHAPTER 13

INTERPOLATION

INTRODUCTION

Finite difference plays an important role in numerical techniques, where tabulated values of the function are available. For instance, consider a function $y = f(x)$. Suppose x takes values $x_0, x_1, x_2, \ldots\ldots x_n$, and the corresponding values of y are $y_0, y_1, y_2, \ldots\ldots, y_n$. For a given table of values, $(x_k, x_k)k = 0, 1, 2, \ldots, n$; the process of estimating any intermediate value of y, for a given value of x lying outside the table of values of x, is known as extrapolation. It may be noted that if the function $f(x)$ is known, the value of y corresponding to any x can be readily computed to the desired accuracy. However, in practice, it may be difficult or sometimes impossible to know the function $y = f(x)$ in its exact form.

To look at a practical example, let us consider the computation of the trajectory of a rocket flight, where we solve Euler's dynamical equations of motion to compute its position and velocity vectors at specific times during the flight. Under the same conditions, suppose that we require the position and velocity vector, at some other intermediate times; we need not compute the trajectory again by solving the dynamical equations.

Instead, we can use the best known interpolation technique to get the desired values.

In general, for the interpolation of a tabulated function, the concept of finite difference is important.

FINITE DIFFERENCE OPERATORS

Forward Differences

For a given table of values $(x_k, y_k), k = 0,1,2,\ldots,n$, with equally spaced abscissas of the function $y = f(x)$, we define the forward difference operator Δ, as follows. The first forward difference is usually expressed as:

$$\Delta y_i = y_{i+1} - y_{i,} \quad i = 0,1, ..,(n-1) \quad (1)$$

To be explicit, we write:

$$\Delta y_0 = y_1 - y_0$$

$$\Delta y_1 = y_2 - y_1$$

$$\vdots$$

$$\Delta y_{n-1} = y_n - y_{n-1}$$

These differences are called the first difference of the function y and are denoted by the symbol, $\Delta y_{i,}$.Here, v is called the forward difference operator.

Similarly, the difference of the first difference is called the second difference, defined by

$$\Delta^2 y_0 = -\Delta y_1 - \Delta y_0, \quad \Delta^2 y_1 = \Delta y_2 - \Delta y_1$$

Thus, in general

$$\Delta^2 y_1 = \Delta y_{i+1} - \Delta y_i \qquad (2)$$

Here, Δ^2 is called the second difference operator. Thus, continuing, we can define the rth difference of y, as:

$$\Delta^r y_i = \Delta^{r-1} y_{i+1} - \Delta^{r-1} y_i \qquad (3)$$

By defining a difference table as a convenient device for displaying various differences, the above defined difference can be written down systematically by constructing a differences table of values $(x_k, y_k), k = 0,1, \ldots, 6$:

x	y	Δy	$\Delta^2 y$	$\Delta^3 y$	$\Delta^4 y$	$\Delta^5 y$	$\Delta^6 y$
x_0	y_0						
x_1	y_1	Δy_0	$\Delta^2 y_0$				
x_2	y_2	Δy_1	$\Delta^2 y_1$	$\Delta^3 y_0$			
x_3	y_3	Δy_2	$\Delta^2 y_2$	$\Delta^3 y_1$	$\Delta^4 y_0$	$\Delta^5 y_0$	
x_4	y_4	Δy_3	$\Delta^2 y_2$	$\Delta^3 y_2$	$\Delta^4 y_1$	$\Delta^5 y_1$	$\Delta^6 y_0$
x_5	y_5	Δy_4	$\Delta^2 y_3$	$\Delta^3 y_3$	$\Delta^4 y_2$		
x_6	y_6	Δy_5	$\Delta^2 y_4$				

This difference table is called a forward difference table or diagonal difference table. Here, each difference is located in its appropriate column, midway between the elements of the previous column. It can be noted that the subscript remains constant along each diagonal of the table. The first term in the table, y_0, is called the leading term, while the differences $\Delta y_0, \Delta^2 y_0, \Delta^3 y_0, \ldots$ are the leading differences.

Example

Construct a forward difference table for the following values of x and y.

x	0.1	0.3	0.5	0.7	0.9	1.1	1.3
y	0.003	0.067	0.148	0.248	0.370	0.518	0.697

Solution

x	y	Δy	$\Delta^2 y$	$\Delta^3 y$	$\Delta^4 y$	$\Delta^5 y$
0.1	0.003					
0.3	0.067	0.064				
0.5	0.148	0.081	0.017			
0.7	0.248	0.100	0.019	0.002		
0.9	0.370	0.122	0.022	0.003	0.001	
1.1	0.518	0.148	0.026	0.004	0.001	0.000
1.3	0.697	0.197	0.031	0.005	0.001	0.000

Example

Express $\Delta^2 y_0$ and $\Delta^3 y_0$ in terms of their values of the function y.

Solution

Noting that each higher order difference is defined in term of the lower order difference, we have:

$$\Delta^2 y_0 = \Delta y_1 - \Delta y_0 = (y_2 - y_1) - (y_1 - y_0) = y_2 - 2y_1 + y_0$$

and

$$\Delta^3 y_0 = \Delta^2 y_1 - \Delta^2 y_0 = (\Delta y_2 - \Delta y_1) - (\Delta y_1 - \Delta y_0)$$

$$= (y_3 - y_2) - (y_2 - y_1) - (y_2 - y_1) + (y_i - y_0)$$

$$= y_3 - 3y_2 + 3y_1 - y_0.$$

Hence, we observe that the coefficients of the values of y, in the expansion of $\Delta^2 y_0, \Delta^3 y_0$ are binomial coefficients. Thus, in general, we arrive at the following result:

$$\Delta^n y_0 = y_n - C_1^n y_{n-1} + C_2^n y_{n-2} - C_3^n y_{n-3} + \ldots\ldots + (-1)^n y_0 \qquad (4)$$

Example

Show that the value of y_n can be expressed in terms of the leading value y_0 and the leading difference

$\Delta y_0, \Delta^2 y_0, \ldots\ldots, \Delta^n y_0$

Solution

We have from the forward difference table:

$$\left.\begin{aligned} y_1 - y_0 &= \Delta y_0 \quad or \quad y_1 = y_0 + \Delta y_0 \\ y_2 - y_1 &= \Delta y_1 \quad or \quad y_2 = y_1 + \Delta y_1 \\ y_3 - y_2 &= \Delta y_2 \quad or \quad y_3 = y_2 + \Delta y_2 \end{aligned}\right\} \tag{5}$$

and so on. Similarly

$$\left.\begin{aligned} \Delta y_1 - \Delta y_0 &= \Delta^2 y_0 \quad or \quad \Delta y_1 = \Delta y_0 + \Delta^2 y_0 \\ \Delta y_2 - \Delta y_1 &= \Delta^2 y_1 \quad or \quad \Delta y_2 = \Delta y_1 + \Delta^2 y_1 \end{aligned}\right\} \tag{6}$$

and so on. Similarly, we can write

$$\left.\begin{aligned} \Delta^2 y_1 - \Delta^2 y_0 &= \Delta^3 y_0 \quad or \quad \Delta^2 y_1 = \Delta^2 y_0 + \Delta^3 y_0 \\ \Delta^2 y_2 - \Delta^2 y_1 &= \Delta^3 y_1 \quad or \quad \Delta^2 y_2 = \Delta^2 y_1 + \Delta^3 y_1 \end{aligned}\right\} \tag{7}$$

and so on. Also from Eqs. 6 and 7, we can rewrite Δy_2 as

$$\Delta y_2 = \left(\Delta y_0 + \Delta^2 y_0\right) + \left(\Delta^2 y_0 + \Delta^3 y_0\right)$$

$$= \Delta y_0 + 2\Delta^2 y_0 + \Delta^3 y_0 \tag{8}$$

From Eqs. $(5)-(8)$, y_3 can be rewritten as:

$$y_3 = y_2 + \Delta y_2$$

$$= \left(y_1 + \Delta y_1\right) + \left(\Delta y_1 + \Delta^2 y_1\right)$$

$$=\left(y_0+\Delta y_1\right)\left(\Delta y_1+\Delta^2 y_1\right)$$

$$=y_0+3\Delta y_0+3\Delta^2 y_0+\Delta^3 y_0$$

$$=\left(1+\Delta\right)^3 y_0 \qquad (9)$$

Similarly, we can symbolically write

$$y_1=\left(1+\Delta\right)y_0,\ y_2=\left(1+\Delta\right)^2 y_0,\quad y_3=\left(1+\Delta\right)^3 y_0$$

Continuing this procedure, we can show in general

$$y_n=\left(1+\Delta\right)^n y_0$$

Hence, we obtain:

$$y_n=y_0+C_1^n\Delta y_0+C_2^n\Delta^2 y_0+\ldots\ldots..+\Delta^n y_0 \qquad (10)$$

Equivalently, we can also write the result as:

$$y_n=\sum_{i=1}^{n}C_1^n\Delta^i y_0. \qquad (11)$$

Backward Difference

For a given table of values $\left(x_k,y_k\right),k=0,1,2,\ldots\ldots,n$ of a function $y=f\left(x\right)$ with equally spaced abscissas, the first backward differences are usually expressed in terms of the backward difference operator ∇:

$$\nabla y_i=y_i-y_{i-1}i=n,\ \left(n-1\right),\ldots..1. \qquad (12)$$

Thus,

$$\nabla y_1=y_1-y_0$$

$$\nabla y_2 = y_2 - y_1$$

$$\vdots \quad \vdots \quad \vdots$$

$$\nabla y_n = y_n - y_{n-1}.$$

The difference of these differences are called the second difference, denoted by $\nabla^2 y_2,\ \nabla^2 y_3, \ldots\ldots \nabla^2 y_n$. That is:

$$\nabla^2 y_2 = \nabla y_2 - \nabla y_1$$

$$\nabla^2 y_3 = \nabla y_3 - \nabla y_2$$

$$\vdots \quad \vdots \quad \vdots$$

$$\nabla^2 y_n = \nabla y_n - \nabla y_{n-1}.$$

Thus, in general, the second backward differences are

$$\nabla^2 y_i = \nabla y_i - \nabla y_{i-1,} \quad i = n, (n-1), \ldots.., 2. \tag{13}$$

While the kth backward differences are

$$\nabla^k y_i = \nabla^{k-i} y_i - \nabla^{k-1} y_{i-1} \quad i = n, (n-1), , \ldots\ldots.., k. \tag{14}$$

These backward differences can be systematically arranged for a table of values $(x_k, y_k), k = 0,1, \ldots.., 6,$ as indicated in the table below.

Backward Difference Table

x	y	∇y	$\nabla^2 y$	$\nabla^3 y$	$\nabla^4 y$	$\nabla^5 y$	$\nabla^6 y$
x_0	y_0						
x_1	y_1	∇y_1	$\nabla^2 y_2$				
x_2	y_2	∇y_2	$\nabla^2 y_3$	$\nabla^3 y_3$	$\nabla^4 y_4$	$\nabla^5 y_5$	
x_3	y_3	∇y_3	$\nabla^2 y_4$	$\nabla^3 y_4$	$\nabla^4 y_5$	$\nabla^5 y_6$	$\nabla^6 y_6$
x_4	y_4	∇y_4	$\nabla^2 y_5$	$\nabla^3 y_5$	$\nabla^4 y_6$		
x_5	y_5	∇y_5	$\nabla^2 y_6$	$\nabla^3 y_6$			
x_6	y_6	∇y_6					

From this table, it can be observed that the subscript remains constant, along every backward diagonal.

Example

Show that any value of y can be expressed in terms of y_n and its backward differences.

Solution

From Eq. (12) we have

$y_{n-1} = y_n - \nabla y_n$ and $y_{n-2} = y_{n-1} - \nabla y_{n-1}$.

Also, from the definition given in Eq (13), we get

$\nabla y_{n-1} = \nabla y_n - \nabla^2 y_n$.

From these equation, we obtain

$y_{n-2} = y_n - 2\nabla y_n + \nabla^2 y_n$.

Similarly, we can show that

$y_{n-3} = y_n - 3\nabla y_n + 3\nabla^2 y_n - \nabla^3 y_n$.

Symbolically, these results can be rewritten as follows:

$$y_{n-1} = (1-\nabla)y_n, \quad y_{n-2} = (1-\nabla)^2 y_n, \quad y_{n-3} = (1-\nabla)^3 y_n.$$

Thus, is general, we can write:

$$y_{n-1} = (1-\nabla)^r y_n$$

That is,

$$y_{n-r} = y_n - C_1^r \nabla y_n + C_2^r \nabla^2 y_n - \ldots\ldots + (-1)^r \nabla^r y_n. \tag{15}$$

Central Differences

In some applications, the central differences notation is found to be more convenient to represent the successive difference of a function. Here, we use the symbol δ to represent the central difference operator and the subscript of δy for any difference as the average of the subscripts of the two members of the difference. Thus, we write:

$$\delta y_{1/2} = y_1 - y_0, \quad \delta y_{3/2} = y_2 - y_1.$$

In general

$$\delta y_i = y_{i+(1/2)} - y_{i-(1/2)}. \tag{16}$$

Higher order differences are define as follows:

$$\delta^2 y_i = \delta y_{i+(1/2)} - \delta y_{i-\left(\frac{1}{2}\right)}. \tag{17}$$

These central differences can be systematically arranged as indicated, in table 3 .

Central Difference Table

x	y	δy	$\delta^2 y$	$\delta^3 y$	$\delta^4 y$	$\delta^5 y$	$\delta^6 y$
x_0	y_0						
x_1	y_1	$\delta y_{1/2}$	$\delta^2 y_1$				
x_2	y_2	$\delta y_{3/2}$	$\delta^2 y_2$	$\delta^3 y_{3/2}$	$\delta^4 y_2$	$\delta^5 y_{5/2}$	
x_3	y_3	$\delta y_{5/2}$	$\delta^2 y_3$	$\delta^3 y_{5/2}$	$\delta^4 y_3$	$\delta^5 y_{7/2}$	$\delta^6 y_3$
x_4	y_4	$\delta y_{7/2}$	$\delta^2 y_4$	$\delta^3 y_{7/2}$	$\delta^4 y_4$		
x_5	y_5	$\delta y_{9/2}$	$\delta^2 y_5$	$\delta^3 y_{9/2}$			
x_6	y_6	$\delta y_{11/2}$					

Thus, we observe that all the odd differences have a fractional suffix and all the even difference with the same subscript lie horizontally.

The following alternative notation may also be adopted to introduce finite difference operators. Let $y = f(x)$ be a functional relationship between x and y, denoted by y_x. Suppose, we are given consecutive values of x differing by h, say $x, x+h, x+2h, x+3h, etc.$ The corresponding values of y are $y_{x,}, y_{x+3h,}$ $etc.$ As before, we can form the differences of these values. Thus

$$\Delta y_x = y_{x+h} - y_x = f(x+h) - f(x) \tag{19}$$

$$\Delta^2 y_x = \Delta y_{x+h} - \Delta y_x$$

Similarly

$$\nabla \mathrm{y_x} = \mathrm{y_x} - \mathrm{y_{x-h}} = \mathrm{f(x) - f(x-h)} \tag{20}$$

And

$$\delta y_x = y_{x+(h/2)} - y_{x-(h/2)} = f\left(x + \frac{h}{2}\right) - f\left(x - \frac{H}{2}\right). \tag{21}$$

Shift Operator, E

Let $y = f(x)$ be a function of x, and let x taken the consecutive values $x, x+h, x+2h, etc$. We then define an operator E having the property

$$E\,f(x) = f(x+h). \tag{22}$$

Thus, when E is applied to $f(x)$, the result is the next value of the function. Here, E is called the shift operator. If we apply the operator E twice to $f(x)$, we get

$$E^2 f(x) = E\left[E\,d(x)\right] = E\left[f(x+h)\right] = f(x+2h).$$

Thus, in general, if we apply the operator E n times to $f(x)$, we arrive at

$$E^n f(x) = f(x+nh).$$

In terms of the new notation, we can write

$$E^n y_x = y_{x+nh}$$

Or

$$E^n f(x) = f(x+nh), \tag{23}$$

for all real values of n. Also, if $y_0, y_1, y_3, \ldots.$ are the consecutive values of the function y_x, then we can also write

$$Ey_0 = y_i,\ E^2 y_0 = y_2,\ \ E^4 y_0 = y_4,\ \ldots\ldots,\ E^2 y_2 = y_4$$

and so on. The inverse operator E^{-1} is define as

$$E^{-1}\ f(x) = f(x-h).$$

And similarly

$$E^{-n}\ f(x) = f(x-nh). \tag{24}$$

Average Operator, μ

The average operator μ is defined as

$$\mu f(x) = \frac{1}{2}\left[f\left(x+\frac{h}{2}\right)+f\left(x-\frac{h}{2}\right)\right] = \frac{1}{2}\left[y_{x-\left(\frac{h}{2}\right)} + y_{x+\left(\frac{h}{2}\right)}\right]. \qquad (25)$$

Differential Operator, D

D represents a differential operator with the following property:

$$\left.\begin{aligned} Df(x) &= \frac{d}{dx}f(x) = f'(x) \\ D^2 f(x) &= \frac{d^2}{dx^2}f(x) = f''(x) \end{aligned}\right\} \qquad (26)$$

Having defined various difference operators Δ, ∇, E, μ and D, we can easily obtain the following relation:

From the definition of the operators Δ and E, we have

$$\Delta y_x = y_{x+h} - y_x = Ey_x - y_x = (E-1)y_x.$$

Therefore,

$$\Delta = E - 1 \qquad (27)$$

Following the definition of operators ∇ and E^{-1}, we have

$$\nabla y_x = y_x - y_{x-h} = y_x - E^{-1}y_x = \left(1-E^{-1}\right)y_x.$$

Therefore,

$$\nabla = 1 - E^{-1} = \frac{E-1}{E}. \qquad (28)$$

The definition of operators δ and E gives us:

$$\delta y_x = y_{x+(h/2)} - y_{x-(h/2)} = E^{1/2} y_x - E^{-1/2} y_x = \left(E^{1/2} - E^{-1/2}\right) y_x.$$

Hence,

$$\delta = E^{1/2} - E^{-\frac{1}{2}}.$$

The definition of μ and E similarly yields:

$$\mu y_x = \frac{1}{2}\left[y_{x+(h/2)} - y_{x-(h/2)}\right] = \frac{1}{2}\left(E^{1/2} - E^{-1/2}\right) y_x.$$

Therefore,

$$\mu = \frac{1}{2}\left(E^{1/2} - E^{-1/2}\right). \quad (30)$$

It is known that

$$E y_x = y_{x+h} = f(x+h).$$

Using the Taylors series expansion, we have:

$$E y_x = f(x) + hf'(x) + \frac{h^2}{2!} f''(x) + \ldots$$

$$= f(x) + hDf(x) + \frac{h^2}{2!} f''(x) + \ldots.$$

$$= \left(1 + \frac{hD}{1!} + \frac{h^2 D^2}{2!} + \ldots..\right) f(x) = E^{hd} y_x$$

Thus,

$$hD = logE. \quad (31)$$

Example

Prove that

$hd = \log(1+\Delta) = -\log(1-\nabla) = \sinh^{-1}(\mu\delta)$.

Solution

Using the standard equations (27) – (31), we have

$$hd = \log E = \log(1+\Delta) = -\log E^{-1} = -\log(1-\nabla) \quad (1)$$

Also,

$$\mu\delta = \frac{1}{2}\left(E^{1/2} + E^{-1/2}\right)\left(E^{1/2} - E^{-1/2}\right) =$$

$$\frac{1}{2}\left(E - E^{-1}\right) = \frac{1}{2}\left(e^{hd} - e^{-hd}\right) = \sinh(hD)$$

Therefore,

$$hD = \sinh^{-1}(\mu\delta).$$

Equation (1) and (2) constitute the required result.

Example

Let Δ, ∇ and δ denote the forward, backward and central difference operators, respectively, and E and μ the shift and average operators, respectively, in the analysis of data with equal spacing h. Show that

$$(i) 1+\delta^2\mu^2 = \left(1+\frac{\delta^2}{2}\right),$$

$$(ii) E^{1/2} = \mu + \frac{\delta}{2}$$

$$(iii) \Delta = \frac{\delta^2}{2} + \delta\sqrt{1+(\delta^2/4}$$

$$(iv) \mu\delta = \frac{\Delta E^{-1}}{2} + \frac{\Delta}{2}$$

$$(v)\ \mu\delta = \frac{\Delta + \nabla}{2}$$

Solutions

(i) From the definition of operators, we have:

$$\mu\delta = \frac{1}{2}\left(E^{1/2} + E^{-1/2}\right)\left(E^{1/2} - E^{-1/2}\right) = \frac{1}{2}\left(E - E^{-1}\right).$$

Therefore,

$$1 + \mu^2\delta^2 = 1 + \frac{1}{4}\left(E^2 - 2 + E^{-2}\right) = \frac{1}{4}\left(E + E^{-1}\right)^2.$$

Also,

$$1 + \frac{\delta^2}{2} = \frac{1}{2}\left(E^{1/2} + E^{-1/2}\right) = \frac{1}{2}\left(E + E^{-1}\right).$$

From Eqs (1) and (2), the first result follows.

(ii) Now

$$\mu + \frac{\delta}{2} = \frac{1}{2}\left(E^{1/2} + E^{-1/2} + E^{1/2} - E^{-1/2}\right) = E^{1/2}.$$

Thus, the second result is proved.

(iii) We can write

$$\frac{\delta^2}{2} + \delta\sqrt{1 + \left(\delta^2/4\right)} = \frac{\left(E^{1/2} - E^{-1/2}\right)^2}{2}$$

$$+ \frac{\left(E^{1/2} - E^{-1/2}\right)^2 \sqrt{1 + \frac{1}{4}\left(E^{1/2} - E^{-1/2}\right)^2}}{1}$$

$$=\frac{E-2+E^{-1}}{2}+\frac{1}{2}\left(E^{\frac{1}{2}}-E^{-\frac{1}{2}}\right)\left(E^{1/2}+E^{-1/2}\right)$$

$$\frac{E-2+E^{-1}}{2}+\frac{E-E^{-1}}{2}$$

$$E=-1.$$

Using Eq. (27), we get

$$E-1=\Delta$$

(iv) We have

$$\mu\delta=\frac{1}{2}\left(E^{1/2}+E^{-1/2}\right)\left(E^{1/2}-E^{-1/2}\right)=\frac{1}{2}\left(E-E^{-1}\right).$$

Now, using Eq. (27), we get

$$=\frac{1}{2}\left(1+\Delta-E^{-1}\right)=\frac{\Delta}{2}+\frac{1}{2}\left(1-E^{-1}\right)$$

$$=\frac{\Delta}{2}+\frac{1}{2}\left(\frac{E-1}{E}\right)=\frac{\Delta}{2}+\frac{\Delta}{2E}$$

(v) We can write

$$\mu\delta=\frac{1}{2}\left(E^{1/2}+E^{-1/2}\right)\left(E^{1/2}-E^{-1/2}\right)=\frac{1}{2}\left(E-E^{-1}\right).$$

Now using Eq. (27) and (28), we have

$$\mu\delta=\frac{1}{2}(1+\Delta-1+\nabla)=\frac{1}{2}(\Delta+\nabla)$$

Example

Show that the operations μ and E commute.

Solution

From the definition of operators μ and E, we have

$$\mu\delta y_0 = \mu y_1 = \frac{1}{2}\left(y_{3/2} + y_{1/2}\right). \tag{1}$$

While

$$E\mu\delta y_0 = \frac{1}{2}E\left(y_{1/2} + y_{-1/2}\right) = \frac{1}{2}\left(y_{3/2} + y_{1/2}\right). \tag{2}$$

Equating (1) and (2), we have

Therefore, the operators μ and E commute.

Newton's Forward Difference Interpolation Formula

Let $y = f(x)$ be a function with values $f(x_0), f(x_0+h), f(x_0+2h), \ldots\ldots,$ corresponding to various equi-spaced values of x with spacing h: $x_0, x_0+h, x_0+2h, \ldots.$ Suppose that we wish to evaluate the function $f(x)$ for a value x_0+ph, where P is any real number. Then, for any real number P, we have an operator E such that $E^p f(x) = f(x+ph)$. Therefore, using Eq. we have

$$f(x_0+ph) = E^p f(x_0) = (1+\Delta)^p (x_0)$$

$$= \left[1 + p\Delta + \frac{p(p-1)}{2!}\Delta^2 + \frac{p(p-1)(p-2)}{3!}\Delta^3 + \ldots\right] f(x_0).$$

That is,

$$f(x_0+ph) = f(x_0) + p\Delta f(x_0) + \frac{p(p-1)}{2!}\Delta^2 f(x_0) + \frac{p(p-1)(p-2)}{3!}$$

$$\Delta^3 f(x_0) + \ldots + \frac{P(p-1)\ldots(p-n+1)}{n!}\Delta^n f(x_0) + Error.$$

This is known as Newton's forward difference formula for interpolation, which gives the value of $f(x_0 + ph)$ in terms of $f(x_0)$ and its leading differences. This formula is also known as Newton-Gregory forward difference interpolation formula: Hence, here $P = (x - x_0)lh.$ The equation can also be written in another alternate form:

$$y_x = y_0 + p\Delta y_0 + \frac{p(p-1)}{2!}\Delta^2 y_0 + \frac{p(p-1)(p-2)}{3!}\Delta^3 y_0 + \ldots\ldots + \frac{P(p-1)\ldots(p-n+1)}{n!}\Delta^n y_0 + Error$$

If we retain $(r+1)$ terms, we obtain a polynomial of degree r agreeing with y_x at $x_0, x_1, \ldots\ldots, x_r$.

This formula is mainly used for interpolating the values of y near the beginning of a set of tabular values, and for extrapolating values of y that are a short distance behind y_0. We shall illustrate these formula by considering the following examples.

Example

Evaluate $f(15)$, given the following table of values:

x	10	20	30	40	50
$y = f(x)$	46	66	81	93	101

Solution

We many note that x=15 is very near to the beginning of the table. Hence, we use Newton's forward difference interpolation formula. The forward differences are calculated and tabulated as given below:

x	$y=f(x)$	Δy	$\Delta^2 Y$	$\Delta^3 y$	$\Delta^4 y$
10	46	20			
20	66	15	-5		
30	81	12	-3	2	
40	93	8	-4	-1	-3
50	101				

We have Newton's forward difference interpolation formula as

$$y = y_0 + p\Delta y_0 + \frac{p(p-1)}{2!}\Delta^2 y_0 + \frac{p(p-1)(p-2)}{3!}\Delta^3 y_0 + \frac{P(p-1)(P-2)(p-3)}{4!}\Delta^4 y_0$$

In this example, from the above table, we have.

$$x_0 = 10, y_0 = 46, \Delta y_0 = 20. \Delta^2 y_0 = -5, \Delta^3 y_0 = 2,\ \Delta^4 y_0 = -3$$

Let y_{15} be the value of y when x = 15, then:

$$P = \frac{x - x_0}{h} = \frac{15-10}{10} = 0.5$$

Substituting these values in Eq. (1), we get:

$$f(15) = y_{15} = 46 + (0.5)(20) + \frac{(0.5)(0.5-1)}{2}(-5) + \frac{(0.5)(0.5-1)(0.5-2)}{6}(2)\frac{(0.5)(0.5-1)(0.5)(0.5-2)}{24}(-3)$$

$$= 46 + 10 + 0.625 + 0.125 + 0.1172.$$

Example

Find the Newton forward difference interpolating polynomial for the following data:

x	0.1	0.2	0.3	0.4	0.5
$y=f(x)$	1.40	1.56	1.76	2.00	2.28

Solution

We shall first constructed the forward difference table to the given data as indicated below:

x	$y = f(x)$	Δy	$\Delta^2 Y$	$\Delta^3 y$	$\Delta^4 y$
0.1	1.40				
0.2	1.56	0.16			
0.3	1.76	0.20	0.04		
0.4	2.00	0.24	0.04	0.00	0.00
0.5	2.28	0.28	0.04	0.00	

Since the third and fourth leading differences are zero, we have Newton's forward difference interpolating formula as

$$y = y_0 + p\Delta y_0 + \frac{p(p-1)}{2}\Delta^2 y_0. \qquad (1)$$

In this problem, $x_0 = 0.1$, $y_0 = 1.40, \Delta y_0 = 0.16, \Delta^2 y_0 = 0.04$, and

$$P = \frac{x-0.1}{0.1} = 10x - 1.$$

Substituting these values into Eq.(1), we obtain

$$y = f(x) = 140 + (10x-1)(0.16) + \frac{(10x-1)(10x-2)}{2}(0.04):$$

That is, $y = 2x^2 + x + 1.28$. This is the required Newton's interpolating polynomial.

Example

Estimate the missing figure in the following table:

x	1	2	3	4	5
$y = f(x)$	2	5	7	–	32

Solution

Since we are given four entries in the table, the function $y = f(x)$ can be represented by a polynomial of degree three. Using Theorem, we have:

$\Delta^3 f(x) = \text{Constant}$ and $\Delta^4 f(x) = 0.$

for all x. In particular, $\Delta^4 f(x_0) = 0$. Equivalently, $(E-1)^4 f(x_0) = 0$. Expanding, we have

$$\left(E^4 - 4E^3 + 6E^2 - 4E + 1\right) f(x_0) = 0.$$

That is,

$$f(x_4) - 4f(x_3) + 6f(x_2) - 4f(x_1)n + f(x_0) = 0.$$

Using the values given in the table, we obtain

$$32 - 4f(x_3) + 6x7 - 4x5 + 2 = 0.$$

Which gives $f(x_3)$, the missing value equal to 14.

Example

Find a cubic polynomial in x which takes on the values –3.3.11.27.57 and 107, when $x = 1, 2, 3, 4$ and 5 respectively.

Solution

Here, the observations are given at equal intervals of unit width. To determine the required polynomial, we first construct the difference table as follows:

x	$f(x)$	$\Delta f(x)$	$\Delta^2 f(x)$	$\Delta^3 f(x)$
0	-3			
1	3	6		
2	11	8	2	
3	27	16	8	6
4	57	30	14	6
5	107	50	20	6

Since the fourth and higher order differences are zero, we require Newton's interpolation formula in the form

$$f(x_0 + ph) = f(x_0) + p\Delta f(x_0) + \frac{p(p-1)}{2}\Delta^2 f(x_0) + \frac{p(p-1)(p-2)}{6}\Delta^3 f(x_0) \quad (1)$$

Here,

$$P = \frac{x - x_0}{h} = \frac{x-0}{1} = x,\ \Delta f(x_0) = 6,\ \Delta^2(x_0) = 2, \Delta^3 f(x_0) = 6.$$

Substituting these values into Eq. (1), we have

$$f(x) = -3 + 6x + \frac{x(x-1)}{2}(2) + \frac{x(x-1)(x-2)}{6}(6).$$

That is, $f(x) = x^3 - 2x^2 + 7x - 3$, is the required cubic polynomial?

Example

For the following table of values, estimate f.

x	1	2	3	4	5	6	7	8
$y = f(x)$	1	8	27	64	125	216	343	512

Solution

The value to be interpolated is at the end of the table. Hence, it is appropriate to use Newton's backward interpolation formula. We shall first construct the difference table for the given data:

x	$y = f(x)$	Δy	$\Delta^2 y$	$\Delta^3 y$	$\Delta^4 y$
1	1				
2	8	7			
3	27	19	12		
4	64	37	18	6	0
5	125	61	24	6	0
6	216	91	30	6	0
7	343	127	36	6	0
8	512				

Since the fourth and higher order differences are zero, the required Newton's backward interpolation formula is

$$y_x = y_n + p\nabla y_n + \frac{p(p+1)}{2!}\nabla^2 y_n + \frac{p(p+1)(p+2)}{3!}\nabla^3 y_n.$$

In this problem,

$$p = \frac{x - x_n}{h} = \frac{7.5 - 8.0}{1} = -0.5$$

and $\nabla y_n = 169, \nabla^2 y_n = 42, \nabla^3 y_n = 6.$

Therefore,

$$y_{7.5} = 512 + (-0.5)(169) + \frac{(-0.5)(0.5)}{2}(42) + \frac{(-0.5)(0.5)(1.5)}{6}(6)$$

$$= 512 - 84.5 - 5.25 - 0.375$$

$$= 421.875.$$

SUMMARY AND DISCUSSION

Finite difference plays an important role in numerical techniques, where tabulated values of the function are available. For instance, consider a function $y = f(x)$. As x takes the values $x_0, x_1, x_2, \ldots.. x_n$, let the corresponding values of y be $y_0, y_1, y_2, \ldots\ldots, y_n$. That is, for a given table of values, $(x_k, x_k)k = 0,1,2,\ldots,n;$ the process of estimating any intermediate value of y, for a given value of x lying outside its table of values, is known as extrapolation. It may be noted that if the function $f(x)$ is known, the value of y corresponding to any x can be readily computed to the desired accuracy. However, in practice, it may be difficult or sometimes impossible to know the function $y = f(x)$ in its exact form. In general, for interpolation of a tabulated function, the concept of finite difference is important.

Forward Differences

For a given table of values $(x_k, y_k), k = 0,1,2,\ldots,n,$ with equally spaced abscissas of the function $y = f(x),$ we define the forward difference operator Δ as follows. The first forward difference is usually expressed as:

$$\Delta y_i = y_{i+1} - y_{i,} \quad i = 0,1, \ldots,(n-1).$$

Explicitly, we write

$$\Delta y_0 = y_1 - y_0$$

$$\Delta y_1 = y_2 - y_1$$

$$\vdots$$

$$\Delta y_{n-1} = y_n - y_{n-1}$$

These differences are called first difference of the function y and are denoted by the symbol, $\Delta y_{i,}$.Here,v is called the forward difference operator.

Backward Difference

For a given table of values $(x_k, y_k), k = 0,1,2,\ldots\ldots,n,$ of a function

$y = f(x)$ with equally spaced abscissas, the first backward differences are usually expressed in terms of the backward difference operator ∇:

$$\nabla y_i = y_i - y_{i-1} i = n, \ (n-1), \ldots..1.$$

Thus,

$$\nabla y_1 = y_1 - y_0$$

$$\nabla y_2 = y_2 - y_1$$

$$\vdots \quad \vdots \quad \vdots$$

$$\nabla y_n = y_n - y_{n-1}.$$

The differences of these differences are called the second difference and denote by $\nabla^2 y_2, \ \nabla^2 y_3, \ldots\ldots \nabla^2 y_n$. That is,

$$\nabla^2 y_2 = \nabla y_2 - \nabla y_1$$

$$\nabla^2 y_3 = \nabla y_3 - \nabla y_2$$

$$\vdots \quad \vdots \quad \vdots$$

$$\nabla^2 y_n = \nabla y_n - \nabla y_{n-1}.$$

Thus, in general, the second backward differences are

$$\nabla^2 y_i = \nabla y_i - \nabla y_{i-1,} \quad i = n, \ (n-1), \ \ldots.., 2.$$

While the kth backward differences are

$$\nabla^k y_i = \nabla^{k-i} y_i - \nabla^{k-1} y_{i-1} \quad i = n, (n-1), , \ldots\ldots.., k.$$

Central Differences

In some applications, the central difference notation is found to be more convenient to represent the successive difference of a function. Here, we use the symbol δ to represent central difference operator and the subscript of δy for any difference as the average of the subscripts of the two members of the difference. Thus, we write:

$$\delta y_{1/2} = y_1 - y_0, \qquad \delta y_{3/2} = y_2 - y_1.$$

In general:

$$\delta y_i = y_{i+(1/2)} - y_{i-(1/2)}.$$

Higher order differences are defined as follows:

$$\delta^2 y_i = \delta y_{i+(1/2)} - \delta y_{i-(1/2)}.$$

Newton's Forward Difference Interpolation Formula

Let $y = f(x)$ be a function that takes values $f(x_0), f(x_0+h), f(x_0+2h),.....$, corresponding to various equi-spaced values of x with spacing h, x_0, x_0+h, x_0+2h ,.... Suppose that we wish to evaluate the function $f(x)$ for a value $x_0 + ph$, where P is any real number. Then for any real number P, we have the operator E such that $E^p f(x) = f(x+ph)$. Therefore, using Eq., we have

$$f(x_0+ph) = E^p f(x_0) = (1+\Delta)^p (x_0)$$

$$= \left[1 + p\Delta + \frac{p(p-1)}{2!}\Delta^2 + \frac{p(p-1)(p-2)}{3!}\Delta^3 + \ldots.\right] f(x_0).$$

That is,

$$f(x_0+ph) = f(x_0) + p\Delta f(x_0) + \frac{p(p-1)}{2!}\Delta^2 f(x_0) + \frac{p(p-1)(p-2)}{3!}\Delta^3 f(x_0) +$$

$$\ldots + \frac{P(p-1)\ldots(p-n+1)}{n!}\Delta^n f(x_0) + Error$$

This is known as the Newton forward difference formula for interpolation, which gives the value of $f(x_0+ph)$ in terms of $f(x_0)$ and its leading differences. This formula is also known as the Newton-Gregory forward difference interpolation formula.

This formula is mainly used for interloping the values of y near the beginning of a set of tabular values and for extrapolating values of y a short distance behind y_0.

CHAPTER 14

PARABOLIC PARTIAL DIFFERENTIAL EQUATIONS

INTRODUCTION

In many applications of science and engineering, there has been a growing desire for the development of numerical new numerical methods. Therefore, in recent years, a greater emphasis has shifted from analytic techniques to computer oriented numerical methods. The principal attraction of numerical methods is that solutions can be obtained for many problems that cannot be solved with analytical methods. The availability of modem digital computers has paved the way for the development of efficient and more general finite difference methods for solving particular partial differential equations. Broad classification details and analytical solutions of partial differential equations are covered in Sankara Rao (1995).

In this chapter, we shall present some of the important and powerful finite difference methods for solving parabolic partial differential equations that model heat and mass transfer in fluids and solids, unsteady behavior of fluid flows and so on. For example, if a fluid is incompressible and has constant thermal conductivity *k,* the law of conservation of energy can be expressed mathematically in the form

$$pC_p \frac{DT}{dt} = k\nabla^2 T,$$

where p is the density, C_p, T is temperature and D/Dt is the material derivative. In Cartesian coordinates, this can be written as:

$$= \frac{\partial T}{\partial t} + \mu \frac{\partial T}{\partial x} + v \frac{\partial T}{\partial y} + w \frac{\partial T}{\partial z} = \alpha \left(\frac{\partial^2 T}{\partial x^2} + \frac{\partial^2 T}{\partial y^2} + \frac{\partial^2 T}{\partial z^2} \right).$$

Here $\alpha = \frac{k}{pC_p}$; μ, v and w are the components of the fluid at a point (x, y, z) and at time t. At time t, it is a non –linear parabolic partial differential equation. For fluids which are at rest, this becomes.

$$\frac{\partial T}{\partial t} = \left(\frac{\partial^2 T}{\partial x^2} + \frac{\partial^2 T}{\partial y^2} + \frac{\partial^2 T}{\partial z^2} \right),$$

which is also called a diffusion equation.

The finite difference method for solving equations of this type, though fairly simple, is also a powerful technique for solving a variety of heat transfer and fluid flow problems in regular geometries. Most of the concepts associated with the numerical solutions of partial differential equations by finite difference methods can be illustrated and understood by considering a simple example; the dimensional diffusion equation in the form:

$$\frac{\partial T}{\partial t} = \alpha \frac{\partial^2 T}{\partial x^2},$$

This holds true in some prescribed region R of the (x, t) space.

BASIC CONCEPTS IN FINITE DIFFERENCE METHODS

For the purpose of illustration, let us consider a two dimensional region as covered by a rectangular grid formed by two sets of lines drawn parallel to the coordinate axes, with grid spacing Δx and Δy in the x and y directions, respectively. The numerical values of the dependent variable are obtained at the points of intersection of the parallel lines, called mesh points, or differential equations over the region of interest to derive approximately equivalent algebraic equations. The discretization consists of replacing each derivate of the partial differential equation at a mesh point by a finite difference approximation in terms of the values of the dependent variable at the mesh point, the immediate neighboring mesh points, and boundary points. In doing so, a set of algebraic equations arise. Let the temperature T at a representation point be a function of two special coordinates x, y at time t. We adopt the following notation. Let the mesh spacing in x and y directions be denoted by Δx and Δy and the time step by Δt. Thus $T(x, y, t)$ can be represented by $T\ (i\Delta x, j\Delta y, n\Delta t) = T_i^n j$. With this notation, let us assume that the function T and its derivates are continuous. Then the finite difference approximations to the derivates can be obtained from Taylor series expansions. For example, the Taylor series expansion of $T_{i+1} j$ about the grid point (i, j) gives us:

$$T_{i+1,j} = T_{i,j} \left[\Delta x \frac{\partial T}{\partial x} + \frac{(\Delta x)^2}{2} \frac{\partial^2 T}{\partial x^2} + \frac{(\Delta x)^3}{6} \frac{\partial^3 T}{\partial x^3} + \frac{(\Delta x)^4}{24} \frac{\partial^4 T}{\partial x^4} + higher\ order\ terms \right]_{i,j}$$

Similarly,

$$T_{i+1,j} = T_{i,j} \left[\Delta x \frac{\partial T}{\partial x} - \frac{(\Delta x)^2}{2} \frac{\partial^2 T}{\partial x^2} + \frac{(\Delta x)^3}{6} \frac{\partial^3 T}{\partial x^3} - \frac{(\Delta x)^4}{24} \frac{\partial^4 T}{\partial x^4} + higher\ order\ terms \right]_{i,j}$$

Solving for $\partial T / \partial x$, gives us:

$$\left(\frac{\partial T}{\partial x} \right)_{i,j} = \frac{T_{i+1} - T_{i,j}}{\Delta x} + O(\Delta x),$$

while Eq. gives

$$\left(\frac{\partial T}{\partial x} \right)_{i,j} = \frac{T_{ij} - T_{i-1,j}}{\Delta x} + O(\Delta x).$$

Equation is the forward difference approximation to the derivative ∂T / ∂x, while Eq. is its backward difference approximation. Both approximations have a truncation error of order Δx, that is they are accurate within the first order. Subtracting Eq. from Eq. we obtain:

$$\left(\frac{\partial T}{\partial x}\right)_{i,j}=\frac{T_{i+1j}-T_{i-1,j}}{2\Delta x}+O\left[(\Delta x)^2\right].$$

This is a central difference approximation of the derivative $\partial T/\partial x$, and is accurate to the second order, in the sense that it has a truncation error of order $0\left(\Delta x^2\right)$.

The central difference approximation of a second order partial derivative $\partial^2 T/\partial x^2$ can be similarly obtained by adding *Eqs.* and. Thus

$$\left(\frac{\partial^2 T}{\partial x^2}\right)_{i,j}=\frac{T_{i+1j}-2T_{i,j}+T_{i-1,j}}{\Delta x^2}+O(\Delta x)^2.$$

Which is accurate to the second order. Similar expressions can be written for the y derivatives.

$$\left(\frac{\partial^2 T}{\partial y^2}\right)_{i,j}=\frac{T_{i+1j}-2T_{i,j}+T_{i-1,j}}{\Delta y^2}+O\left(\Delta y^2\right).$$

The expression for mixed derivatives can be obtained by differentiating with respect to each variable. Thus, for example,

$$\left(\frac{\partial^2 T}{\partial x\partial y}\right)_{i,j}=\frac{\partial}{\partial x}\left(\frac{\partial T}{\partial y}\right)_{i,j}=\frac{\left(\frac{\partial T}{\partial y}\right)_{i+1,j}-\left(\frac{\partial T}{\partial y}\right)_{i-1,j}}{2\Delta x}$$

$$=\frac{\left(T_{i+1,j+1}-T_{i+1,j-1}\right)/(2\Delta y)-\left(T_{i-1,j+1}-T_{i-1,j-1}\right)/(2\Delta y)}{2\Delta y}.$$

Therefore,

$$\left(\frac{\partial^2 T}{\partial x\partial y}\right)_{i,j}=\frac{T_{i+1,j+1}-T_{i+1,j-1}-T_{i-1,j+1}-T_{i-1,j-1}}{4\Delta x\Delta y}$$

Proceeding in a similar manner, the central difference approximation to the third derivative is found to be

$$\left(\frac{\partial^3 T}{\partial x^3}\right)_{i,j} = \frac{T_{i+2,j} - 2T_{i+1,j} + 2T_{i-1,j} - T_{i-2,j}}{2(\Delta x)^3}.$$

Similar approximations can be obtained for even higher order derivatives. It may be observed that, as the other derivatives increase, the number of adjacent nodes involved in the approximation also increase. When we are computing adjacent values to the boundaries, the usefulness of these approximations become restricted. In fact, while computing near the boundaries, one sided formulae involving more adjacent points become necessary. For instance, to implement derivative boundary conditions, a second order accurate difference formula can be derived involving simultaneously the nodes $(i-2, j), (i-1, j)$ and (i, j) in the form:

$$\left(\frac{\partial T}{\partial x^2}\right)_{i,j} = \frac{\alpha T_{ij} + bT_{i-1,j} + cT_{i-2j}}{\Delta x}.$$

Where the coefficients a, b, c can be obtained from the Taylor series expansion of $T_{i-2,j}$ *and* T_{i-1j} as shown below:

$$T_{i-2,j} = T_{ij} - (2\Delta x)\left(\frac{\partial T}{\partial x}\right)_{i,j} + \frac{(2\Delta x)^2}{2}\left(\frac{\partial^2 T}{\partial x^2}\right)_{i,j} - \frac{(2\Delta x)^3}{6}\left(\frac{\partial^3 T}{\partial x^3}\right) + H.O.T$$

$$T_{i-1,j} = T_{ij} - \Delta x\left(\frac{\partial T}{\partial x}\right)_{i,j} + \frac{(\Delta x)^2}{2}\left(\frac{\partial^2 T}{\partial x^2}\right)_{i,j} - \frac{(\Delta x)^3}{6}\left(\frac{\partial^3 T}{\partial x^3}\right) + H.O.T$$

Thus, we have

$$aT_{ij} + bT_{i-1j} + cT_{i-2,j} = (a+b+c)T_{ij} - \Delta x(2c+b)\left(\frac{\partial T}{\partial x}\right)_{i,j} + \frac{(\Delta x)^2}{2}(4c+b)\left(\frac{\partial^2 T}{\partial x^2}\right)_{i,j} + O(\Delta x^3).$$

Comparing *Eqs*. , we get

$$a+b+c=0, \quad b+2c=-1, \quad b+4c=0.$$

Solving these equations, we obtain:

$$a = \frac{3}{2}, b = -\frac{4}{2}, c = \frac{1}{2}.$$

Thus,

$$\left(\frac{\partial T}{\partial x}\right)_{i,j} = \frac{3T_{ij} - 4T_{i-1,j} + T_{i-2j}}{2\Delta x} O\left(\Delta x^2\right)$$

Similarly, one –sided formulae can be derived for even higher order derivatives

We have defined various finite difference operators $\Delta, \nabla \delta, E, \mu$ and D and obtained various relationship connecting them. These relationships can also be used to derive finite difference approximations to derivatives of any order to the desired accuracy.

Thus, the application of a finite difference method to a given physical problems consists of the following three basic steps:

(i) Division of the spatial domain into an orthogonal computational grid.

(ii) Discretization of the governing equations and boundary conditions in space and time to derive approximately equivalent algebraic equations for each node.

(iii) Solving the resulting equations by a suitable matrix inversion or iterative technique.

These steps are further explained in the next section with more detail.

Example

Apply the Taylor series expansion to the form

$$\left(\frac{\partial^4 T}{\partial x^4}\right)_{i,j} = aT_{i+2,j} + bT_{i+1,j} + cT_{ij} + dT_{i-1,j} + eT_{i-2,j}.$$

And hence show that

$$\left(\frac{\partial^4 T}{\partial x^4}\right)_{i,j} = \frac{T_{i+2,j} - 4T_{i+1,j} + 6T_{ij} + 4T_{i-1,j} + T_{i-2,j}}{\Delta x^4} - \frac{(\Delta x)^2}{6}\left(\frac{\partial^6 T}{\partial x^6}\right)_{i,j}$$

Solution

Writing down the Taylor series expansions:

$$T_{i\pm 2,j} = T_{ij} \pm (2\Delta x)\left(\frac{\partial T}{\partial x}\right)_{i,j} + \frac{(2\Delta x)^2}{2}\left(\frac{\partial^2 T}{\partial x^2}\right)_{i,j}$$

$$\pm \frac{(2\Delta x)^3}{6}\left(\frac{\partial^3 T}{\partial x^3}\right)_{i,j} + \frac{(2\Delta x)^4}{24}\left(\frac{\partial^4 T}{\partial x^4}\right)_{i,j}$$

$$\pm \frac{(2\Delta x)^5}{120}\left(\frac{\partial^5 T}{\partial x^5}\right)_{i,j} + \frac{(2\Delta x)^6}{720}\left(\frac{\partial^6 T}{\partial x^6}\right)_{i,j} \pm \ldots..$$

and

$$T_{i\pm 1,j} = T_{ij} \pm (\Delta x)\left(\frac{\partial T}{\partial x}\right)_{i,j} + \frac{(\Delta x)^2}{2}\left(\frac{\partial^2 T}{\partial x^2}\right)_{i,j}$$

$$\pm \frac{(\Delta x)^3}{6}\left(\frac{\partial^3 T}{\partial x^3}\right)_{i,j} + \frac{(\Delta x)^4}{24}\left(\frac{\partial^4 T}{\partial x^4}\right)_{i,j}$$

$$\pm \frac{(\Delta x)^5}{120}\left(\frac{\partial^5 T}{\partial x^5}\right)_{i,j} + \frac{(\Delta x)^6}{720}\left(\frac{\partial^6 T}{\partial x^6}\right)_{i,j}$$

Substituting these expression into the form

$$\left(\frac{\partial^4 T}{\partial x^4}\right)_{i,j} = aT_{i+2j} + bT_{i+1j} + cT_{ij} + dT_{i-1,j} + eT_{i-2,j}$$

We obtain the following equations:

$$a+b+c+d+e=0$$

$$2a+b-d-2e=0$$

$$2a+\frac{b}{2}+\frac{d}{2}+2e=0$$

$$\frac{4}{3}a+\frac{b}{6}-\frac{d}{6}-\frac{4}{3}e=0$$

$$\frac{2}{3}a+\frac{b}{24}+\frac{d}{24}+\frac{2}{3}e=1$$
,

whose solution is found to be

$$a=1,\ b=-4\ ,\ c=6d=-4,\ e=1$$

$$T_{i+2,j}-4T_{i+2,j}+6T_{i,j}-4T_{i-1,j}+T_{i-2\ ,j}=(\Delta x)^4\left(\frac{\partial^4 T}{\partial x^4}\right)_{i,j}+\frac{\Delta x^6}{6}\left(\frac{\partial^6 T}{\partial x^6}\right)_{i,j}+\ldots.$$

This immediately gives

$$\left(\frac{\partial^4 T}{\partial x^4}\right)_{i,j}=\frac{T_{i+2,j}-4T_{i+2,j}+6T_{i,j}-4T_{i-1,j}+T_{i-2\ ,j}}{\Delta x^4}-\frac{(\Delta x)^2}{6}\left(\frac{\partial^6 T}{\partial x^6}\right)_{i,j}$$

Explicit Methods

These are basic finite difference methods for solving partial differential equations. Here, our approach in solving such equations is to replace the derivates of the given differential equation by their finite difference formulae.

A formula which expresses one unknown value at a given node in terms of the known preceding values is called an explicit formula. Some of the well-known explicit methods for solving parabolic partial differential equations are presented below.

Schmidt Method

This is a simple two- level explicit method. For illustration, let us consider a one-dimensional diffusion equation at the (i,n) grid point. Thus, we have:

$$\left(\frac{\partial T}{\partial t}\right)_i^n = \alpha\left(\frac{\partial^2 T}{\partial x^2}\right)_i^n .$$

Introducing the forward difference approximation to the time derivative and the central difference approximation to space derivative (FTCS), the above equation becomes:

$$\frac{T_i^{n+1} - T_i^1}{\Delta t} = \alpha \frac{T^n{}_{i+1} - 2T_i^n + T_{i-1}^n}{(\Delta x)^2}.$$

That is,

We now define

$$\frac{\alpha \Delta t}{(\Delta x)^2} = r.$$

Then the above equation becomes

$$T_i^{n+1} = rT_{i-1}^n + (1-2r)T_i^n + rT_{i+1.}^{.n}$$

This is the Schmidt explicit formula which gives the approximate value rT_i^{n+1} at the $(n+1)$ time level. This is also called a two-level explicit formula.

For $n=0$, Eq. gives approximate values of T_i^1 at grid points along the first time row $t = \Delta t$, in terms of the known initial and boundary value T_i^0 $(i = 1, 2, \ldots, I)$. Taking $n = 1$ min., we can compute approximate values of T_i^2 at the grid points along the second time level, in terms of just the computed values T_i^1, and so on. This process is continued until a predetermined time is reached. We shall illustrate this method through the following examples.

Example

Solve the following initial boundary value problem using an explicit finite difference method:

$$\frac{\partial T}{\partial t} = \frac{\partial^2 T}{\partial x^2} \quad 0 \le x \le 1,$$

given

$$T = \sin \pi x \quad when\, t = 0, 0 \le x \le 1$$

$$T = 0 \quad at\, x = 0 \;and\; x = 1 \;for\, t > 0.$$

And hence examine the accuracy of the numerical solution at t = 0.006 with its analytical solution.

Solution

Let us choose $\Delta x = 0.1$. To get a numerically stable solution, we choose Δt in such a way as to satisfy the stability condition:

$$r = \frac{\Delta t}{\Delta x^2} \le 0.5,$$

which gives $\Delta t \le 0.005$. In this example, we take $\Delta t = 0.002$, so that r = 0.2. Thus, using the FTCS approximation, the given differential equation reduces to

$$T_i^{n+1} = rT_{i-1}^n + (1-2r)T_i^n + rT_{i+1}^n.$$

Note that $n=0$ corresponds to t at $i=0$ and the end values of T are given by the boundary conditions. Thus,

$$T_1^0 = 0.3090, \qquad T_2^0 = \sin\frac{\pi}{5} = 0.5878, \quad T_3^0 = 0.8090$$

$T_4^0 = 0.9511, \quad T_5^0 = \sin\frac{\pi}{2} = 1.0, \quad T_6^0 = 0.9511$

$T_7^0 = 0.8090, \quad T_8^0 = 0.5878, \quad T_9^0 = 0.3090$

The boundary conditions give $T_0^0 = T_{10}^0 = 0.$ Now the plan of computation, the initial and boundary conditions are shown below.

We observe the symmetry of the solution on either side of the center line $x=0.5$ and therefore the value of T and $x=0.6$ is the same as x = 0.4 and so on. In other words, $T_6^0 = T_4^0$ etc. These entries are shown as the first row for $i=1, n=0.$ Using the numerical scheme given by Eq. we get:

$T_1^1 = 0.2T_0^0 + 0.6T_1^0 + 0.2T_2^0 - 0.3030.$

Similarly, for $i = 2,3,4$ and $5, n = 0,$ we get T_2^1 , T_3^1, T_4^1 and T_5^1, which is the solution of the given initial boundary value problem for $t = \Delta t = 0.002.$ Thus we can use the sample algorithm described by Eq. and further compute the values of T for subsequent time steps at all the nodes, which are conveniently tabulated in the table below.

		x										
n	t	0.0	0.1	0.2	0.3	0.4	0.5	0.6	0.7	0.8	0.9	1.0
0	0.0	0.0	0.3090	0.5878	0.8090	0.9511	1.0	0.9511	0.8090	0.5878	0.3090	0.0
1	0.002	0.0	0.3030	0.5763	0.7932	0.9325	0.9804	0.9325	0.7932	0.5763	0.3030	0.0
2	0.004	0.0	0.2970	0.5650	0.7777	0.9142	0.9612	0.9142	0.7777	0.5650	0.2970	0.0
3	0.006	0.0	0.2912	0.5540	0.7625	0.8963	0.9424	0.8963	0.7625	0.5540	0.2912	0.0
		0.0	0.2912	0.5520	0.7625	0.8964	0.9425	0.8964	0.7625	0.5540	0.2912	0.0

The analytical solution to the present problem can be easily found to be see (Sankara Rao):

$$T(x,t) = e^{-\pi^2 t} \sin \pi x\,.$$

Thus,

$$T(x, 0.006) = 0.9425 \sin \pi x.$$

The exact solution and the computed solution at t = 0.006 are shown in the table. Both solutions found to agree up to the third decimal place. This result demonstrates the power of even simple explicit numerical methods.

Example

Solve the differential equation

$$\frac{\partial T}{\partial t}=\frac{\partial^2 T}{\partial x^2}, \qquad 0 \le x \le \frac{1}{2},$$

given that $T=0$ when $t=0, 0\le x \le \frac{1}{2}$, with boundary conditions

$$\frac{\partial T}{\partial x}=0 \ at\, x=0 \ and \frac{\partial T}{\partial x}=1 \ at\, x=\frac{1}{2} \text{ for } t>0.$$

Taking $\Delta x = 0.1, \Delta t = 0.001$.Give the solution for three time steps.

Solution

Using FTCS approximation, the discredited equation of the given partial differential equation for each node i is given by

$$T_i^{n+1} = rT_{i-1}^n + (1-2r)T_i^n + T_{i+1}^n,$$

where $r = \Delta t / \Delta x^2$. The data of the present problem gives us $r = 0.01$. In this problem, the derivative boundary conditions are specified, which require special care. The implementation of which are specified as follows:

The left and side, that is at $x = 0, \partial T / \partial x = 0 \text{ for } t > 0$, can be equivalently written as $\frac{T_{i+1}^n - T_i^n}{\Delta x} = 0$ using the forward difference approximation, implying $T_1^n = T_{i+1}^n$. Therfore, $T_0^n = T_1^n$.

At the right hand side, that is, at x = ½, the boundary condition is $\partial T / \partial x = 1, t > 0$. Using the backward difference approximation, its equivalent is:

$\frac{T_1^n - T_{i-1}^n,}{\Delta x} = 0, \text{implying}\, T_i^n = T_{i-1}^n + \Delta x, \text{ therefore, } T_5^n = T_4^n + \Delta x$.

When $i = 1$,

$$T_1^{n+1} = rT_0^n + (1-2r)T_1^n + rT_2^n.$$

Eliminating t_0^n from Eqs. we get

$$T_1^{n+1} = (1-r)T_1^n + rT_2^n.$$

Thus, we use the following equations from $x = 0 \,\text{to}\, 0.3$, that is, $\text{for}\, i = 01, 2 \,\text{and}\, 3$

$$\left.\begin{aligned} T_1^{n+1} &= (1-r)T_1^n + rT_2^n \\ T_0^{n+1} &= T_1^{n+1} \\ T_i^{n+1} &= rT_{1-l}^n + (1-2r)T_1^n + rT_{1+l}^n, \quad for\, i = 2, 3 \end{aligned}\right\}$$

Also, $for\, i = 4, Eq.(1)$ becomes

$$T_4^{n+1} = rT_3^n + (1-2r)T_4^n + rT_5^n.$$

From *Eqs.* , we eliminate T_5^n to get:

$$T_4^{n+1} = rT_3^n + (1-r)T_4^n + \frac{\Delta t}{\Delta x}, \quad \text{for } i = 4$$

and Eq. gives us:

$$T_5^{n+1} = T_4^{n+1}.$$

Thus, we use Eqs. and to compute T at $x=0.4$ and $x=0.5$, that is, for $i=4$ and 5 and Eqs. for internal nodes.

Following the procedure, the solution T at each node for three time steps are computed and depicted in the table below.

		0.0	0.1	0.2	0.3	0.4	0.5
0	0.0	0.0	0.0	0.0	0.0	0.0	0.0
1	0.001	0.0	0.0	0.0	0.0	0.01	0.11
2	0.002	0.0	0.0	0.0	0.001	0.019	0.119
3	0.003	0.0	0.0	0.0001	0.0027	0.0272	0.1272

Dulfort- Frankel Method (1953)

This is another explicit method developed for solving the diffusion equation:

$$T_t = \alpha T_{xx}$$

In this method, both the time and space derivative are replaced by their central difference approximations. Thus:

$$\frac{T_i^{n+1} - T_1^{n-1}}{2\Delta t} = \alpha \frac{T_{i-1}^n - 2T_i^n + T_{i+1}^n}{\Delta x^2}.$$

On rearranging, this simplifies to:

$$T_i^{n+1} - T_1^{n-1} + 2r\left(T_{i-1}^n - 2T_i^n + T_{i+1}^n\right),$$

where

$$r = \alpha \frac{\Delta t}{\Delta x^2}.$$

Further, in Eq., T_i^n is replaced by its average value at the $(n+1)^{\text{th}}$ and $(n-1)^{\text{th}}$ time levels. Thus, if we substitute

$$T_i^n = \frac{T_i^{n+1} + T_i^{n-1}}{2}$$

into Eq. , we obtain

$$T_i^{n+1} = \frac{2r}{1+2r}\left(T_{i-1}^n + T_{i+1}^n\right) + \frac{1-2r}{1+2r} T_i^{n-1}.$$

Since it involves three-time levels, it is called the explicit three-level finite difference scheme. To initiate computation using the Durfort-Frankle Scheme, we should know the solution at two time levels, that is when $n = 0$ and $n = 1$ or $t =$ and t $= \Delta t$.

To implement this scheme, from the given initial and boundary conditions, we shall compute the solution at t $= \Delta$, using the two-level explicit method and then using Eq. to obtain the numerical solution at subsequent time steps, at all nodes covering the region of computation.

Implicit Methods

These are also basic finite difference methods for solving parabolic partial differential equations. A method in which the computation of many present unknown values requires the solution of a set of simultaneous equations is called an implicit method. For a given problem, the choice of a particular method depends on the computational accuracy required, the affordable computational time and the storage memory. Some of the well-known implicit methods are discussed in the following sections.

Classical Implicit Method

This method was developed by O'Brien et al. to solve the diffusion equation. Here, the time derivative is replaced by its central different approximation. Thus, using the BTCS approximation, the diffusion equation evaluated at $(i, n+1)$ grid point, namely

$$\left(\frac{\partial T}{\partial t}\right)_t^{n+1} = \alpha \left(\frac{\partial^2 T}{\partial x^2}\right)_t^{n+1}$$

Gives

$$\frac{T_i^{n+1} - T_i^n}{\Delta T} = á\left[\frac{\mathrm{T}_{\mathrm{i-1}}^{\mathrm{n+1}} - 2\mathrm{T}_{\mathrm{i}}^{\mathrm{n+1}} + \mathrm{T}_{\mathrm{i+1}}^{\mathrm{n+1}}}{(\Delta \mathrm{x})^2}\right]$$

$$-rT_{i-1}^{n+1} + (1+2r)_i^{n+1} T_i^{n+1} - rT_{i-1}^{n+1} = T_1^n,$$

for $i = 1.2, \ldots N-1$, the number of unknowns in $x-$direction. Thus the equation is implicit in the sense that there are three unknown values of T at the $(n+1)^{\text{th}}$ time level. For $n=0$ and for $i = 1,2,\ldots, N-1, Eq.$ gives $(N-1)$ simultaneous linear algebraic equations for $(N-1)$ unknown values of T at internal grid points along the first time level in terms of known initial and boundary values. These equations can be solved to get the values of T_i^1, for $i = 1,2,\ldots, N-1$. Similarly, by taking $n = 1$, we get $(N-1)$ equations for $(N-1)$ unknown values of T at the second time level, in terms of the boundary values and the recently computed values of T at the first time level, and so on. Thus, for example, if the values of T are known at the $(n)^{\text{th}}$ time level. By subsisting $i = 1, 2, \ldots N-1$ in Eq. we get the following set of simultaneous linear algebraic equations in tridiagonal from, as given below:

$$\left.\begin{array}{r} (1+2r)T_1^{n+1} - rT_2^{n+1} = d_1^n \\ -rT_1^{n+1} + (1+2r)T_2^{n+1} - rT_3^{n+1} = d_2^n \\ -rT_2^{n+2} + (1+2r)T_3^{n+1} - rT_4^{n+1} = d_3^n \\ -rT_{N-2}^{n+1} + (1+2r)T_{N-1}^{n+1} = d_{N-1}^n \end{array}\right\}$$

Where,

$$d_1^n = T_1^n + rT_0^{n+1}, \quad d_i^n = T_i^n. \quad i = 2, 3, \ldots, N-2$$

and

$$d_{N-1}^n = T_{N-1}^n + rT_N^{n+1}.$$

The subscripts 0 and N correspond to the value. This tridiagonal system can be solved exactly using Crout's LU reduction technique, as explained in the following section.0

Crank-Nicolson Method

In this method, the diffusion equation is approximated by replacing both the time and space derivative by their central difference approximations

(CTCS) at a point.

$$\left[i\Delta x, \left(n + \frac{1}{2} \right) \Delta t \right],$$

which is half way between the grid points $(i,n)$and$(i,n+1)$. Thus, the diffusion equations can be written as

$$\left(\frac{\partial T}{\partial t} \right)_t^{n+1(1/2)} = \alpha \left[\left(\frac{\partial^2 T}{\partial x^2} \right)_i^n + \left(\frac{\partial^2 T}{\partial x^2} \right)_i^{n+1} \right].$$

Now, using the central difference approximation on both the space and time derivatives, we get

$$\frac{T_i^{n+1} - T_i^n}{\Delta T} = \frac{á}{2} \left(\frac{T_{i+1}^n - 2T_i^n + T_{i-1}^n}{\Delta x^2} + \frac{T_{i+1}^{n+1} - 2T_i^{n+1} + T_{i-1}^{n+1}}{\Delta x^2} \right)$$

Or

$$-\frac{r}{2} T_{i-1}^{n+1} + (1+r) - \frac{r}{2} T_{i-1}^n + (1-r) T_i^n + \frac{r}{2} T_{i+1}^n,$$

for $i = 0,1,2,\ldots..,N-1,N$, where N is the number of division into which the interval of integration is subdivided. Thus, $i = 0$ and $i = N$ correspond to the left and right boundaries respectively.

It can be observed from Eq., that the new value of temperature T_i^{n+1} is not given directly in terms of known T_i^n at one time step, but is also a function of unknown temperatures at adjacent positions, and is therefore called an implicit method.

Thus, assuming that the values of T are known at the nth time level and taking $i = 1, 2\ldots, N-1$, Eq yields a set of linear simultaneous equations in terms of unknowns T_i^{n+1}, as given below in a tridiagonal form.

Where

$$\left.\begin{aligned}
d_1^n &= \frac{r}{2}T_0^n + (1+r)T_1^n + \frac{r}{2}T_2^n + \frac{r}{2}T_0^{n+1} \\
d_1^n &= \frac{r}{2}T_{i-1}^n + (1-r)T_i^n + \frac{r}{2}T_{i+1,}^n \qquad i = 2,3,\ldots, N-2 \\
d_{N-1}^n &= \frac{r}{2}T_{N-2}^n + (1-r)T_{N-1}^n + \frac{r}{2}T_N^n + \frac{r}{2}T_N^{n+1}
\end{aligned}\right\}$$

l of which are known quantities. This resulting tridiagonal system can be solved exactly using Crout's reduction technique.

In this method, the computational effort and the required computer memory space can be minimized. Also, since we use the same coefficient tridiagonal matrix to compute the solution at subsequent time steps, a considerable amount of computational time will be saved, particularly while handling higher order systems. An example follows for the illustration of this method.

Example

Use the Crank-Nicolson method to find the numerical solution of the following parabolic partial differential equation after one time step:

$$T_t = T_{xx,} \qquad 0 < x < 1,$$

subject to the initial condition

$$T(x,0) = 1, \quad 0 < x < 1$$

and the boundary conditions

$$T(0,t) = T(1,t) = 0, \;\; t \geq 0$$

Compute the solution by taking $\Delta x = 1/4$ and $\Delta t = 1/32$.

Solution

The Crank-Nicolson implicit finite difference scheme for the given differential equation is given by Eq., that is:

$$-\frac{r}{2}T_{i-1}^{n+1}+(1+r)T_i^{n+1}-\frac{r}{2}T_{i+1}^{n+1}=\frac{r}{2}T_{i-1}^{n+1}+(1-r)T_i^{n}-\frac{r}{2}T_{i+1}^{n}.$$

Noting that N = 1 / $\Delta x = 4, \Delta t$ / $\Delta x^2 = 1$ /2 and $N = 0,4$ correspond to the left and right boundaries respectively, for $i = 1,2,3$ and n = 0, Eq yields:

$$-\frac{1}{4}T_0^1+\frac{3}{2}T_1^1=\frac{1}{4}T_2^1=\frac{1}{4}T_0^0+\frac{1}{2}T_1^0+\frac{1}{4}T_2^0$$

$$-\frac{1}{4}T_1^1+\frac{3}{2}T_2^1=\frac{1}{4}T_3^1=\frac{1}{4}T_1^0+\frac{1}{2}T_2^0+\frac{1}{4}T_3^0$$

$$-\frac{1}{4}T_2^1+\frac{3}{2}T_3^1=\frac{1}{4}T_4^1=\frac{1}{4}T_2^0+\frac{1}{2}T_3^0+\frac{1}{4}T_4^0.$$

From the boundary conditions, we get

$$T_0^n = T_4^n = 0 \quad for\, all \;\; n = 0\,,1\,,2\ldots.$$

Thus, using the boundary conditions, we get

$$0+6T_1^1-T_2^1=0+2+1$$

$$-T_1^1+6T_2^1-T_3^1=1+2+1$$

$$-T_2^1+6T_3^1-0=1+2+0,$$

which in matrix notation reduces to the tridiagonal system

$$\begin{bmatrix} 6\ -1 & \\ -16-1 \\ -1\ 6 \end{bmatrix}\begin{pmatrix} T_1^1 \\ T_2^1 \\ T_3^1 \end{pmatrix}=\begin{pmatrix} 3 \\ 4 \\ 3 \end{pmatrix}$$

To solve this system, we use Crout's reduction technique, where we decompose the coefficient matrix of $Eq.$ into the form:

$$A = LU = \begin{bmatrix} l_{11} & 0 & 0 \\ l_{21} & l_{22} & 0 \\ l_{21} & l_{32} & l_{33} \end{bmatrix} \begin{bmatrix} 1 & u_{12} & u_{13} \\ 0 & 1 & u_{23} \\ 0 & 0 & 1 \end{bmatrix}.$$

Thus, we get:

$$l_{11} = 6,\ l_{21} = -1,\ l_{31} = 0$$

$$l_{11}u_{12} = -1\ , l_{11}u_{13} = 0,\ \text{implying}\, u_{12} = -\frac{1}{6},\ u_{13} = 0$$

$$l_{21}u_{12} + l_{22} = 6\,\text{gives}\, l_{22} = 6 - \frac{1}{6} = \frac{35}{6}$$

$$l_{31}u_{12} + l_{32} = -1\,\text{gives}\, l_{32} = -1$$

$$l_{31}u_{13} + l_{32}u_{23} + l_{33} = 6 \text{ or } l_{33} = \frac{204}{35}.$$

Now, the system can be written as

$$\begin{bmatrix} 6 & 0 & 0 \\ -1 & \frac{35}{6} & 0 \\ 0 & -1 & \frac{204}{35} \end{bmatrix} \begin{bmatrix} 1 & -\frac{1}{6} & 0 \\ 0 & 1 & -\frac{6}{35} \\ 0 & 0 & 1 \end{bmatrix} \begin{pmatrix} T_1^1 \\ T_2^1 \\ T_3^1 \end{pmatrix} = \begin{pmatrix} 3 \\ 4 \\ 3 \end{pmatrix}$$

The solution of which is given by

$$\left(T_1^1, T_2^1, T_3^1\right) = \left(\frac{11}{17}, \frac{15}{17}, \frac{11}{17}\right).$$

Similarly, we can advance the solution one time step at a time, by solving the same tridiagonal ∂T. To solve this problem by an explicit method and to get a result of the same accuracy, we should chose Δt such that:

$$\Delta t < \frac{1}{32}$$

Example

Solve the unsteady heat equation $T_t = T_{xx,}$ $0 \le x \le 4$, subject to the initial conditions $T(x,0)=1000$, and the boundary conditions

$$\frac{\partial T}{\partial x}(0,t)=0.15(T-70),\ \frac{\partial T}{\partial x}(4,t)=0.25(T-70).$$

Using the Crank-Nicolson finite difference method by taking $\alpha = 0.132, \Delta x = 1.0$ and $\Delta t = 0.893$. Give solution after one time step.

Solution

Form the given data, we find that

$$r = \frac{\alpha \Delta t}{\Delta x^2} = 0.25$$

and the initial conditions are shown below:

$i=1$ $i=2$ $i=3$ $i=4$ $i=5$

$x=0$ $x=1$ $x=2$ $x=3$ $x=4$

$T_1^0 = 1000$ $T_2^0 = 1000$ $T_3^0 = 1000$ $T_4^0 = 1000$ $T_5^0 = 1000$.

Using the Crank-Nicolson implicit scheme, we have from Eq.

$$-0.257_{i-1}^{n+1} + 2.57_i^{n+1} - 0.257_{i-1}^{n+1} = 0.257_{i-1}^n + 1.57_i^n + 0.257_{i+1}^n.$$

Let $n = 0, i = 1, 2, \ldots, 5,$ then $Eq(1)$ gives

In this system of equation, are fictitious points which we shall eliminate using the boundary conditions. Using the central difference approximation to the left and the boundary conditions, we have

$$\frac{T_2^N - T_0^n}{\Delta x} = 0.15\left(T_1^n - 70\right).$$

Thus,

$$T_0^1 = T_2^1 - 0.15\left(T_1^1 - 70\right) \text{ and } T_0^0 = T_2^0 - 0.15\left(T_1^0 - 70\right).$$

Therefore,

$$0.25\left(T_0^0 + T_0^1\right) = 0.25\left(T_2^0 + T_2^1\right) - 0.0375\left(T_1^0 + T_1^1\right) + 5.25.$$

Similarly, using the central difference approximation to the right-end boundary condition, we have

$$\frac{T_5^N - T_4^n}{\Delta x} = 0.25\left(T_5^n - 70\right)$$

Thus,

$$T_6^1 = T_4^1 - 0.25\left(T_5^1 - 70\right) \ and\ T_6^0 = T_4^0 - 0.25\left(T_5^0 - 70\right).$$

Therefore,

$$0.25\left(T_6^0 + T_4^1\right) = 0.25\left(T_4^0 + T_4^1\right) - 0.0625\left(T_5^0 + T_5^1\right) - 8.75.$$

Hence, the last of becomes

$$-0.5T_4^1 + 2.4375T_5^1 = 0.5T_4^0 + 1.5625_5^0 - 8.75.$$

Using , it can be written in matrix notation, after using the initial conditions in the form:

$$\begin{bmatrix} 2.5375 & -0.5 & 0 & 0 & 0 \\ -0.25 & 2.5 & 0.25 & 0 & \\ 0 & 0 & -0.25 & 2.5 & -0.25 \\ 0 & 0 & 0 & -0.5 & 2.4375 \end{bmatrix} \begin{pmatrix} T_1^1 \\ T_2^1 \\ T_3^1 \\ T_4^1 \\ T_5^1 \end{pmatrix} = \begin{pmatrix} 1967.75 \\ 2000 \\ 2000 \\ 2000 \\ 2053.75 \end{pmatrix}$$

Applying Crout's reduction technique, this matrix equation can be re-written as

Now, the solution of Eq. can be easily found as

$$\left(T_1^1 T_2^1 T_3^1 T_4^1 T_5^1\right) = \left(971.8725 \quad 996.9774997.8171\ 983.2880\ 1044.2641\right).$$

This is the heat distribution after one-time step. At subsequent time steps, the right hand side values of *Eq*, change while the coefficient matrix remains unchanged, and hence the solution of the given problem can be obtained at subsequent time steps with little effort.

Weighted Average Implicit Method

To explain this method, we shall recall the heat condition equation

$$T_t = \alpha T_{xx}.$$

In this method, we shall replace the time derivative by its backward difference approximation and the space derivative by its central difference approximation. Thus, using BTCS approximation, Eq. is first evaluated at the grid point $(i, n+1)$ to get

$$T_i^{n+1} - T_i^n = r\left(T_{i-1}^{n+1} - 2T_i^{n+1} + T_{i+1}^{n+1}\right).$$

Where

$$r = \frac{\alpha \Delta t}{x^2}.$$

Now, we introduce the weight λ such that $0 \le \lambda \le 1$, and rewrite the above expression in the form

$$T_i^{n+1} - T_i^n = r\left[\lambda\left(T_{i-1}^{n+1} - 2T_i^{n+1} + T_{i+1}^{n+1}\right) + (1-\lambda)\left(T_{i-1}^n - 2T_{i-1}^n + T_{i+1}^n\right)\right]$$

as a weighted average at the (n+1)[th] and n[th] levels. On rewriting the above equations, we obtain

$$-r\lambda T_{i-1}^{n+1} + (1+2r\lambda)T_i^{n+1} - r\lambda T_{i+1}^{n+1} = r(1-\lambda)T_{i-1}^n.$$

From Eq. it can be observed that when $\lambda = 1$, we can recover the classical implicit formula; when $\lambda = 1/2$, we obtain the Crank-Nicolson formula and when $\lambda = 0$, we get the explicit formula represented by Eq. Hence, this weighted average method can be considered as the most general implicit method, which combines many methods so far discussed.

THE CONCEPT OF STABILITY

We have introduced various finite difference methods for solving parabolic partial differential equations that model the heat conduction equation. All these methods essentially reduce the solution of the partial differential equation to the solution of a set of algebraic equations. We are aware that numerical computations carried out on any computer can be accurate only to a finite number of decimal places. At each stage of the calculation, some round-off error will be introduced, no matter how small. Let e be an error introduced into the calculations at each mesh point (i,n) and $\overline{u}_i^n$, the resulting numerical solution of the differential equation,

$$u_t = \alpha u_{xx}$$

with finite difference approximation. Let u_i^n be the corresponding solution of the differential equation with a finite difference method, assume no round-off error is involved in each calculation. Then, we define

$$e_i^n = u_i^n - \overline{u}_i^n .$$

Stability considerations are very important in deriving numerical solutions of a differential equation using finite difference methods. The solution of the finite difference equation is a finite difference method. The solution of the exponentially as we progress from one time step to another. That is, e_i^n should remain uniformly bounded a $n \to \infty$, while $\Delta t, \Delta x \to 0$. There are two standard methods for investigating the growth of error in solving the finite difference equations: the matrix method and the Von-Neumann Fourier series expansion method. However, we shall discuss only the matrix method in the following section to show the important aspects of the theoretical study of numerical stability. Consistency and convergence of the finite difference method are other important theoretical aspects, for which the reader is advised to consult advances books for further reading.

The Matrix Method

Let us consider a one- dimensional heat conduction problem described by the differential equation

$$u_{1=}\alpha u_{xx} \quad 0 \le x \le L$$

subject to the boundary conditions

$$u = 0\, at \;\; x = 0 \;\; \text{and}\, x = L, t \ge 0$$

and the initial condition

$$u = f(x),\; 0 < x < L,\; t = 0.$$

Use the simple explicit finite difference scheme, FTCS to find the given differential equation for $n = 0,1,2\; ,\ldots,$and for $i = 0,1,2,\ldots.,N$, such that $\Delta x = L/$N. The subscripts 0 and N correspond to x = 0 and L, the left and right-end boundaries, where u is known for all t. Expanding Eq. for $i = 1,2\ldots.,(N-1)$, using the boundary conditions $U_0^n = u_N^n = 0$ for all t, the above system written in matrix notation takes the following form:

$$\begin{pmatrix} u_1^{n+1} \\ u_2^{n+1} \\ u_3^{n+1} \\ \vdots \\ u_{N+1}^{n+1} \end{pmatrix} = \begin{bmatrix} 1-2r & r & & \\ r & 1-2r & r & \\ & r & 1-2r & r \\ & & & \\ & & r & 1-2r \end{bmatrix} \begin{pmatrix} u_1^{n} \\ u_2^{n} \\ u_3^{n} \\ \vdots \\ u_{N-1}^{n} \end{pmatrix}$$

Equivalently

$$U^{n+1} = [A]U^n.$$

Where U is a column vector, [A] is an (N-1) x (N-1) square matrix, which happens to be a tridiagonal matrix. Here, [A] is called an amplification matrix. U^0 is a vector of initial values, with which we have initiated our computation. The Successive rows of computation with each time step are given as

$$U^1 = [A]U^0$$

$$U^2 = [A]U^1 = [A][A]U^0 = \left[A^2\right]U^0$$

$$U^{n+1} = [A]U^n = \left[A^2\right]U^{n-1} = \ldots. = \left[A^{n+1}\right]U^0.$$

Now, suppose we have introduced errors at every node along the n = 0 row or at t = 0, and start the computation with the vector of values $\overline{U^0}$, we then compute the successive rows as follow:

$$\overline{U^1} = [A]\overline{U^0}$$

$$\overline{U^2} = [A]\overline{U^1} = \left[A^2\right]U^0$$

$$\overline{U^{n+1}} = [\mathrm{A}]\mathrm{U}^{\mathrm{n}} = \left[\mathrm{A}^2\right]\mathrm{U}^{\mathrm{n}-1} = \ldots. = \left[\mathrm{A}^{\mathrm{n}+1}\right]\mathrm{U}^0$$

Then the error vector e can be written as

$$e^{n+1} = U^{n+1} - \overline{U^{n+1}} = \left[A^{n+1}\right]\left(U^0 - \overline{U^0}\right) = \left[A^{n+1}\right]e^0,$$

indicating that the formula for error propagation is similar to the one used in the computation of U. Since the difference equation is linear, the superposition principle is applicable and hence it is enough to consider the propagation of only one line of errors. A finite difference scheme is known to be stable if e^n remains bounded as $n \to \infty$, which is ensured, provided that the eigenvalue with the largest modulus of the amplification matrix [A] is less than or equal to unity.

Now, from Eq. the matrix [A] can rewritten as

$$[A] = \begin{bmatrix} 100 & & \\ 010 & & \\ & & \\ & & 0\,1\,0 \\ & & 0\,1 \end{bmatrix} + r \begin{bmatrix} -2\,1 & & \\ 1-21 & & \\ & & \\ & & 1-2\,1 \\ & & 1\ -2 \end{bmatrix}$$

Alternatively

$$[A] = [l_{n-1}] + r[D_{N-1}],$$

where $[l_{N-1}]$ isa unit matrix of order (N-1) and $[D_{N-1}]$ isa n$(N-1)$ tridiagonal matrix. We may recall from linear algebra that the eigenvalues of an $N\,xN$ tridiagonal matrix

$$\begin{bmatrix} bc & & \\ abc & & \\ abc & & \\ & & \\ & & a\ b\ c \\ & & a\ b \end{bmatrix}$$

are given by the formula

$$\lambda_s = B + 2\sqrt{ca}\cos\frac{s\pi}{N+1}, \quad s = 1,2.....,N$$

Thus, the eigenvalues of $[D_{N-1}]$ are

$$\lambda_s = -2 + 2\cos\frac{s\pi}{N} = -4\sin^2\frac{s\pi}{2N}, \; s = 1,2,\ldots,N$$

Thus, the eigenvalue are

$$1 + r\left(-4\sin^2\frac{s\pi}{2N}\right).$$

Hence, the condition for the stability of the simple explicit scheme to solve the heat conduction equation is

$$\left|1 - 4r\sin^2\frac{s\pi}{2N} \le 1\right|.$$

Therefore,

$$-1 \le -4r\sin^2\frac{s\pi}{2N} \le 1,$$

which gives the limiting value of r as

$$-1 \le 1 - 4r\sin^2\frac{s\pi}{2N}$$

$$r \le \frac{1}{2\sin^2\frac{s\pi}{2N}}.$$

Thus, the explicit scheme is stable if

$$\left(r = \alpha\frac{\Delta t}{\Delta x^2} \le \frac{1}{2}\right)$$

Example

Show that the Crank-Nicolson scheme for the diffusion equation

$T_t = T_{xx}$, $0 \le x \le L$

satisfy the conditions

$T = 0$ at $x = 0$, $T = L$ for $t > 0$

and $T = f(x)$, $0 \le x \le L$ for $t = 0$

has unrestricted stability.

Solution

We may recall the Crank-Nicolson finite difference scheme for the given diffusion equation, given by:

$$-rT_{i-1}^{n+1}+2(1+r)T_{i+1}^{n+1}-rT_{i+1}^{n+1}=T_{i-1}^{n}+2(1-r)T_{i}^{n}+rT_{i+1,}^{n}\quad i=0,1,2,\ldots\ldots,N$$

$i=0,N$ correspond to the left and right boundaries respectively. For $i=1,2,\ldots,(N-1),Eq$ can be expanded and written in the following matrix form:

$$\begin{bmatrix} 2(1+r) & -r & & \\ -r & 2(1+r) & -r & \\ & & & \\ & -r & 2(1+r) & -r \\ & & -r & 2(1+r) \end{bmatrix}\begin{pmatrix} T_1^{n+1} \\ T_2^{n+1} \\ \vdots \\ T_{N-2}^{n+1} \\ T_{N-1}^{n+1} \end{pmatrix}$$

$$=\begin{bmatrix} 2(1-r) & & & \\ r & 2(1-r) & r & \\ & & & \\ & r & 2(1-r) & r \\ & & r & 2(1-r) \end{bmatrix}\begin{pmatrix} T_1^{n} \\ T_2^{n} \\ \vdots \\ T_{N-2}^{n} \\ T_{N-1}^{n} \end{pmatrix}$$

That is,

$$(2l-rD_{N-1})\overline{T}^{n+1}=(2l+rD_{n-1})\overline{T}^{n}$$

Where,

$$D_{N-1}=\begin{bmatrix} -2 & 1 & & \\ 1 & -2 & 1 & \\ & \vdots & & \\ & 1 & -2 & 1 \\ & & 1 & -2 \end{bmatrix}$$

Thus,

$$\overline{T}^{n+1} = \left(2l - rD_{N-1}\right)^{-1}\left(2l + rD_{N-1}\right)\overline{T}^{n}.$$

Alternatively, it can be expressed in the form $\overline{T}^{n+1} = [A]\overline{T}^{n}$, where

$$[A] = \left(2l - rD_{N-1}\right)^{-1}\left(2l + r\,D_{N-1}\right).$$

For stability, the modulus of the eigenvalue of $[A]$ should each be less than unity. However, the eigenvalues of D_{N-1} are known to be given by

$$\lambda_s = -2 + 2\cos\frac{s\pi}{N} = -4\sin^2\frac{s\pi}{2n}.$$

Hence, the eigenvalues of $[A]$ are

$$\frac{2 - 4r\sin^2\dfrac{s\pi}{2N}}{2 + 4r\sin^2\dfrac{s\pi}{2N}},\ s = 1,2.\ldots,N-1$$

These are clearly less than one for all positive values of r. This proves that the Crank-Nicolson scheme for the given problem is stable for all positive values of r, so has unrestricted stability.

METHOD FOR TWO-DIMENSIONAL EQUATIONS

Consider the finite difference methods of the solution for the equation

$$\frac{\partial u}{\partial t} = \alpha\left(\frac{\partial^2 u}{\partial x^2} + \frac{\partial^2 u}{\partial y^2}\right)$$

in a rectangular region bounded by $\overline{R} = R\,x[0 \le t \le T]$, where R = $[0 \le x, y \le 1]$, with appropriate initial and boundary conditions. Here again, the region $\overline{R}$ in the (x, y, t) space is covered by a rectangular grid with sides parallel to the axes, with $\Delta x = \Delta y = h$ and $\Delta t = k$, with grid spacing in the space and time directions. The grid points (x, y, t) are given by $x = i\Delta x, y = j\Delta y, t = n\Delta t$; where i, j, n are all integers and $i = j = 0$ corrospond to the origin. The function u satisfying the difference equation at the grid points is denoted by $u^n_{i,j}$. With this notation, we shall extend to the two dimensional parabolic.

Implicit Method

Taking $\Delta x = y\Delta = h,$ if we apply the Crank=Nicolson scheme to, we obtain

$$u^{n+1}_{i,j} = u^n_{i,j} + \frac{r}{2}\left(u^{n+1}_{i-1,j} + u^{n+1}_{i+1,j} + u^{n+1}_{ij-1} + u^{n+1}_{ij+1} - 4u^{n+1}_{ij}\right)$$

$$+\frac{r}{2}\left(u^n_{i-1,j} + u^n_{i+1,j} + u^n_{i\,j-1} + u^n_{i\,j+1} - 4u^{n+1}_{i,j}\right),$$

where, $r = \alpha\,\Delta t / h^2$. Further, if we define $a = -r/2,\ c = (1+2r)$, and

$$b^n_{i,j} = u^n_{i,j} + \frac{r}{2}\left(u^n_{i-1,j} + u^n_{i+1,j} + u^n_{i\,j-1} + u^n_{i\,j+1} - 4u^{n+1}_{i,j}\right),$$

we can rewrite Eq. as

$$\left(au^{n+1}_{i-1,j} + au^{n+1}_{i,j} + au^{n+1}_{i\,j-1} + au^{n+1}_{i\,j+1} = b^n_{i,j}\right).$$

This resulting linear algebraic system is no longer tridiagonal, in view of the unknowns $u^{n+1}_{ij}, u^{n+1}_{i-1j}, u^{n+1}_{i+j}, u^{n+1}_{i\,j-1} = u^{n+1}_{ij+1}$ in *Eq*. Such a system is in general solved by an iterative method.

Alternate Direction Implicit Method

While solving time-dependent partial differential equations in more than one space variable, it is advantageous to break the numerical process into simpler sequentially solvable problems. Such a process is called splitting.

For example, we will discover that much is to be gained by splitting a two- dimensional problem into a sequence of two one dimensional problems. This procedure leads to the development of the so called alternate direction implicit (ADI) method.

This method is a two-step process involving the solution of a tridiagonal system of equations along lines parallel to the x and y axes at the first and second steps respectively. When the (ADI) method is applied to Eq., firstly, we use row-wise traverses. That is, we advance from $m\Delta t$ to $(n+1)\Delta t$ by approximating the space derivatives using the central difference approximation at the th time level in the x-direction and at n^{th} time level in the y-direction. After the row wise transverse, we proceed to the column-wise traverse as a second step. In this case, we advance from the $(n+1)^{\text{th}}$ time level to the $(n+2)^{\text{th}}$ time level, using the central difference approximation on both the space derivatives, this time at the $(n+1)^{\text{th}}$ level in the x-direction and at $((n+2)^{\text{th}}$ time level in the y-direction. Thus, the resulting expression, by taking

$\Delta x = \Delta y = h = r = \dfrac{\alpha \Delta t}{h^2}$, can be written as:

$$u_{i,j}^{n+1} = u_{i,j}^{n} = r\left(u_{i-1,j}^{n+1} - 2u_{i,j}^{n+1} + u_{i+1,j}^{n+1}\right) + r\left(u_{i\,j-1}^{n} + u_{i,j+1}^{n} - 2u_{ij}^{n}\right)$$

$$u_{i,j}^{n+1} = u_{i,j}^{n} = r\left(u_{i-1,j}^{n+1} - 2u_{i,j}^{n+1} + u_{i+1,j}^{n+1}\right) + r\left(u_{ij-1}^{n+2} - 2u_{i,j}^{n+2} + u_{ij+1}^{n+2}\right).$$

Let us apply this formula to a two-dimensional problem as follows.

Example

A rectangular plate is of 8 units. Initially, all the points on the plate are at 〖50〗^0 C. The edges are suddenly brought to the temperatures and held at these temperatures. Use the ADI method to trace the history of temperature at the internal nodes spaced two units apart, assuming that heat flows only in the x and y directions.

Solution

The governing equations of the heat transfer are given by $u_1 = u_{xx} + u_{yy}$.

Taking the discredited form of the above equation is

$$\frac{u_{i,j}^{n+1} = u_{i,j}^{n}}{\Delta t} = \frac{u_{i-1j} - 2u_{ij} + u_{i+1j}}{h^2} + \frac{u_{ij-1} - 2u_{ij} + u_{ij+1}}{h^2}.$$

Let then we apply the ADI method, whereas a first step we use row- wise traverses. Thus we can rearrange:

$$u_{i,j}^{n+1} = u_{i,j}^{n} = r\left(u_{i-1,j}^{n+1} - 2u_{i,j}^{n+1} + u_{i+1,j}^{n+1}\right) + r\left(u_{i\,j-1}^{n} + u_{i,j+1}^{n} - 2u_{ij}^{n}\right).$$

After the row-wise traverse, we proceed to the column-wise traverse as a second step and the corresponding finite difference form is:

$$u_{i,j}^{n+1} = u_{i,j}^{n} = r\left(u_{i-1,j}^{n+1} - 2u_{i,j}^{n+1} + u_{i+1,j}^{n+1}\right) + r\left(u_{ij-1}^{n+2} - 2u_{i,j}^{n+2} + u_{ij+1}^{n+2}\right).$$

For a row-wise traverse, we consider $j = 1, 2$ and vary $i = 1, 2, 3$. Thus. Eq. gives us:

$$u_{1,1}^{1} - u_{1,1}^{0} = r\left(u_{0,1}^{1} - 2u_{1,1}^{1} + u_{2,1}^{1}\right) + r\left(u_{1,0}^{1} - 2u_{1,1}^{0} + u_{1,2}^{1}\right)$$

$$u_{2,1}^{1} - u_{2,1}^{0} = r\left(u_{1,1}^{1} - 2u_{2,1}^{1} + u_{3,1}^{1}\right) + r\left(u_{2,0}^{0} - 2u_{2,1}^{0} + u_{2,2}^{0}\right)$$

$$u_{3,1}^{1} - u_{3,1}^{0} = r\left(u_{1,1}^{1} - 2u_{3,1}^{1} + u_{4,1}^{1}\right) + r\left(u_{3,0}^{0} - 2u_{3,1}^{0} + u_{3,2}^{0}\right)$$

$$u_{1,2}^{1} - u_{1,2}^{0} = r\left(u_{0,2}^{1} - 2u_{1,2}^{1} + u_{2,2}^{1}\right) + r\left(u_{1,1}^{0} - 2u_{1,2}^{0} + u_{1,3}^{0}\right)$$

$$u_{2,2}^{1} - u_{2,2}^{0} = r\left(u_{1,2}^{1} - 2u_{2,2}^{1} + u_{3,2}^{1}\right) + r\left(u_{2,1}^{0} - 2u_{2,2}^{0} + u_{2,3}^{0}\right)$$

$$u_{3,2}^{1} - u_{3,2}^{0} = r\left(u_{2,2}^{1} - 2u_{3,2}^{1} + u_{4,2}^{1}\right) + r\left(u_{3,1}^{0} - 2u_{3,2}^{0} + u_{3,3}^{0}\right)$$

This can be written in matrix notation as:

Taking r = 1 and using the computed values in the first step, the above equations can be written in matrix notation:

$$\begin{bmatrix} 3-1 \\ -1\ 3 \\ 3-1 \\ -1\ 3 \\ 3\ -1 \\ -1\ 3 \end{bmatrix} \begin{pmatrix} u_{1,1}^2 \\ u_{1,2}^2 \\ u_{2,1}^2 \\ u_{3,1}^2 \\ u_{3,2}^2 \end{pmatrix} = \begin{pmatrix} -56.1904+65+53.5714+100 \\ -40.4762+25+46.4286+10 \\ -53.5714+56.1904+54.5238+90 \\ -46.4286+40.4762+48.8095+20 \\ 54.5238+53.5714+60+80 \\ -48.8095+46.4286+50+30 \end{pmatrix} = \begin{pmatrix} 162381 \\ 40.9524 \\ 147.1428 \\ 62.8571 \\ 139.0476 \\ 77.6191 \end{pmatrix}$$

whose solution is given by

$$u_{1,1}^2 = 66.0119, \quad u_{2,2}^2 = 41.6943$$

$$u_{1,2}^2 = 35.6548 \quad u_{3,1}^2 = 61.8452$$

$$u_{2,1}^2 = 63.0357, \quad u_{3,2}^2 = 46.4881$$

This is the ADI method of solution at t = 8.

This procedure can be continued alternatively between row-wise and column-wise traverses and a steady safe solution is obtained This technique is more convenient for computer adaptation rather than for hand computation.

INDEX